CHARLES PROTEUS
STEINMETZ

Lectures on Electrical Engineering

EDITED BY PHILIP L. ALGER
CONSULTING PROFESSOR OF ELECTRICAL ENGINEERING
RENSSELAER POLYTECHNIC INSTITUTE

IN
THREE VOLUMES

III

Transient Electrical Phenomena

DOVER PUBLICATIONS, INC.
NEW YORK

Published in Canada by General Publishing Company, Ltd., 30 Lesmill Road, Don Mills, Toronto, Ontario.

Published in the United Kingdom by Constable and Company, Ltd., 10 Orange Street, London WC 2.

This Dover edition, first published in 1971, is a reprinting, in three volumes, of selected portions of the Steinmetz Electrical Engineering Library, originally published between 1915 and 1920 in nine volumes by the McGraw-Hill Book Company, Inc. The correspondence between the present edition and the original set is indicated on page vi of the General Preface.

International Standard Book Number: 0-486-62516-8
Library of Congress Catalog Card Number: 70-137004

Manufactured in the United States of America
Dover Publications, Inc.
180 Varick Street
New York, N. Y. 10014

GENERAL PREFACE

Steinmetz is well described by his adopted middle name, Proteus, for the range of his interests and accomplishments was vast. To his friends he was known simply as "the Doctor," but by others he was referred to variously as the wizard of Schenectady, forger of thunderbolts, and latter-day Vulcan. Harvard University called him "the foremost expert in applied electricity of this country," and he was a prolific writer on both technical and general subjects.

But it is as a teacher that he is best known, and with good reason. Steinmetz had a remarkable ability to explain things. He liked to talk with children, telling them why the stars twinkle, what makes the wind blow, and what makes flowers bloom. He was able to explain engineering problems in the same way, clearly and simply; he would resolve them into distinct, readily understood elements, and then solve them step by step, using approximations freely in order to emphasize method rather than detailed calculation.

In Steinmetz's day alternating-current phenomena were considered very mysterious, and there were very few textbooks available to explain them. Thus, a whole generation of engineers eagerly read the nine volumes of his Union College lectures and monographs published by McGraw-Hill, some of which went into five editions. Steinmetz realized that many engineers did not have the time or inclination for a rigorous mathematical presentation, and therefore endeavored in this series "to give a descriptive exposition of the physical nature and meaning, the origin and effects, of these phenomena, with the use of very little and only the simplest form of mathematics."

In view of the remarkable clarity of Steinmetz's explanations, and the breadth of his treatment of fundamental concepts, his writings of fifty years ago still offer one of the best ways of getting a basic electrical engineering education. They are also of historical interest: Steinmetz's use of the complex number $a + jb$ for alternating-current circuit calculations was just one of many unique contributions to electrical engineering recorded in these books.

This Dover edition of the lectures, in three volumes, includes all the material still of use to present-day students of engineering, with the exception of Volume 2 of the McGraw-Hill set, which is already available in both hardcover and paperback in a new edition revised by the present writer (*Mathematics for Science and Engineering*, by Philip L. Alger, McGraw-Hill, Second Edition, 1969). The relationship between the new Dover edition and the 9-volume McGraw-Hill edition of the lectures is:

Dover Edition	McGraw-Hill Edition
Volume I: *Elements of Electrical Engineering*	Volume 3, pp. 1–120; Volume 4, pp. 1–206; Volume 5, pp. 1–228
Volume II: *Electric Waves and Impulses*	Volume 4, pp. 341–465; Volume 7, complete; Volume 8, pp. 3–219; Volume 4, pp. 466–473 (Appendix)
Volume III: *Transient Electric Phenomena*	Volume 8, pp. 223–683

Omitted from the Dover edition are Volume 1, which consists of general lectures too diffuse to be of present-day interest, Volumes 6 and 9, and parts of the other volumes which deal with rotating machines, with more complex theories, or with apparatus now obsolete.

These lectures present concepts usually studied by electrical engineers in the first three years of college. They should be readily understood by any intelligent technician or engineering apprentice, so that by his own efforts he can gain a working knowledge of electrical engineering.

Some necessary changes have been made in the original text: a number of misprints have been silently emended, and cross references have been altered so as to be useful to the reader of these three new volumes. The cross reference "III, ii, 25," for example, refers to volume III, Part II, Section 25 of the Dover edition. Where footnotes have been added they are differentiated from Steinmetz's own by the use of brackets.

Steinmetz occasionally refers the reader to his book *Engineering Mathematics* for a more detailed mathematical explanation than that given in the text. As mentioned above, this book is now available in a revised edition.

The reader should take note of the few changes in terminology that have occurred since Steinmetz's day:

1. Condensers are now called *capacitors*, condensance is

called *capacitance*, and condensive reactance is known as *capacitive reactance.*

2. A vector which indicates the phase position of a sinusoidally varying quantity is now known as a *phasor*; Steinmetz refers to these simply as vectors.

3. Wave screens are now called *wave filters.*

Philip L. Alger

Schenectady, New York
March, 1969

The books listed below will be useful as supplements to this work.

Alger, Philip L.: *Mathematics for Science and Engineering* (Based on *Engineering Mathematics* by C. P. Steinmetz), Second Edition; McGraw-Hill, 1969.

———: *The Nature of Induction Machines*; Gordon and Breach Science Publishers, 1965.

Bewley, L. V.: Alternating Current Machinery; The Macmillan Co., 1949.

Clement, P. R. and W. C. Johnson: *Electrical Engineering Science*; McGraw-Hill, 1960.

Concordia, C.: *Synchronous Machines*; John Wiley, 1951.

Dow, W. G.: *Fundamentals of Engineering Electronics*; John Wiley, 1952.

Fano, Chu, and Adler: *Electromagnetic Fields, Energy and Forces*; John Wiley, 1960.

Fitzgerald, A. E. and C. Kingsley, Jr.: *Electric Machinery*, Second Edition; McGraw-Hill, 1961.

Guillemin, E. A.: *Introductory Circuit Theory*; John Wiley, 1953.

Ham, J. M. and G. R. Slemon: *Scientific Basis of Electrical Engineering*; John Wiley, 1961.

Ku, Y. H.: *Electric Energy Conversion*; Ronald Press, 1959.

Lewis, W. W. and C. F. Goodheart: *Basic Electric Circuit Theory*; Ronald Press, 1958.

Majmudar, Harit: *Electromechanical Energy Converters*; Allyn and Bacon, 1965.

Messerle, H. K.: *Dynamic Circuit Theory*; Pergamon Press, 1965.

Scott, W. T.: *The Physics of Electricity and Magnetism*; John Wiley, 1959.

Slemon, G. R.: *Magnetoelectric Devices*; John Wiley, 1966.

Veinott, C. G.: *Fractional Horsepower Electric Motors*, Second Edition; McGraw-Hill, 1948.

Zimmerman, H. J. and S. J. Mason: *Electric Circuit Theory*; John Wiley, 1959.

The following books provide more information about the life and thought of C. P. Steinmetz.

Alger, Philip L. and E. Caldecott: *Steinmetz—The Philosopher*; Mohawk Development Service, Inc., P.O. Box 7, Bellevue Station, Schenectady, New York 12306.

Miller, Floyd: *The Electrical Genius of Liberty Hall*; McGraw-Hill, 1962.

Miller, J. A.: *Modern Jupiter*; The American Society of Mechanical Engineers, 1958.

PREFACE TO VOLUME III

In this third volume of Steinmetz's lectures, a great many kinds of transient electromagnetic phenomena are explained in physical terms and analyzed with the aid of mathematics.

The first part deals with periodic transients, such as those in voltage regulators and rectifiers, both mechanical and electrical. The second part is devoted to transients in space, particularly those occurring on long transmission lines. The variations in circuit constants due to high frequency are considered, including the depth of penetration of magnetic flux in laminations and of current in conductors. Also, the effects of the finite velocity of electric field propagation are studied, especially the resistance and inductance due to radiation.

The third part considers transients that occur in both time and space. The equations of the electric circuit are developed in their most general form, and from these simpler formulas are derived for a large number of special cases. Periodic and nonperiodic impulse currents, standing and traveling waves, free oscillations, and inductive discharges are considered in turn. In the final chapters wave attenuation and the flattening of steep wave fronts on transmission lines are studied.

Tables of the velocity functions used in the study of wave propagation in space are given in an appendix.

Philip L. Alger

Schenectady, New York
March, 1969

CONTENTS FOR VOLUME III

PART I
PERIODIC TRANSIENTS

CHAPTER I. INTRODUCTION

CHAPTER II. CIRCUIT CONTROL BY PERIODIC TRANSIENT PHENOMENA

CHAPTER III. MECHANICAL RECTIFICATION

CHAPTER IV. ARC RECTIFICATION

PART II
TRANSIENTS IN SPACE

PART I

PERIODIC TRANSIENTS

CHAPTER I

INTRODUCTION

1. Whenever in an electric circuit a sudden change of the circuit conditions is produced, a transient term appears in the circuit, that is, at the moment when the change begins, the circuit quantities, as current, voltage, magnetic flux, etc., correspond to the circuit conditions existing before the change, but do not, in general, correspond to the circuit conditions brought about by the change, and therefore must pass from the values corresponding to the previous condition to the values corresponding to the changed condition. This transient term may be a gradual approach to the final condition, or an approach by a series of oscillations of gradual decreasing intensities.

Gradually — after indefinite time theoretically, after relatively short time practically — the transient term disappears, and permanent conditions of current, of voltage, of magnetism, etc., are established. The numerical values of current, of voltage, etc., in the permanent state reached after the change of circuit conditions, in general, are different from the values of current, voltage, etc., existing in the permanent state before the change, since they correspond to a changed condition of the circuit. They may, however, be the same, or such as can be considered the same, if the change which gives rise to the transient term can be considered as not changing the permanent circuit conditions. For instance, if the connection of one part of a circuit, with regard to the other part of the circuit, is reversed, a transient term is produced by this reversal, but the final or permanent condition after the reversal is the same as before, except that the current, voltage, etc., in the part of the circuit which has been reversed, are now in opposite direction. In this latter case, the same change can be produced again and again after equal intervals of time t_0, and thus the transient term made to recur periodically. The electric quantities i, e, etc., of the circuit, from time $t = 0$ to $t = t_0$, have the same values as from time $t = t_0$ to $t = 2\,t_0$, from $t = 2\,t_0$ to $t = 3\,t_0$, etc., and it is sufficient to investigate one cycle, from $t = 0$ to $t = t_0$.

In this case, the starting values of the electrical quantities during each period are the end values of the preceding period, or, in other words, the terminal values at the moment of start of the transient term, $t = 0$, $i = i_0$ and $e = e_0$, are the same as the values at the end of the period $t = t_0$, $i = i'$ and $e = e'$; that is, $i_0 = \pm\, i'$, $e_0 = \pm\, e'$, etc.; where, the plus sign applies for the unchanged, and the minus sign for the reversed part of the circuit.

2. With such periodically recurrent changes of circuit conditions, the period of recurrence t_0 may be so long, that the transient term produced by a change has died out, the permanent conditions reached, before the next change takes place. Or, at the moment where a change of circuit conditions starts a transient term, the transient term due to the preceding change has not yet disappeared, that is, the time, t_0, of a period is shorter than the duration of the transient term.

In the first case, the terminal or starting values, that is, the values at the moment when the change begins, are the same as the permanent values, and periodic recurrence has no effect on the character of the transient term, but the phenomenon is calculated as discussed in "Transients in Time" [Dover Volume II, Part III], as single transient term, which gradually dies out.

If, however, at the moment of change, the transient term of the preceding change has not yet vanished, then the starting or terminal values of the electric quantities, as i_0 and e_0, also contain a transient term, namely, that existing at the end of the preceding period. The same term then exists also at the end of the period, or at $t = t_0$. Hence in this case, the terminal conditions are given, not as fixed numerical values, but as an equation between the electric quantities at time $t = 0$ and at time $t = t_0$; or, at the beginning and at the end of the period, and the integration constants, thus, are calculated from this equation.

3. In general, the permanent values of electric quantities after a change are not the same as before, and therefore at least two changes are required before the initial condition of the circuit is restored, and the cycle can be repeated. Periodically recurring transient phenomena, thus usually consist of two or more successive changes, at the end of which the original condition of the circuit is reproduced, and therefore the series of changes can be repeated. For instance, increasing the resistance of a circuit brings about a change. Decreasing this resistance again to its original value brings about a second change, which restores the condition existing before the first change, and thus completes the cycle. In this case, then, the starting values of the electric quantities during the first part of the period equal the end values during the second part of the period, and the starting values of the second part of the period equal the end values of the first part of the period. That is, if a resistor is inserted at time $t = 0$, short circuited at time $t = t_1$, and inserted again at time $t = t_0$, and e and i are voltage and current respectively during the first, e_1 and i_1 during the second part of the period, we have

$$/e/_{t=0} = /e_1/_{t=t_0}; \quad /e_1/_{t=t_1} = /e/_{t=t_1},$$

and

$$/i/_{t=0} = /i_1/_{t=t_0}; \quad /i_1/_{t=t_1} = /i/_{t=t_1}.$$

If during the times t_1 and $t_0 - t_1$ the transient terms have already vanished, and permanent conditions established, so that the transient terms of each part of the period depend only upon the permanent values during the other part of the period, the length of time t_1 and t_0 has no effect on the transient term, that is, each change of circuit conditions takes place and is calculated independently of the other change, or the periodic recurrence. A number of such cases have been discussed in "Transients in Time" [Dover Volume II, Part III], as for instance, the effect of cutting a resistor in and out of a divided inductive circuit, paragraph 75, Fig. 33 (page 403). In this case, four successive changes are made before the cycle recurs: a resistor is cut in, in two steps, and cut out again in two steps, but at each change, sufficient time elapses to reach practically permanent condition.

In general, and especially in those cases of periodic transient phenomena, which are of engineering importance, successive changes occur before the permanent condition is reached, or even approximated after the preceding change, so that frequently the values of the electric quantities are very different throughout the whole cycle from the permanent values which they would gradually assume; that is, the transient term preponderates in the values of current, voltage, etc., and the permanent term occasionally is very small compared with the transient term.

4. Periodic transient phenomena are of engineering importance mainly in three cases: (1) in the control of electric circuits; (2) in the production of high frequency currents, and (3) in the rectification of alternating currents.

1. In controlling electric circuits, etc., by some operating mechanism, as a potential magnet increasing and decreasing the resistance of the circuit, or a clutch shifting brushes, etc., the main objections are due to the excess of the friction of rest over the friction while moving. This results in a lack of sensitiveness, and an overreaching of the controlling device. To overcome the friction of rest, the deviation of the circuit from normal must become greater than necessary to maintain the motion of the operating mechanism, and when once started, the mechanism overreaches. This objection is eliminated by never allowing the operating mechanism to come to rest, but arranging it in unstable equilibrium, as a "floating system," so that the condition of the circuit is never normal, but continuously and periodically varies between the two extremes, and the resultant effect is the average of the transient terms, which rapidly and periodically succeed each other. By changing the relative duration of the successive transient terms, any resultant intermediary between the two extremes can thus be produced. On this principle, for instance, operated the controlling solenoid of the Thomson-Houston arc machine, and also numerous automatic potential regulators.

2. Production of high frequency oscillating currents by periodically recurring condenser discharges has been discussed under "oscillating current generator," in paragraph 44, Dover edition, Volume II, Part III, page 339.

High frequency alternating currents are produced by an arc, when made unstable by shunting it with a condenser, as discussed before.

The Ruhmkorff coil or inductorium also represents an application of periodically recurring transient phenomena, as also does Prof. E. Thomson's dynamostatic machine.

3. By reversing the connections between a source of alter-

nating voltage and the receiver circuit, synchronously with the alternations of the voltage, the current in the receiver circuit is made unidirectional (though more or less pulsating) and therefore rectified.

In rectifying alternating voltages, either both half waves of voltage can be taken from the same source, as the same transformer coil, and by synchronous reversal of connections sent in the same direction into the receiver circuit, or two sources of voltage, as the two secondary coils of a transformer, may be used, and the one half wave taken from the one source, and sent into the receiver circuit, the other half wave taken from the other source, and sent into the receiver circuit in the same direction as the first half wave. The latter arrangement has the disadvantage of using the alternating current supply source less economically, but has the advantage that no reversal, but only an opening and closing of connections, is required, and is therefore the method commonly applied in stationary rectifying apparatus.

5. In rectifying alternating voltages, the change of connections between the alternating supply and the unidirectional receiving circuit can be carried out as outlined below:

(*a*) By a synchronously moving commutator or contact maker, in mechanical rectification. Such mechanical rectifiers may again be divided, by the character of the alternating supply voltage, into single phase and polyphase, and by the character of the electric circuit, into constant potential and constant current rectifiers. Mechanical rectification by a commutator driven by a separate synchronous motor has not yet found any extensive industrial application. Rectification by a commutator driven by the generator of the alternating voltage has found very extended and important industrial use in the excitation of the field, or a part of the field (the series field) of alternators and synchronous motors, and especially in the constant-current arc machine. The Brush arc machine is a quarter-phase alternator connected to a rectifying commutator on the armature shaft, and the Thomson-Houston arc machine is a star-connected three-phase alternator connected to a rectifying commutator on the armature shaft. The reason for using rectification in these machines, which are intended to produce constant direct current at very high voltage, is that the ordinary commutator of the continuous-current machine cannot safely commutate, even at limited current, more than 30 to 50 volts per commutator segment, while the rectifying commutator of the constant-current arc machine can control from 2000 to 3000 volts per segment, and therefore rectification is superior to commutation for very high voltages at limited current, as explained by the character of this phenomenon, discussed in Chapter III.*

(*b*) The synchronous change of circuit connection required by the rectification of alternating e.m.fs. can be brought about without any mechanical motion in so-called "arc rectifiers," by the characteristic properties of the electric arc, to be a good conductor in one, an insulator in the opposite direction. By thus inserting an arc in the path of the alternating circuit, current can exist and thus a circuit be established for that half wave of alternating voltage, which sends the current in the same direction as the current in the arc, while for the reversed half wave of voltage the arc acts as open circuit. As seen, the arc cannot reverse, but only open and close the circuit, and so can rectify only one half wave, that is, two separate sources of alternating voltage, or two rectifiers with the same source of voltage, are required to rectify both half waves of alternating voltage.

(*c*) Some electrolytic cells, as those containing aluminum as one terminal, offer a low resistance to the passage of current in one direction, but a very high resistance, or practically interrupt the current, in opposite direction, due to the formation of a non-conducting film on the aluminum, when it is the positive terminal. Such electrolytic cells can therefore be used for rectification in a similar manner as arcs.

The three main classes of rectifiers thus are: (*a*) mechanical rectifiers; (*b*) arc rectifiers; (*c*) electrolytic rectifiers.

Still other methods of rectification, as by the unidirectional character of vacuum discharges, of the conduction in some crystals, etc., are not yet of industrial importance.

* Dover, Volume II, Part III.

CHAPTER II

CIRCUIT CONTROL BY PERIODIC TRANSIENT PHENOMENA

6. As an example of a system of periodic transient phenomena, used for the control of electric circuits, may be considered an automatic potential regulator operating in the field circuit of the exciter of an alternating current system.

Let, $r_0 = 40$ ohms = resistance and $L = 400$ henrys = inductance of the exciter field circuit.

A resistor, having a resistance, $r_1 = 24$ ohms, is inserted in series to r_0, L in the exciter field, and a potential magnet, controlled by the alternating current system, is arranged so as to short circuit resistance, r_1, if the alternating potential is below, to throw resistance r_1 into circuit again, if the potential is above normal.

With a single resistance step, r_1, in the one position of the regulator, with r_1 short circuited, and only r_0 as exciter field winding resistance, the alternating potential would be above normal, that is, the regulator cannot remain in this position, but as soon after short circuiting resistance r_1 as the potential has risen sufficiently, the regulator must change its position and cut resistance r_1 into the circuit, increasing the exciter field circuit resistance to $r_0 + r_1$. This resistance now is too high, would lower the alternating potential too much, and the regulator thus cuts resistance r_1 out again. That is, the regulator continuously oscillates between the two positions, corresponding to the exciter field circuit resistances r_0 and $(r_0 + r_1)$ respectively, at a period depending on the momentum of the moving mass, the force of the magnets, etc., that is, approximately constant. The time of contact in each of the two positions, however, varies: when requiring a high field excitation, the regulator remains a longer time in position r_0, hence a shorter time in position $(r_0 + r_1)$, before the rising potential throws it over into the next position; while at light load, requiring low field excitation, the duration of the period of high resistance,

$(r_0 + r_1)$, is greater, and that of the period of low resistance, r_0, less.

7. Let, t_1 = the duration of the short circuit of resistance r_1; t_2 = the time during which resistance r_1 is in circuit, and $t_0 = t_1 + t_2$.

During each period t_0, the resistance of the exciter field, therefore, is r_0 for the time t_1, and $(r_0 + r_1)$ for the time t_2.

Furthermore, let, i_1 = the current during time t_1, and i_2 = the current during time t_2.

During each of the two periods, let the time be counted anew from zero, that is, the transient current i_1 exists during the time $0 < t < t_1$, through the resistance r_0, the transient current, i_2, during the time $0 < t < t_2$, through the resistance $(r_0 + r_1)$.

This gives the terminal conditions:

$$\left.\begin{aligned} /i_1/_{t=o} &= /i_2/_{t=t_2} \\ \text{and} \quad /i_2/_{t=o} &= /i_1/_{t=t_1}; \end{aligned}\right\} \qquad (1)$$

that is, the starting point of the current, i_1, is the end value of the current, i_2, and inversely.

If now, e = voltage impressed upon the exciter field circuit, the differential equations are:

$$\left.\begin{aligned} e &= r_0 i_1 + L\frac{di_1}{dt} \\ \text{and} \quad e &= (r_0 + r_1)\, i_2 + L\frac{di_2}{dt}; \end{aligned}\right\} \qquad (2)$$

or,

$$\left.\begin{aligned} \frac{di_1}{i_1 - \frac{e}{r_0}} &= -\frac{r_0}{L}\,dt, \\ \frac{di_2}{i_2 - \frac{e}{r_0 + r_1}} &= -\frac{r_0 + r_1}{L}\,dt. \end{aligned}\right\} \qquad (3)$$

Integrated,

$$\left.\begin{aligned} i_1 &= \frac{e}{r_0} + c_1 \varepsilon^{-\frac{r_0}{L} t} \\ \text{and} \qquad & \\ i_2 &= \frac{e}{r_0 + r_1} + c_2 \varepsilon^{-\frac{r_0 + r_1}{L} t}. \end{aligned}\right\} \qquad (4)$$

Substituting the terminal conditions (1) in equations (4), gives for the integration constants c_1 and c_2 the equations,

$$\left.\begin{aligned} \frac{e}{r_0} + c_1 &= \frac{e}{r_0 + r_1} + c_2 \varepsilon^{-\frac{r_0+r_1}{L} t_2} \\ \text{and} \qquad & \\ \frac{e}{r_0 + r_1} + c_2 &= \frac{e}{r_0} + c_1 \varepsilon^{-\frac{r_0}{L} t_2}; \end{aligned}\right\}$$

herefrom,

$$\left.\begin{aligned} c_1 &= -\frac{e r_1 \left\{1 - \varepsilon^{-\frac{r_0+r_1}{L} t_2}\right\}}{r_0 (r_0 + r_1) \left\{1 - \varepsilon^{-\frac{r_0}{L} t_1 - \frac{r_0+r_1}{L} t_2}\right\}} \\ \text{and} \qquad & \\ c_2 &= +\frac{e r_1 \left\{1 - \varepsilon^{-\frac{r_0}{L} t_1}\right\}}{r_0 (r_0 + r_1) \left\{1 - \varepsilon^{-\frac{r_0}{L} t_1 - \frac{r_0+r_1}{L} t_2}\right\}}. \end{aligned}\right\} \qquad (5)$$

Substituting (5) in (4),

$$\left.\begin{aligned} i_1 &= \frac{e}{r_0}\left\{1 - \frac{r_1 \left\{1 - \varepsilon^{-\frac{r_0+r_1}{L} t_2}\right\}}{(r_0 + r_1) \left\{1 - \varepsilon^{-\frac{r_0}{L} t_1 - \frac{r_0+r_1}{L} t_2}\right\}} \varepsilon^{-\frac{r_0}{L} t}\right\} \\ \text{and} \qquad & \\ i_2 &= \frac{e}{r_0 + r_1}\left\{1 + \frac{r_1 \left\{1 - \varepsilon^{-\frac{r_0}{L} t_1}\right\}}{r_0 \left\{1 - \varepsilon^{-\frac{r_0}{L} t_1 - \frac{r_0+r_1}{L} t_2}\right\}} \varepsilon^{-\frac{r_0+r_1}{L} t}\right\}. \end{aligned}\right\} \qquad (6)$$

If, $e = 250$ volts; $t_0 = 0.2$ sec., or 5 complete cycles per sec. $t_1 = 0.15$, and $t_2 = 0.05$ sec.; then

$$\left.\begin{aligned} i_1 &= 6.25 \left\{1 - 0.128\, \varepsilon^{-0.1\, t}\right\} \\ \text{and} \qquad & \\ i_2 &= 3.91 \left\{1 + 0.391\, \varepsilon^{-0.16\, t}\right\}. \end{aligned}\right\} \qquad (7)$$

8. The mean value of current in the circuit is

$$i = \frac{1}{t_1 + t_2} \left\{ \int_0^{t_1} i_1 dt + \int_0^{t_2} i_2 dt \right\}. \tag{8}$$

This integrated gives,

$$i = \frac{t_1 \dfrac{e}{r_0} + t_2 \dfrac{e}{r_0 + r_1}}{t_1 + t_2}; \tag{9}$$

and, if

$$i_1' = \frac{e}{r_0}$$

and

$$i_2' = \frac{e}{r_0 + r_1} \tag{10}$$

are the two extreme values of permanent current, corresponding respectively to the resistances r_0 and $(r_0 + r_1)$, we have

$$i = \frac{t_1 i_1' + t_2 i_2'}{t_1 + t_2} = \frac{t_1}{t_0} i_1' + \frac{t_2}{t_0} i_2^1; \tag{11}$$

that is, the current, i, varies between i_1' and i_2' as linear function of the durations of contact, t_1 and t_2.

The maximum variation of current during the periodic change is given by the ratio of maximum current and minimum current; or,

$$\left|\frac{i_2}{i_1}\right|_{t=o} = q, \tag{12}$$

and is

$$q = \frac{r_0 (1 - \varepsilon^{-s_1 - s_2}) + r_1 (1 - \varepsilon^{-s_1})}{r_0 (1 - \varepsilon^{-s_1 - s_2}) + r_1 \varepsilon^{-s_2} (1 - \varepsilon^{-s_1})}; \tag{13}$$

where,

$$s_1 = \frac{r_0}{L} t_1$$

and

$$s_2 = \frac{r_0 + r_1}{L} t_2. \tag{14}$$

Substituting

$$1 - \varepsilon^{-x} = x - \frac{x^2}{2} + \frac{x^3}{6} - + \ldots, \tag{15}$$

by using only term of first order;

gives

$$\left.\begin{aligned} 1 - \varepsilon^{-x} &= x, \\ q &= 1; \end{aligned}\right\} \tag{16}$$

that is, the primary terms eliminate, and the difference between i_1 and i_2 is due to terms of secondary order only, hence very small.

Substituting

$$1 - \varepsilon^{-x} = x - \frac{x^2}{2}; \tag{17}$$

that is, using also terms of second order, gives

$$q = \frac{\{r_0(s_1 + s_2) + r_1 s_1\} - \frac{1}{2}\{r_0(s_1 + s_2)^2 + r_1 s_1{}^2\}}{\{r_0(s_1 + s_2) + r_1 s_1\} - \frac{1}{2}\{r_0(s_1 + s_2)^2 + r_1 s_1{}^2 + 2 r_1 s_1 s_2\}}; \tag{18}$$

or, approximately,

$$q = 1 + \frac{r_1 s_1 s_2}{r_0(s_1 + s_2) + r_1 s_1}, \tag{19}$$

and, substituting (14),

$$q = 1 + \frac{r_1 t_1 t_2}{L(t_1 + t_2)}; \tag{20}$$

that is, the percentage variation of current is

$$q - 1 = \frac{r_1 t_1 t_2}{L(t_1 + t_2)}. \tag{21}$$

Equation (21) is a maximum for

$$t_1 = t_2 = \frac{t_0}{2}, \tag{22}$$

and, then, is

$$q - 1 = \frac{r_1 t_0}{4L}; \tag{23}$$

or, in the above example, $(r_1 = 24;\ L = 400;\ t_0 = 0.2)$;

$$q - 1 = 0.003;$$

that is, 0.3 per cent.

The time t_0 of a cycle, which gives 1 per cent variation of current, $q - 1 = 0.01$, is

$$t_0 = \frac{4L}{r_1}(q - 1), \tag{24}$$

$$= \tfrac{2}{3} \text{ sec.}$$

The pulsation of current, 0.3 per cent respectively 1 per cent, thus is very small compared with the pulsation of the resistance, $r_1 = 24$ ohms, which is 46 per cent of the average resistance $r_0 + \frac{r_1}{2} = 52$ ohms.

CHAPTER III.

MECHANICAL RECTIFICATION.

9. If an alternating-current circuit is connected, by means of a synchronously operated circuit breaker or rectifier, with a second circuit in such a manner, that the connection between the two circuits is reversed at or near the moment when the alternating voltage passes zero, then in the second circuit current and voltage are more or less unidirectional, although they may not be constant, but pulsating.

If i = instantaneous value of alternating current, and i_0 = instantaneous value of rectified current, then we have, before reversal, $i_0 = i$, and after reversal, $i_0 = -i$; that is, during the reversal of the circuit one of the currents must reverse. Since, however, due to the self-inductance of the circuits, neither current can reverse instantly, the reversal occurs gradually, so that for a while during rectification the instantaneous value of the alternating and of the rectified current differ from each other. Thus means have to be provided either to shunt the difference between the two currents through a non-inductive bypath, or, the difference of the two currents exists as arc over the surface of the rectifying commutator.*

The general phenomenon of single-phase rectification thus is: The alternating and the rectified circuit are in series. Both circuits are closed upon themselves at the rectifier, by the resistances, r and r_0, respectively. The terminals are reversed. The shunt-resistance circuits are opened, leaving the circuits in series in opposite direction.

Special cases hereof are:

1. If $r = r_0 = 0$, that is, during rectification both circuits are short circuited. Such short-circuit rectification is feasible only in limited-current circuits, as on arc lighting machines, or circuits of high self-inductance, or in cases where the voltage of the recti-

* If the circuit is reversed at the moment when the alternating current passes zero, due to self-inductance of the rectified circuit its current differs from zero, and an arc still appears at the rectifier.

fied circuit is only a small part of the total voltage, and thus the current not controlled thereby, as when rectifying for the supply of series fields of alternators.

2. $r = r_0 = \infty$, or open circuit rectification. This is feasible only if the rectified circuit contains practically no self-inductance, but a constant counter e.m.f., e, (charging storage batteries), so that in the moment when the alternating impressed e.m.f. falls to e, and the current disappears, the circuit is opened, and closed again in opposite direction when after reversal the alternating impressed e.m.f. has reached the value, e.

In polyphase rectification, the rectified circuit may be fed successively by the successive phases of the system, that is shifted over from a phase of falling e.m.f. to a phase of rising e.m.f., by shunting the two phases with each other during the time the current changes from the one to the next phase. Thus the Thomson-Houston arc machine is a star-connected three-phase constant-current alternator with rectifying commutator. The Brush arc machine is a quarter-phase machine with rectifying commutator.

In rectification frequently the sine wave term of the current is entirely overshadowed by the transient exponential term, and thus the current in the rectified circuit is essentially of an exponential nature.

As examples, three cases will be discussed:

1. Single-phase constant-current rectification; that is, a rectifier is inserted in an alternating-current circuit, and the voltage consumed by the rectified circuit is small compared with the total circuit voltage; the current thus is not noticeably affected by the rectifier. In other words, a sine wave of current is sent over a rectifying commutator.

2. Single-phase constant-potential rectification; that is, a constant-potential alternating e.m.f. is rectified, and the impedance between the alternating voltage and the rectifying commutator is small, so that the rectified circuit determines the current wave shape.

3. Quarter-phase constant-current rectification as occurring in the Brush arc machine.

I. Single-phase constant-current rectification.

10. A sine wave of current, $i_0 \sin \theta$, derived from an e.m.f. very large compared with the voltage consumed in the rectified circuit, feeds, after rectification, a circuit of impedance $Z = r - jx$. This circuit is permanently shunted by a circuit of resistance r_1.

Rectification takes place over short-circuit from the moment $\pi - \theta_2$ to $\pi + \theta_1$; that is, at $\pi - \theta_2$ the rectified and the alternating circuit are closed upon themselves at the rectifier, and this short-circuit opened, after reversal, at $\pi + \theta_1$, as shown by the diagrammatic representation of a two-pole model of such a rectifier in Fig. 1. In this case the space angles $\pi + \tau_1$ and $\pi - \tau_2$ and the time angles $\pi + \theta_1$ and $\pi - \theta_2$ are identical.

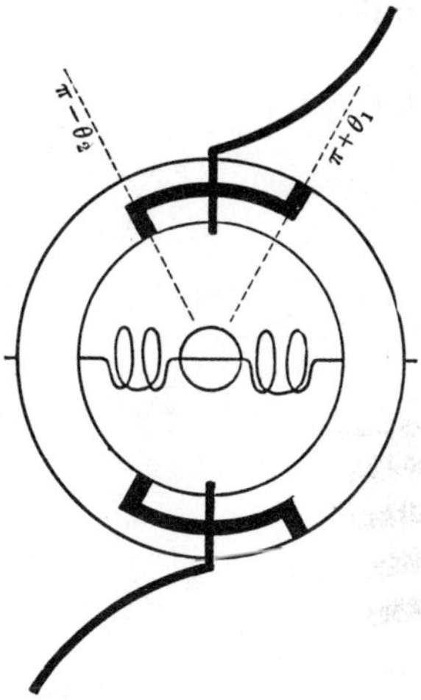

FIG. 1. Single-phase current rectifier commutator.

This represents the conditions existing in compound-wound alternators, that is, alternators feeding a series field winding through a rectifier.

Let, during the period from θ_1 to $\pi - \theta_2$, i = current in impedance Z, and i_1 = current in resistance r_1, then:

$$i + i_1 = i_0 \sin \theta. \tag{1}$$

However,

$$i_1 r_1 = ir + x \frac{di}{d\theta} \tag{2}$$

and substituting (1) in (2) gives the differential equation:

$$i\,(r + r_1) + x \frac{di}{d\theta} - i_0 r_1 \sin \theta = 0, \tag{3}$$

which is integrated by the function:

$$i = A\varepsilon^{-a\theta} + B \sin (\theta - \delta). \tag{4}$$

Substituting (4) in (3) and arranging, gives:

$$A\,(r + r_1 - ax)\,\varepsilon^{-a\theta} + [B\,([r + r_1] \cos \delta + x \sin \delta) - i_0 r_1] \sin \theta - [(r + r_1) \sin \delta - x \cos \delta]\,B \cos \theta = 0, \tag{5}$$

which equation must be an identity, thus:

$$\left.\begin{aligned} r + r_1 - ax &= 0, \\ B\left([r + r_1]\cos\delta + x\sin\delta\right) - i_0 r_1 &= 0 \\ \text{and} \qquad (r + r_1)\sin\delta - x\cos\delta &= 0, \end{aligned}\right\}$$

and herefrom:

$$\left.\begin{aligned} a &= \frac{r + r_1}{x}, \\ \tan\delta &= \frac{x}{r + r_1} \\ \text{and} \qquad B &= i_0 \frac{r_1}{\sqrt{(r + r_1)^2 + x^2}} = \frac{i_0 r_1}{z}, \end{aligned}\right\} \tag{6}$$

where

$$z = \sqrt{(r + r_1)^2 + x^2}; \tag{7}$$

hence:

$$i = A\varepsilon^{-\frac{r + r_1}{x}\theta} + i_0 \frac{r_1}{z}\sin(\theta - \delta). \tag{8}$$

During the time of short-circuit, from $\pi - \theta_2$ to $\pi + \theta_1$, if i' = current in impedance Z, we have

$$i'r + x\frac{di'}{d\theta} = 0, \tag{9}$$

hence:

$$i' = A'\varepsilon^{-\frac{r}{x}\theta}. \tag{10}$$

The condition of sparkless rectification is, that no sudden change of current occur anywhere in the system. In consequence hereof we must have:

$$i = i' = i_0 \sin\theta \text{ at the moment } \theta = \pi - \theta_2,$$

and, at the moment $\theta = \pi + \theta_1$, i' must have reached the same value as i and $i_0 \sin\theta$ at the moment $\theta = \theta_1$.

This gives the two double equations:

$$\left.\begin{aligned} i_{\pi-\theta_2} &= i'_{\pi-\theta_2} = i_0 \sin(\pi - \theta_2) \\ \text{and} \qquad i_{\theta_1} &= i'_{\pi+\theta_1} = i_0 \sin \theta_1; \end{aligned}\right\} \tag{11}$$

or, substituting (8) and (9),

$$A\varepsilon^{-\frac{r+r_1}{x}(\pi-\theta_2)} + i_0 \frac{r_1}{z} \sin(\delta + \theta_2) = A'\varepsilon^{-\frac{r}{x}(\pi-\theta_2)} = i_0 \sin \theta_2 \tag{12}$$

and

$$A\varepsilon^{-\frac{r+r_1}{x}\theta_1} - i_0 \frac{r_1}{z} \sin(\delta - \theta_1) = A'\varepsilon^{-\frac{r}{x}(\pi+\theta_1)} = i_0 \sin \theta_1. \tag{13}$$

These four equations (12) (13) determine four of the five quantities, A, A', θ_1, θ_2, r_1, leaving one indeterminate.

Thus, one of these five quantities can be chosen. The determination of the four remaining quantities, however, is rather difficult, due to the complex character of equations (12) (13), and is feasible only by approximation, in a numerical example.

11. Example: Let an alternating current of effective value of 100 amp., that is, of maximum value $i_0 = 141.4$, be rectified for the supply of a circuit of impedance $Z = 0.2 - 2j$, shunted by a non-inductive circuit of resistance r_1.

Let the series connection of the rectified and alternating circuits be established 30 time-degrees after the zero value of alternating current, that is, $\theta_1 = 30 \text{ deg.} = \frac{\pi}{6}$ chosen.

Then, from equation (13), we have

$$A'\varepsilon^{-\frac{r}{x}(\pi+\theta_1)} = i_0 \sin \theta_1,$$

hence, substituting r, x, θ_1, i_0, gives

$$A' = 102.$$

From equation (12),

$$A'\varepsilon^{-\frac{r}{x}(\pi-\theta_2)} = i_0 \sin \theta_2,$$

and, substituting,

$$\sin \theta_2 = 0.527\, \varepsilon^{0.1\theta_2};$$

approximately

$$\sin \theta_2 = 0.527 \text{ and } \theta_2 = 32°;$$

thus, more closely

$$\sin \theta_2 = 0.527\ \epsilon^{3.20°} = 0.558, \text{ and } \theta_2 = 34°;$$

thus, more closely

$$\sin \theta_2 = 0.527\ \varepsilon^{3.4°} = 0.559, \quad \text{and} \quad \theta_2 = 34°.$$

From equations (12) and (13) it follows:

$$A\varepsilon^{-\frac{r+r_1}{x}(\pi-\theta_2)} + i_0 \frac{r_1}{z} \sin(\delta + \theta_2) = i_0 \sin \theta_2,$$

$$A\varepsilon^{-\frac{r+r_1}{x}\theta_1} - i_0 \frac{r_1}{z} \sin(\delta - \theta_1) = i_0 \sin \theta_1;$$

eliminating A gives

$$\varepsilon^{-\frac{r}{x}(\pi-\theta_1-\theta_2)}\varepsilon^{-\frac{r_1}{x}(\pi-\theta_1-\theta_2)} = \frac{z \sin \theta_2 - r_1 \sin(\delta + \theta_2)}{z \sin \theta_1 + r_1 \sin(\delta - \theta_1)};$$

substituting $\sin \delta = \frac{x}{z}$, $\cos \delta = \frac{r + r_1}{z}$, $z^2 = (r + r_1)^2 + x^2$, and substituting for r, x, θ_1, θ_2, gives after some changes:

$$\varepsilon^{-58 r_1°} = \frac{1.5 - 1.04\, r_1}{1.1 - r_1};$$

calculating by approximation,

assuming $r_1 = 0.5$,

$$0.603 = 0.612;$$

assuming $r_1 = 0.51$,

$$0.597 = 0.602;$$

assuming $r_1 = 0.52$,

$$0.591 = 0.592;$$

hence, $r_1 = 0.52$,

and $z = 2.124$,

$$\delta = 70°.$$

Substituting these values in (12) or (13) gives

$$A = 114;$$

hence, as final equations, we have

$$i = 112\,\varepsilon^{-0.36\,\theta} + 34.6 \sin(\theta - 70°),$$
$$i' = 102\,\varepsilon^{-0.1\,\theta},$$
$$i_0 = 141.4 \sin \theta,$$

and

$$i_1 = i_0 - i;$$

which gives the following results:

Quantity.	Instantaneous Values.										Effective Value.	Arithmetic Mean Value.
$\theta° =$	30	50	70	90	110	130	146	170	190	210	......	.....
$i =$	70.8	70.5	72.6	74.9	79.0	80.4	79.0	......	......	......	75.2	75.2
$i' =$							79.0	75.8	73.2	70.8		
$i_0 \sin \theta =$	70.8	108	133	141.4	133	108	79.0	24.7	−24.7	−70.8	100.0	.....
$i_1 =$	0	37.5	60.4	66.5	54.0	27.6	0	(−51.1	−48.5)	0	38.2	27.3
	....	....	....	.....	....	....	....	....	....	......	(44.9)	

Curves of these quantities are plotted in Fig. 2, for $i_0 = 100 \sin \theta$.

The effective value of the rectified current is 75.2 amp., and this current is fairly constant, pulsating only between 70.5 and 80.4 amp., or by 6.6 per cent from the mean; that is, due to the self-inductance, the fluctuations of current are practically suppressed, and taken up by the non-inductive shunt, and the arithmetic mean value of this current is therefore equal to its effective value. The effective value of the shunt current is 38.2 amp., and this current is unidirectional also, but very fluctuating. Its arithmetic mean value is only 27.3 amp.; that is, in this circuit a continuous-current ammeter would record 27.3, an alternating ammeter 38.2 amperes. The effective value of the total difference between alternating and rectified current (shunt plus short-circuit current) is 44.9 amp.

The current divides between the inductive rectified circuit and its non-inductive shunt, not in proportion to their respective impedances, but more nearly, though not quite, in proportion to the resistances; that is, in a rectified circuit, self-inductance does not greatly affect the intensity of the current, but only its character as regards fluctuations.

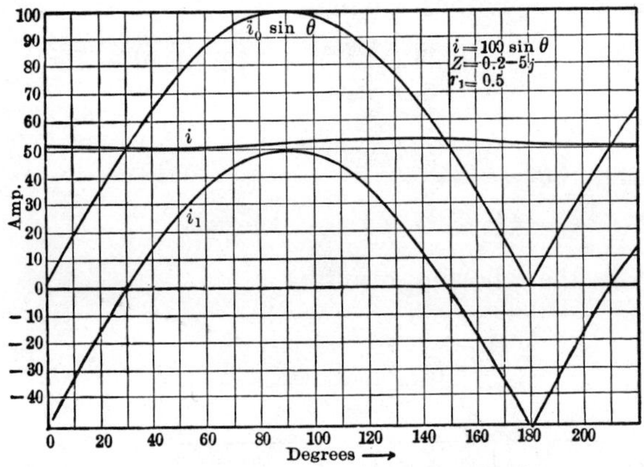

FIG. 2. Single-phase current rectification.

2. Single-phase constant-potential rectification.

12. Let the alternating e.m.f. $e_0 \sin\theta$ of the alternating circuit of impedance $Z_0 = r_0 - jx_0$ be rectified by connecting it at the moment θ_1 with the direct-current receiver circuit of impedance $Z = r - jx$ and continuous counter e.m.f. e, disconnecting it therefrom at the moment $\pi - \theta_2$, and closing during the time from $\pi - \theta_2$ to $\pi + \theta_1$ the alternating circuit by the resistance r_1, the direct-current circuit by the resistance r_2, then connecting the circuits again in series in opposite direction, at $\pi + \theta_1$, etc., as shown diagrammatically by Fig. 3, where

$$r_1 = \frac{1}{\dfrac{1}{r' + r''} + \dfrac{1}{r''' + r''''}};$$

$$r_2 = \frac{1}{\dfrac{1}{r' + r''''} + \dfrac{1}{r'' + r'''}}.$$

1. Then, during the time from θ_1 to $\pi - \theta_2$, if i_1 = current, the differential equation is

$$e_0 \sin\theta - e - i_1 (r + r_0) - (x + x_0) \frac{di_1}{d\theta} = 0, \tag{1}$$

which is integrated by

$$i_1 = A_1 + B_1\varepsilon^{-a_1\theta} + C_1 \sin(\theta - \delta_1) \qquad (2)$$

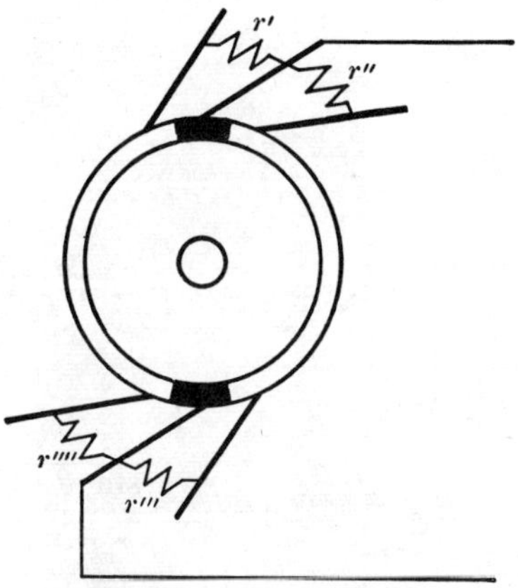

FIG. 3. Single-phase constant-potential rectifying commutator.

Equation (2) substituted in (1) gives

$$e_0 \sin\theta - e - (r + r_0)[A_1 + B_1\varepsilon^{-a_1\theta} + C_1 \sin(\theta - \delta_1)] - (x + x_0)[-a_1B_1\varepsilon^{-a_1\theta} + C_1 \cos(\theta - \delta_1)] = 0;$$

or, transposing,

$$-[+e + (r + r_0)A_1] + B_1\varepsilon^{-a\theta}[a_1(x + x_0) - (r + r_0)] + \sin\theta[e_0 - (r + r_0)C_1 \cos\delta_1 - (x + x_0)C_1 \sin\delta_1] + C_1 \cos\theta[(r + r_0)\sin\delta_1 - (x + x_0)\cos\delta_1] = 0;$$

herefrom it follows that

$$e + (r + r_0)A_1 = 0,$$
$$a_1(x + x_0) - (r + r_0) = 0,$$
$$e_0 - (r + r_0)C_1 \cos\delta_1 - (x + x_0)C_1 \sin\delta_1 = 0,$$

and

$$(r + r_0)\sin\delta_1 - (x + x_0)\cos\delta_1 = 0;$$

hence

$$\left.\begin{aligned} A_1 &= -\frac{e}{r+r_0}, \\ a_1 &= \frac{r+r_0}{x+x_0}, \\ \tan\delta_1 &= \frac{x+x_0}{r+r_0}, \\ \text{and} \qquad C_1 &= \frac{e_0}{\sqrt{(r+r_0)^2+(x+x_0)^2}}, \end{aligned}\right\} \tag{3}$$

and, substituting in (2),

$$\left.\begin{aligned} i_1 &= -\frac{e}{r+r_0} + B_1\varepsilon^{-\frac{r+r_0}{x+x_0}\theta} + \frac{e_0\sin(\theta-\delta_1)}{\sqrt{(r+r_0)^2+(x+x_0)^2}}, \\ &= -\frac{e}{r+r_0} + B_1\varepsilon^{-\frac{r+r_0}{x+x_0}\theta} + \frac{e_0[(r+r_0)\sin\theta-(x+x_0)\cos\theta]}{(r+r_0)^2+(x+x_0)^2}, \\ \tan\delta_1 &= \frac{x+x_0}{r+r_0}. \end{aligned}\right\} \tag{4}$$

2. During the time from $\pi - \theta_2$ to $\pi + \theta_1$, if i_2 = current in the direct circuit, i_3 = current in alternating circuit, we have

Alternating-current circuit:

$$e_0 \sin\theta - i_3(r_0+r_1) - x_0\frac{di_3}{d\theta} = 0, \tag{5}$$

which is integrated the same as in (1), by

$$\left.\begin{aligned} i_3 &= B_3\varepsilon^{-\frac{r_0+r_1}{x_0}\theta} + \frac{e_0\sin(\theta-\delta_3)}{\sqrt{(r_0+r_1)^2+x_0^2}} \\ &= B_3\varepsilon^{-\frac{r_0+r_1}{x_0}\theta} + \frac{e_0[(r_0+r_1)\sin\theta - x_0\cos\theta]}{(r_0+r_1)^2+x_0^2}, \\ \tan\delta_3 &= \frac{x_0}{r_0+r_1}. \end{aligned}\right\} \tag{6}$$

Direct-current circuit:

$$-e - i_2(r + r_2) - x\frac{di_2}{d\theta} = 0, \tag{7}$$

integrated by

$$i_2 = -\frac{e}{r + r_2} + B_2\varepsilon^{-\frac{r + r_2}{x}\theta} \tag{8}$$

At $\theta = \pi - \theta_2$, however, we must have

$$\left.\begin{aligned} &i_1 = i_2 = i_3 \\ \text{and}\quad &i_2 \text{ at } \theta = \pi + \theta_1 \text{ must be equal to} \\ &i_1 \text{ at } \theta = \theta_1, \text{ and opposite to } i_3 \text{ at } \theta = \pi + \theta_1; \\ &i_1[\theta = \theta_1] = i_2[\theta = \pi + \theta_1] = -i_3[\theta = \pi + \theta_1]. \end{aligned}\right\} \tag{9}$$

These terminal conditions represent four equations, which suffice for the determination of the three remaining integration constants, B_1, B_2, B_3, and one further constant, as θ_1 or θ_2, or r_1 or r_2, or e; that is, with the circuit conditions Z_0, Z, r_1, r_2, e_0, e chosen, the moment θ_1 depends on θ_2 and inversely.

13. *Special case:*

$$Z_0 = 0, \quad r_2 = 0, \quad e = 0; \tag{10}$$

that is, the alternating e.m.f. $e_0 \sin\theta$ is connected to the circuit of impedance $Z = r - jx$ during time θ_1 to $\pi - \theta_2$, and closed by resistance r_1, while the rectified circuit is short-circuited, during time $\pi - \theta_2$ to $\pi + \theta_1$.

The equations are:

$$\left.\begin{aligned} &1.\ \text{Time } \theta_1 \text{ to } \pi - \theta_2: \\ &\qquad i_1 = B_1\varepsilon^{-\frac{r}{x}\theta} + \frac{e_0}{r^2 + x^2}[r\sin\theta - x\cos\theta]. \\ &2.\ \text{Time } \pi - \theta_2 \text{ to } \pi + \theta_1: \\ &\qquad i_2 = B_2\varepsilon^{-\frac{r}{x}\theta} \\ &\qquad i_3 = \frac{e_0\sin\theta}{r_1}. \end{aligned}\right\} \tag{11}$$

The terminal conditions now assume the following forms:

At
$$\theta = \pi - \theta_2,$$
$$i_1 = i_2 = i_3$$

$$\left.\begin{aligned} B_1\varepsilon^{-\frac{r}{x}(\pi-\theta_2)} + \frac{e_0}{r^2+x^2}(r\sin\theta_2 + x\cos\theta_2) &= B_2\varepsilon^{-\frac{r}{x}(\pi-\theta_2)} \\ &= \frac{e_0}{r_1}\sin\theta_2; \\ \text{at } \theta = \pi + \theta_1 \text{ and } \theta_1 \text{ respectively} \\ B_1\varepsilon^{-\frac{r}{x}\theta_1} + \frac{e_0}{r^2+x^2}(r\sin\theta_1 - x\cos\theta_1) &= B_2\varepsilon^{-\frac{r}{x}(\pi+\theta_1)} \\ &= \frac{e_0}{r_1}\sin\theta_1. \end{aligned}\right\} \quad (12)$$

These four equations suffice for the determination of the two integration constants B_1 and B_2, and two of the three rectification constants, θ_1, θ_2, r_1, so that one of the latter may be chosen.

Choosing θ_2, the moment of beginning reversal, the equations (12) transposed and expanded give

$$\left.\begin{aligned} \varepsilon^{-\frac{r}{x}(\theta_1+\theta_2)} &= \frac{\sin\theta_1}{\sin\theta_2}, \\ \cot\theta_1 + \varepsilon^{\frac{r\pi}{x}}\cot\theta_2 &= \left(\frac{r^2+x^2}{r_1x} - \frac{r}{x}\right)\left(\varepsilon^{\frac{r\pi}{x}} - 1\right), \\ B_2 &= \frac{e_0\sin\theta_2}{r_1}\varepsilon^{+\frac{r}{x}(\pi-\theta_2)}, \\ \text{and} \quad B_1 &= B_2 - \frac{e_0}{r^2+x^2}(r\sin\theta_2 + x\cos\theta_2)\varepsilon^{\frac{r}{x}(\pi-\theta_2)}, \end{aligned}\right\} \quad (13)$$

which give θ_1, r_1, B_2, B_1: θ_1 is calculated by approximation.

Assuming, as an example,

$$\left.\begin{aligned} e_0 &= 156\sin\theta \text{ (corresponding to 110 volts effective)}, \\ Z &= 10 - 30\,j, \\ \text{and} \quad \theta_2 &= \frac{\pi}{6} = 30^\circ, \end{aligned}\right\} \quad (14)$$

by equations (13) we have:

$$\log \sin \theta_1 = -0.3765 - 0.1448\, \theta_1,$$

$$\left.\begin{aligned} \text{and} \quad \theta_1 &= 21.7°, \\ r_1 &= 7.63, \\ B_2 &= 24.4, \\ \text{and} \quad B_1 &= 12.8; \end{aligned}\right\} \tag{15}$$

thus

$$\left.\begin{aligned} i_1 &= 12.8\, \varepsilon^{-\frac{\theta}{3}} + 1.56\, (\sin \theta - 3 \cos \theta), \\ i_2 &= 24.4\, \varepsilon^{-\frac{\theta}{3}}, \\ \text{and} \quad i_3 &= 20.5 \sin \theta, \end{aligned}\right\} \tag{16}$$

which gives:

$\theta°$.	i_1.	i_2.	i_3.	$\theta°$.	i_1.	i_2.	i_3.
21.7	7.55	....	......	135	10.27	....	....
30	7.47	....	......	150	10.20	10.2	10.2
45	7.7	....	......	165	....	9.4	5.3
60	8.02	....	......	180	....	8.6	0
75	8.56	....	......	195	....	7.9	−5.3
90	9.18	....	......	201.7	....	7.55	−7.55
105	9.67	....	......	....	....	....	....
120	10.09	....	......	....	....	....	....

The mean value of the rectified current is derived herefrom as 8.92 amp., while without rectification the effective value of alternating current would be $\frac{110}{\sqrt{r^2 + x^2}} = 3.48$. 110 volts effective corresponds to $\frac{2\sqrt{2}}{\pi}\, 110 = 99$ volts mean, which in $r = 10$ would give the current as 9.9 amp.

Thus, in a rectified circuit, self-inductance has little effect besides smoothing out the fluctuations of current, which in this case varies between 7.47 and 10.27, with 8.92 as mean, while without self-inductance it would vary between 0 and 15.6, with 9.9 as mean, and without rectification the current would be $4.95 \sin (\theta - 71.6°)$.

As seen, in this case the exponential or transient term of current largely preponderates over the permanent or sinusoidal term.

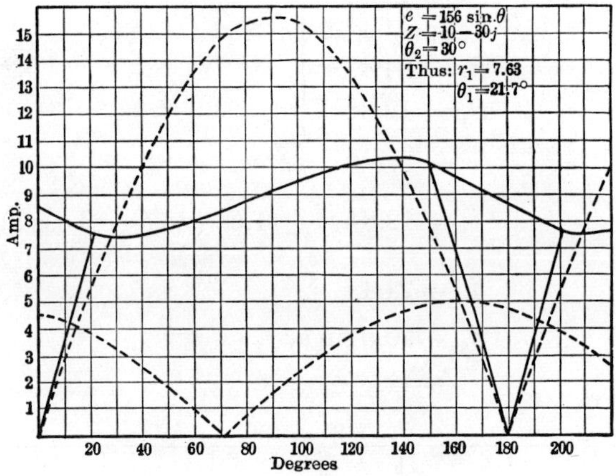

FIG. 4. Single-phase e.m.f. rectification.

In Fig. 4 is shown the rectified current in drawn line, the value it would have without self-inductance, and the value the alternating current would have, in dotted lines.

3. Quarter-phase constant-current rectification.

14. In the quarter-phase constant-current arc machine, as the Brush machine, two e.m.fs., $E_1 = e \cos \theta$ and $E_2 = e \sin \theta$, are connected to a rectifying commutator, so that while the first E_1 is in circuit E_2 is open-circuited. At the moment θ_1, E_2 is connected in parallel, as shown diagrammatically in Fig. 58, with E_1, and the rising e.m.f. in E_2 gradually shifts the current i_0 away from E_1 into E_2, until at the moment θ_2, E_1 is disconnected and E_2 left in circuit.

Assume that, due to the superposition of a number of such quarter-phase e.m.fs., displaced in time-phase from each other, and rectified by a corresponding number of commutators offset against each other, and due to self-inductance in the external circuit, the rectified current is practically steady and has the value i_0. Thus up to the moment θ_1 the current in E_1 is i_0, in

E_2 is 0. From θ_1 to θ_2 the current in E_2 may be i; thus in E_1 it is $i_2 = i_0 - i$. After θ_2, the current in E_1 is 0, in E_2 it is i_0.

A change of current occurs only during the time from θ_1 to θ_2, and it is only this time that needs to be considered.

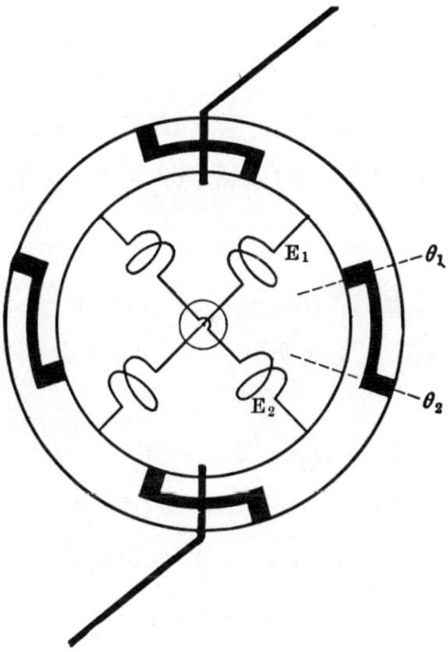

FIG. 5. Quarter-phase constant-current rectifying commutator.

Let $Z = r - jx$ = impedance per phase, where $x = 2\pi fL$; then at the time t and the corresponding angle $\theta = 2\pi ft$ the difference of potential in E_1 is

$$\left.\begin{array}{l} e\cos\theta - (i_0 - i)\,r - L\dfrac{d\,(i_0 - i)}{dt} \\ \quad = e\cos\theta - (i_0 - i)\,r + x\dfrac{di}{d\theta}; \\ \text{the difference of potential in } E_2 \text{ is} \\ e\sin\theta - ir - x\dfrac{di}{d\theta}; \end{array}\right\} \qquad (1)$$

and, since these two potential differences are connected in parallel, they are equal

$$e\,(\sin\theta - \cos\theta) + i_0 r - 2\,ir - 2\,x\frac{di}{d\theta} = 0. \qquad (2)$$

The differential equation (2) is integrated by

$$i = A + B\varepsilon^{-a\theta} + C\cos(\theta - \delta); \tag{3}$$

thus

$$\frac{di}{d\theta} = -aB\varepsilon^{-a\theta} - C\sin(\theta - \delta),$$

and substituting in (2),

$$e(\sin\theta - \cos\theta) + i_0 r - 2Ar - 2Br\varepsilon^{-a\theta} - 2Cr\cos(\theta - \delta) + 2aBx\varepsilon^{-a\theta} + 2Cx\sin(\theta - \delta) = 0\cdot$$

or, transposed,

$$(i_0 - 2A)r + 2B\varepsilon^{-a\theta}(ax - r) + \sin\theta[e - 2Cr\sin\delta + 2Cx\cos\delta] - \cos\theta[e + 2Cr\cos\delta + 2Cx\sin\delta] = 0;$$

thus

$$i_0 - 2A = 0,$$

$$ax - r = 0,$$

$$e - 2Cr\sin\delta + 2Cx\cos\delta = 0,$$

and

$$e + 2Cx\sin\delta + 2Cr\cos\delta = 0,$$

and herefrom, letting $\frac{x}{r} = \tan\sigma$, we have

$$\left.\begin{aligned} e &= -2Cz\sin(\sigma - \delta), \\ e &= -2Cz\cos(\sigma - \delta), \\ A &= \frac{i_0}{2}, \\ a &= \frac{r}{x}, \\ \tan\tau &= \frac{x - r}{x + r}, \\ C &= -\frac{e}{\sqrt{2(x^2 + r^2)}}, \\ \tan(\sigma - \delta) &= 1, \\ \text{and} \quad C &= \frac{e}{\sqrt{2}z}\cdot \end{aligned}\right\} \tag{4}$$

These values substituted in (3) give

$$i = \frac{i_0}{2} + B\varepsilon^{-\frac{r}{x}\theta} - \frac{e}{\sqrt{2(x^2 + r^2)}} \cos(\theta - \delta),$$

$$\tan \delta = \frac{x - r}{x + r}\cdot \tag{5}$$

At $\theta = \theta_1$, $i = 0$, and we have

$$0 = \frac{i_0}{2} + B\varepsilon^{-\frac{r}{x}\theta_1} - \frac{e}{\sqrt{2(x^2 + r^2)}} \cos(\theta_1 - \delta);$$

hence,

$$B\varepsilon^{-\frac{r}{x}\theta_1} = \frac{e}{\sqrt{2(x^2 + r^2)}} \cos(\theta_1 - \delta) - \frac{i_0}{2}; \tag{6}$$

substituting in (5), we have the equations of current in the two coils as follows:

$$\left.\begin{aligned} i &= \frac{i_0}{2} + \left(\frac{e}{\sqrt{2(x^2 + r^2)}} \cos(\theta_1 - \delta) - \frac{i_0}{2}\right)\varepsilon^{-\frac{r}{x}(\theta - \theta_1)} \\ &\quad - \frac{e}{\sqrt{2(x^2 + r^2)}} \cos(\theta - \delta) \\ &= \frac{i_0}{2}\left(1 - \varepsilon^{-\frac{r}{x}(\theta - \theta_1)}\right) \\ &\quad - \frac{e}{\sqrt{2(x^2 + r^2)}}\left(\cos(\theta - \delta) - \cos(\theta_1 - \delta)\,\varepsilon^{-\frac{r}{x}(\theta - \theta_1)}\right)\theta_1; \\ \text{and} \qquad & \\ i_0 - i &= \frac{i_0}{2} - \left(\frac{e}{\sqrt{2(x^2 + r^2)}} \cos(\theta_1 - \delta) - \frac{i_0}{2}\right)\varepsilon^{-\frac{r}{x}(\theta - \theta_1)} \\ &\quad + \frac{e}{\sqrt{2(x^2 + r^2)}} \cos(\theta - \delta). \end{aligned}\right\} \tag{7}$$

At $\theta = \theta_2$, $i = i_0$; thus

$$\frac{i_0}{2} - \left(\frac{e}{\sqrt{2\,(x^2 + r^2)}} \cos\,(\theta_1 - \delta) - \frac{i_0}{2}\right) \varepsilon^{-\frac{r}{x}(\theta_2 - \theta_1)} + \frac{e}{\sqrt{2\,(x^2 + r^2)}} \cos\,(\theta_2 - \delta) = 0;$$

or, multiplied by $\varepsilon^{-\frac{r}{x}\theta_1}$ and rearranged, we have the condition connecting moments θ_1 and θ_2, as follows:

$$\frac{i_0}{2}\left(\varepsilon^{-\frac{r}{x}\theta_1} + \varepsilon^{-\frac{r}{x}\theta_2}\right) + \frac{e}{\sqrt{2\,(x^2 + r^2)}} \left\{ \varepsilon^{-\frac{r}{x}\theta_1} \cos\,(\theta_2 - \delta) - \varepsilon^{-\frac{r}{x}\theta_2} \cos\,(\theta_1 - \delta) \right\} = 0$$

and

$$\frac{i_0}{2}\left(\varepsilon^{\frac{r}{x}\theta_1} + \varepsilon^{\frac{r}{x}\theta_2}\right) = \frac{e}{\sqrt{2\,(x^2 + r^2)}} \left\{ \varepsilon^{\frac{r}{x}\theta_1} \cos\,(\theta_1 - \delta) - \varepsilon^{\frac{r}{x}\theta_2} \cos\,(\theta_2 - \delta) \right\}. \tag{8}$$

Rearranged equation (8) gives

$$f\,(\theta_2) = \varepsilon^{\frac{r}{x}\theta_2}\left[1 + \frac{2\,e}{i_0\sqrt{2\,(x^2 + r^2)}} \cos\,(\theta_2 - \delta)\right] = \varepsilon^{\frac{r}{x}\theta_1}\left[\frac{2\,e}{i_0\sqrt{2\,(x^2 + r^2)}} \cos\,(\theta_1 - \delta) - 1\right], \tag{9}$$

where $$\tan\,\delta = \frac{x - r}{x + r}.$$

By approximation, from this equation the value of θ_2, corresponding to a given θ_1, is derived.

15. Example:

$$e = 2000, \quad i_0 = 10, \quad \text{and } Z = 10 - 40\,j.$$

Thus $\delta = 31° = 0.54$ radians

and $$i = 5 + [34.3 \cos(\theta_1 - 31^\circ) - 5]\,\varepsilon^{-0.25(\theta-\theta_1)} - 34.3 \cos(\theta - 31^\circ),$$

$$f(\theta_2) = \varepsilon^{0.25\theta_2}[1 + 6.86 \cos(\theta_2 - 31^\circ)] = \varepsilon^{0.25\theta_1}[6.86 \cos(\theta_1 - 31^\circ) - 1].$$

Substituting for θ_1, $30^\circ = \frac{\pi}{6}$, $45^\circ = \frac{\pi}{4}$, and $60^\circ = \frac{\pi}{3}$, respectively, gives:

$$\theta_1 = \frac{\pi}{6},\; i = 5 + 29.3\,\varepsilon^{-0.25\left(\theta - \frac{\pi}{6}\right)} - 34.3 \cos(\theta - 31^\circ)$$

$$\frac{\pi}{4},\; i = 5 + 28.3\,\varepsilon^{-0.25\left(\theta - \frac{\pi}{4}\right)} - 34.3 \cos(\theta - 31^\circ)$$

$$\frac{\pi}{3},\; i = 5 + 25.1\,\varepsilon^{-0.25\left(\theta - \frac{\pi}{3}\right)} - 34.3 \cos(\theta - 31^\circ)$$

and

	$\theta_1 = \frac{\pi}{6}$		$\theta_1 = \frac{\pi}{4}$		$\theta_1 = \frac{\pi}{3}$			
θ°	i	i_2	i	i_2	i	i_2	e_2	e_1
25	......	.....	......	....	......	......	1810	850
30	0	10	......	......	......	......	......	......
35	......	......	......	......	......	......	1640	1150
40	−0.9	10.9	......	......	......	......	......	......
45	......	......	0	10	......	......	1410	1410
50	−0.6	10.6	......	......	......	......	......	......
55	......	......	.8	9.2	......	......	1150	1640
60	+0.6	9.4	......	......	0	10	......	......
65	......	......	2.5	7.5	......	......	850	1810
70	3.0	7.0	......	......	2.2	7.8	......	......
75	......	......	5.1	4.9	......	......	520	1930
80	6.0	4.0	......	......	5.4	4.6	......	......
85	......	......	8.6	1.4	......	......	170	1990
90	9.9	0.1	......	......	9.3	0.7	......	......
95	......	......	12.8	−2.8	......	......	−170	1990
100	14.3	−4.3	......	......	13.8	−3.8	......	......
105	......	......	17.3	−7.3	......	......	−520	1930
110	19.1	−9.1	......	......	18.6	−8.6	......	......
115	......	......	22.2	−12.2	......	......	−850	1810

These values are plotted in Fig. 6, together with e_1 and e_2. It follows then,

$$\theta_2 = \quad 90.2^\circ \qquad 88.6^\circ \qquad 91.7^\circ$$

The actual curves of an arc machine differ, however, very greatly from those of Fig. 6. In the arc machine, inherent regulation for constant current is produced by opposing a very high armature reaction to the field excitation, so that the resultant m.m.f., or m.m.f. which produces the effective magnetic flux, is

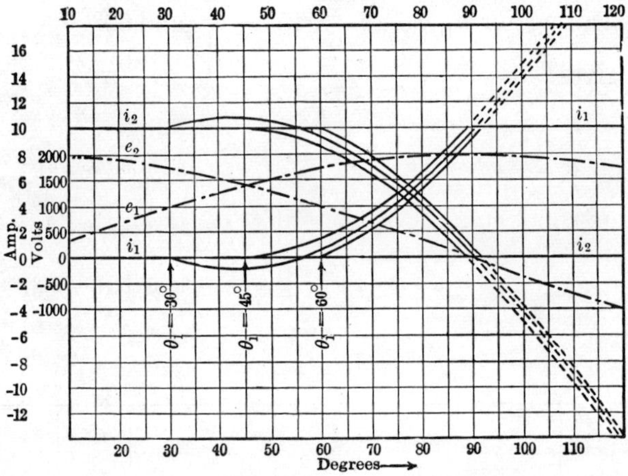

FIG. 6. Quarter-phase rectification.

small compared with the total field m.m.f. and the armature reaction, and so greatly varies with a small variation of armature current. As result, a very great distortion of the field occurs, and the magnetic flux is concentrated at the pole corner. This gives an e.m.f. wave which has a very sharp and high peak, with very long flat zero, and so cannot be approximated by an equivalent sine wave, but the actual e.m.f. curves have to be used in a more exact investigation.

CHAPTER IV

ARC RECTIFICATION

I. The Arc.

16. The operation of the arc rectifier is based on the characteristic of the electric arc to be a good conductor in one direction but a non-conductor in the opposite direction, and so to permit only unidirectional currents.

In an electric arc the current is carried across the gap between the terminals by a bridge of conducting vapor consisting of the material of the negative or the cathode, which is produced and constantly replenished by the cathode blast, a high velocity blast issuing from the cathode or negative terminal towards the anode or positive terminal.

An electric arc, therefore, cannot spontaneously establish itself. Before current can exist as an arc across the gap between two terminals, the arc flame or vapor bridge must exist, i.e., energy must have been expended in establishing this vapor bridge. This can be done by bringing the terminals into contact and so starting the current, and then by gradually withdrawing the terminals derive the energy of the arc flame by means of the current, from the electric circuit, as is done in practically all arc lamps. Or by increasing the voltage across the gap between the terminals so high that the electrostatic stress in the gap represents sufficient energy to establish a path for the current, i.e., by jumping an electrostatic spark across the gap, this spark is followed by the arc flame. An arc can also be established between two terminals by supplying the arc flame from another arc, etc.

The arc therefore must be continuous at the cathode, but may be shifted from anode to anode. Any interruption of the cathode blast puts out the arc by interrupting the supply of conducting vapor, and a reversal of the arc stream means stopping the cathode blast and producing a reverse cathode blast, which, in general, requires a voltage higher than the electrostatic striking voltage (at arc temperature) between the electrodes. With an alternating impressed e.m.f. the arc if established goes out at the end of the half wave, or if a cathode blast is maintained continuously by a second arc (excited by direct current or overlapping sufficiently with the first arc), only alternate half waves can pass, those for which that terminal is negative from which the continuous blast issues. The arc, with an alternating impressed voltage, therefore rectifies, and the voltage range of rectification is the range between the arc voltage and the electrostatic spark voltage through the arc vapor, or the air or residual gas which may be mixed with it. Hence it is highest with the mercury arc, due to its low temperature.

The mercury arc is therefore almost exclusively used for arc rectification. It is enclosed in an evacuated glass vessel, so as to avoid escape of mercury vapor and entrance of air into the arc stream. Due to the low temperature of the boiling point of mercury, enclosure in glass is feasible with the mercury arc.

II. Mercury Arc Rectifier.

17. Depending upon the character of the alternating supply, whether a source of constant alternating potential or constant alternating current, the direct-current circuit receives from the rectifier either constant potential or constant current. Depending on the character of the system, thus constant-potential rectifiers and constant-current rectifiers can be distinguished. They differ somewhat from each other in their construction and that of the auxiliary apparatus, since the constant-potential rectifier operates at constant voltage but varying current, while the constant-current rectifier operates at varying voltage. The general character of the phenomenon of arc rectification is, however, the same in either case, so that only the constant-current rectifier will be considered more explicitly in the following paragraphs.

The constant-current mercury arc rectifier system, as used for the operation of constant direct-current arc circuits from an alternating constant potential supply of any frequency, is sketched diagrammatically in Fig. 7. It consists of a constant-current transformer with a tap C brought out from the middle of the secondary coil AB. The rectifier tube has two graphite anodes *a*, *b*, and a mercury cathode *c*, and usually two auxiliary mercury anodes near the cathode *c* (not shown in diagram, Fig. 7), which are used for excitation, mainly in starting, by establishing between the cathode *c* and the two auxiliary mercury anodes, from a small low voltage constant-potential transformer, a pair of low current rectifying arcs. In the constant-potential rectifier, generally one auxiliary anode only is used, connected through a resistor *r* with one of the main anodes, and the constant-

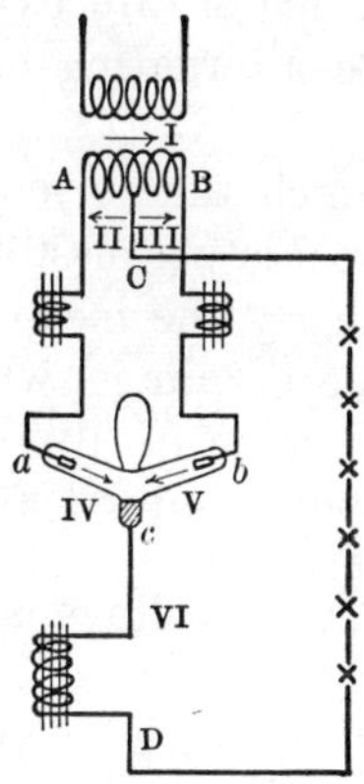

FIG. 7. Constant-current mercury arc rectifier.

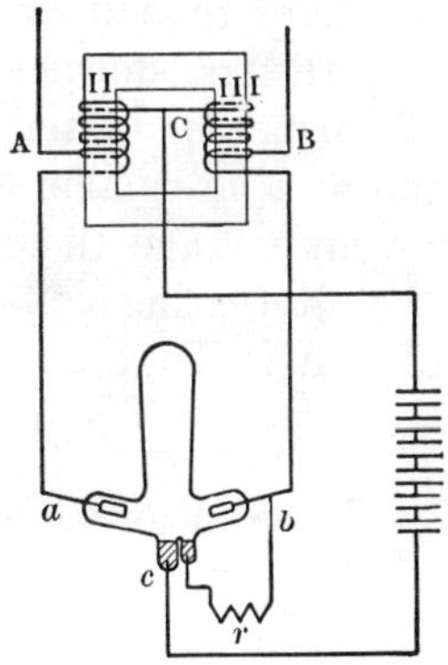

FIG. 8. Constant-potential mercury arc rectifier.

current transformer is replaced by a constant-potential transformer or compensator (auto-transformer) having considerable inductance between the two half coils II and III, as shown in Fig. 8. Two reactive coils are inserted between the outside terminals of the transformer and rectifier tube respectively, for the purpose of producing an overlap between the two rectifying arcs, *ca* and *cb*, and thereby the required continuity of the arc stream at *c*. Or instead of separate reactances, the two half coils II and III may be given sufficient reactance, as in Fig. 8. A reactive coil is inserted into the rectified or arc circuit, which connects between transformer neutral *C* and rectifier neutral *c*, for the purpose of reducing the fluctuation of the rectified current to the desired amount.

In the constant-potential rectifier, instead of the transformer *ACB* and the reactive coils *Aa* and *Ba*, generally a compensator or auto-transformer is used, as shown in Fig. 8., in which the two halves of the coil, AC and BC, are made of considerable self-inductance against each other, as by their location on different magnet cores, and the reactive coil at c frequently omitted. The modification of the equations resulting herefrom is obvious. Such auto-transformer also may raise or lower the impressed voltage, as shown in Fig. 8.

The rectified or direct voltage of the constant-current rectifier is somewhat less than one-half of the alternating voltage supplied by the transformer secondary AB, the rectified or direct current somewhat more than double the effective alternating current supplied by the transformer.

In the constant-potential rectifier, in which the currents are larger, and so a far smaller angle of overlap θ is permissible, the direct-current voltage therefore is very nearly the mean value of half the alternating voltage, minus the arc voltage, which is about 13 volts. That is, if e = effective value of alternating voltage between rectifier terminals ab of compensator (Fig. 8), hence $\frac{2\sqrt{2}}{\pi} e$ = mean value, the direct current voltage is

$$e_0 = \frac{\sqrt{2}}{\pi} e - 13.$$

III. Mode of Operation.

18. Let, in Figs. 9 and 10, the impressed voltage between the secondary terminals AB of an alternating-current transformer be shown by curve I. Let C be the middle or center of the transformer secondary AB. The voltages from C to A and from C to B then are given by curves II and III.

If now A, B, C are connected with the corresponding rectifier terminals a, b, c and at c a cathode blast maintained, those currents will exist for which c is negative or cathode, i.e., the current through the rectifier from a to c and from b to c, under the impressed e.m.fs. II and III, are given by curves IV and V, and the current derived from c is the sum of IV and V, as shown in curve VI.

Such a rectifier as shown diagrammatically in Fig. 9 requires some outside means for maintaining the cathode blast at c, since the current in the half wave 1 in curve VI goes down to zero at the zero value of e.m.f. III before the current of the next half wave 2 starts by the e.m.f. II.

It is therefore necessary to maintain the current of the half wave 1 beyond the zero value of its propelling impressed e.m.f. III until the current of the next half wave 2 has started, i.e., to overlap the currents of the successive half waves. This is done by inserting reactances into the leads from the transformer to the rectifier, i.e., between *A* and *a*, *B* and *b* respectively, as shown in Fig. 7. The effect of this reactance is that the current of half wave 1, V, continues beyond the zero of its impressed e.m.f. III i.e., until the e.m.f. III has died out and reversed, and the current of the half wave 2, IV, started by e.m.f. II; that is, the two half waves of the current overlap, and each half wave lasts for more than half a period or 180 degrees.

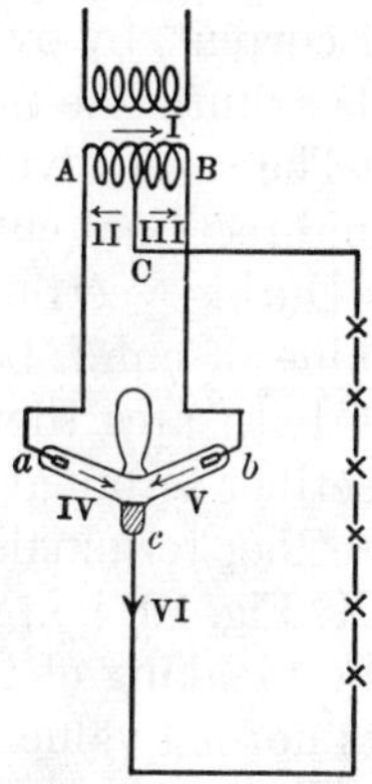

FIG. 9.. Constant-current mercury arc rectifier.

The current waves then are shown in curve VII. The current half wave 1 starts at the zero value of its e.m.f. III, but rises more slowly than it would without react-

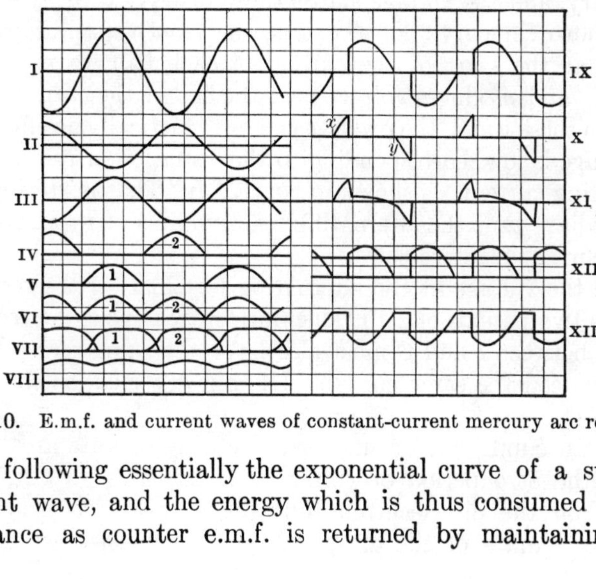

FIG. 10. E.m.f. and current waves of constant-current mercury arc rectifier.

ance, following essentially the exponential curve of a starting current wave, and the energy which is thus consumed by the reactance as counter e.m.f. is returned by maintaining the current half wave 1 beyond the e.m.f. wave, i.e., beyond 180 degrees, by θ_0 time-degrees, so that it overlaps the next half wave 2 by θ_0 time-degrees.

Hereby the rectifier becomes self-exciting, i.e., each half wave of current, by overlapping with the next, maintains the cathode blast until the next half wave is started.

The successive current half waves added give the rectified or unidirectional current curve VIII.

During a certain period of time in each half wave from the zero value of e.m.f. both arcs *ca* and *cb* exist. During the existence of both arcs there can be no potential difference between the rectifier terminals *a* and *b*, and the impressed e.m.f. between the rectifier terminals *a* and *b* therefore has the form shown in curve IX, Fig. 10, i.e., remains zero for θ_0 time-degrees, and then with the breaking of the arc of the preceding half wave jumps up to its normal value.

The generated e.m.f. of the transformer secondary, however, must more or less completely follow the primary impressed e.m.f. wave, that is, has a shape as shown in curve I, and the difference between IX and I must be taken up by the reactance. That is, during the time when both arcs exist in the rectifier, the a. c. reactive coils consume the generated e.m.f. of the transformer secondary, and the voltage across these reactive coils, therefore, is as shown in curve X. That is, the reactive coil consumes voltage at the start of the current of each half wave, at x in curve X, and produces voltage near the end of the current, at y. Between these times, the reactive coil has practically no effect and its voltage is low, corresponding to the variation of the rectified alternating current, as shown in curve XI. That is, during this intermediary time the alternating reactive coils merely assist the direct-current reactive coil.

Since the voltage at the alternating terminals of the rectifier, *a*, *b*, has two periods of zero value during each cycle, the rectified voltage between *c* and *C* must also have the same zero periods, and is indeed the same curve as IX, but reversed, as shown in curve XII.

Such an e.m.f. wave cannot satisfactorily operate arcs, since during the zero period of voltage XII the arcs go out. The voltage on the direct-current line must never fall below the "counter e.m.f." of the arcs, and since the resistance of this circuit is low, frequently less than 10 per cent, it follows that the total variation of direct-current line voltage must be below 10 per cent, i.e., the voltage practically constant, as shown by the straight line in curve XII. Hence a high reactance is inserted into the direct-current circuit, which consumes the excess voltage during that part of curve XII where the rectified voltage is above line voltage, and supplies the line voltage during the period of zero rectified voltage. The voltage across this reactive coil, therefore, is as shown by curve XIII.

IV. Constant-Current Rectifier.

19. The angle of overlap θ_0 of the two arcs is determined by the desired stability of the system. By the angle θ_0 and the impressed e.m.f. is determined the sum total of e.m.fs. which has to be consumed and returned by the a. c. reactive coil, and herefrom the size of the a. c. reactive coil.

From the angle θ_0 also follows the wave shape of the rectified voltage, and therefrom the sum total of e.m.f. which has to be given by the d. c. reactive coil, and hereby the size of the d. c. reactive coil required to maintain the d. c. current fluctuation within certain given limits.

The efficiency, power factor, regulation, etc., of such a mercury arc rectifier system are essentially those of the constant-current transformer feeding the rectifier tube.

Let f = frequency of the alternating-current supply system, i_0 = mean value of the rectified direct current, and a = the pulsation of the rectified current from the mean value, i.e., $i_0\,(1 + a)$ the maximum and $i_0\,(1 - a)$ the minimum value of direct current. A pulsation from a mean of 20 to 25 per cent is permissible in an arc circuit. The total variation of the rectified current then is $2\,ai_0$, i.e., the alternating component of the direct current has the maximum value ai_0, hence the effective value $\frac{a}{\sqrt{2}}\,i_0$ (or for $a = 0.2$, $0.141\,i_0$) and the frequency $2\,f$. Hysteresis and eddy losses in the direct-current reactive coil, therefore, correspond to an alternating current of frequency $2\,f$ and effective value $\frac{a}{\sqrt{2}}\,i_0$, or about $0.141\,i_0$, i.e., are small even at relatively high densities.

In the alternating-current reactive coils the current varies, unidirectionally, between 0 and $i_0 (1 + a)$, i. e., its alternating component has the maximum value $\frac{1 + a}{2} i_0$ and the effective value $\frac{1 + a}{2\sqrt{2}} i_0$ (or, for $a = + 0.2$, $0.425\ i_0$) and the frequency f. The hysteresis loss, therefore, corresponds to an alternating current of frequency f and effective value $\frac{1 + a}{2\sqrt{2}} i_0$, or about $0.425\ i_0$.

With decreasing load, at constant alternating-current supply, the rectified direct current slightly increases, due to the increasing overlap of the rectifying arcs, and to give constant direct current the transformer must therefore be adjusted so as to regulate for a slight decrease of alternating-current output with decrease of load.

V. Theory and Calculation.

20. In the constant-current mercury-arc rectifier shown diagrammatically in Fig. 11, let $e \sin \theta$ = sine wave of e.m.f. impressed between neutral and outside of alternating-current supply to the rectifier; that is, $2 e \sin \theta$ = total secondary generated e.m.f. of the constant-current transformer; $Z_1 = r_1 - jx_1$ = impedance of the reactive coil in each anode circuit of the rectifier ("alternating-current reactive coil"), inclusive of the internal self-inductive impedance between the two halves of the transformer secondary coil; i_1 and i_2 = anode currents, counted in the direction from anode to cathode; e_a = counter e.m.f. of rectifying arc, which is constant; $Z_0 = r_0 - jx_0$ = impedance of reactive coil in rectified circuit ("direct-current reactive coil"); $Z_2 = r_2 - jx_2$ = impedance of load or arc-lamp circuit; e_0' = counter e.m.f. in rectified circuit, which is con-

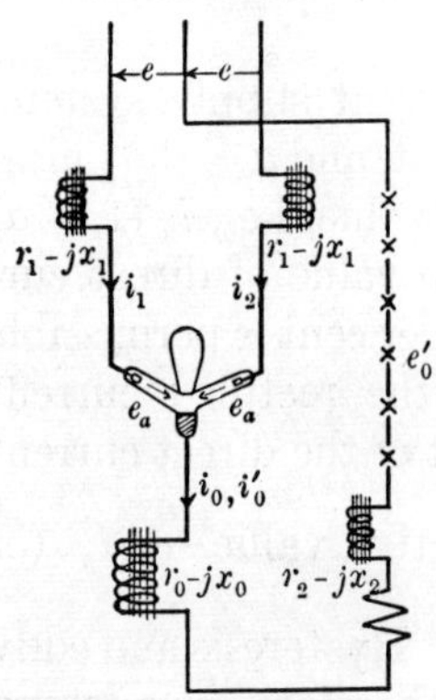

Fig. 11. Constant-current mercury arc rectifier.

stant (equal to the sum of the counter e.m.fs. of the arcs in the lamp circuit); θ_0 = angle of overlap of the two rectifying arcs, or overlap of the currents i_1 and i_2; i_0 = rectified current during the period, $0 < \theta < \theta_0$, where both rectifying arcs exist, and $i_0' =$ rectified current during the period, $\theta_0 < \theta < \pi$, where only one arc or one anode current i_1 exists.

Let $e_0 = e_0' + e_a$ = total counter e.m.f. in the rectified circuit and $Z = r - jx = (r_1 + r_0 + r_2) - j(x_1 + x_0 + x_2)$ = total impedance per circuit; then we have

(*a*) During the period when both rectifying arcs exist,

$$0 < \theta < \theta_0,$$

$$i_0 = i_1 + i_2. \tag{1}$$

In the circuit between the e.m.f. $2e \sin \theta$, the rectifier tube, and the currents i_1 and i_2, according to Kirchhoff's law, it is, **Fig. 11**,

$$2e \sin \theta - r_1 i_1 - x_1 \frac{di_1}{d\theta} + r_1 i_2 + x_1 \frac{di_2}{d\theta} = 0. \tag{2}$$

In the circuit from the transformer neutral over e.m.f. $e \sin \theta$, current i_1, rectifier arc e_a and rectified circuit i_0, back to the transformer neutral, we have

$$e \sin \theta - r_1 i_1 - x_1 \frac{di_1}{d\theta} - e_a - r_0 i_0 - x_0 \frac{di_0}{d\theta} - r_2 i_0 - x_2 \frac{di_0}{d\theta} - e_0' = 0;$$

or,

$$e \sin \theta - r_1 i_1 - x_1 \frac{di_1}{d\theta} - (r_0 + r_2)\, i_0 - (x_0 + x_2) \frac{di_0}{d\theta} - e_0 = 0. \tag{3}$$

(*b*) During the period when only one rectifying arc exists,

$$\theta_0 < \theta < \pi,$$

$$i_1 = i_0';$$

hence, in this circuit,

$$e \sin \theta - r_1 i_0' - x_1 \frac{di_0'}{d\theta} - (r_0 + r_2)\, i_0' - (x_0 + x_2) \frac{di_0'}{d\theta} - e_0 = 0. \tag{4}$$

Substituting (1) in (2) and combining the result (5) of this substitution with (3) gives the *differential equations of the rectifier:*

$$2\,e \sin\theta + r_1\,(i_0 - 2\,i_1) + x_1 \frac{d}{d\theta}\,(i_0 - 2\,i_1) = 0, \tag{5}$$

$$2\,e_0 + (2\,r - r_1)\,i_0 + (2\,x - x_1)\,\frac{di_0}{d\theta} = 0, \tag{6}$$

and

$$e \sin\theta - e_0 - r i_0' - x \frac{di_0'}{d\theta} = 0. \tag{7}$$

In these equations, i_0 and i_1 apply for the time, $0 < \theta < \theta_0$, i_0' for the time, $\theta_0 < \theta < \pi$.

21. These differential equations are integrated by the functions

$$i_0 - 2\,i_1 = A\varepsilon^{-a\theta} + A' \sin(\theta - \beta), \tag{8}$$

$$i_0 = B\varepsilon^{-b\theta} + B', \tag{9}$$

and

$$i_0' = C\varepsilon^{-c\theta} + C' + C'' \sin(\theta - \gamma). \tag{10}$$

Substituting (8), (9), and (10) into (5), (6), and (7) gives three identities:

$$2\,e \sin\theta + A'\,[r_1 \sin(\theta - \beta) + x_1 \cos(\theta - \beta)] + A\varepsilon^{-a\theta}\,(r_1 - a x_1) = 0.$$

$$2\,e_0 + B'(2\,r - r_1) + B\varepsilon^{-b\theta}\,[(2\,r - r_1) - b\,(2\,x - x_1)] = 0,$$

and

$$e \sin\theta - e_0 - C''[r \sin(\theta - \gamma) + x \cos(\theta - \gamma)] - C'r - C\varepsilon^{-c\theta}(r - cx) = 0;$$

hence,

$$\left.\begin{aligned}
r_1 - a x_1 &= 0,\\
(2\,r - r_1) - b\,(2\,x - x_1) &= 0,\\
\text{and} \qquad r - cx &= 0.\\
2\,e_0 + B'\,(2\,r - r_1) &= 0,\\
e_0 + C'r &= 0,\\
2\,e + A'\,(r_1 \cos\beta + x_1 \sin\beta) &= 0,\\
A'\,(r_1 \sin\beta - x_1 \cos\beta) &= 0,\\
e - C''\,(r \cos\gamma + x \sin\gamma) &= 0,\\
C''\,(r \sin\gamma - x \cos\gamma) &= 0.
\end{aligned}\right\} \tag{11}$$

Writing

$$\left.\begin{aligned} z_1 &= \sqrt{r_1^2 + x_1^2}, \\ \tan \alpha_1 &= \frac{x_1}{r_1}, \end{aligned}\right\} \tag{12}$$

and

$$\left.\begin{aligned} z &= \sqrt{r^2 + x^2}, \\ \tan \alpha &= \frac{x}{r} \cdot \end{aligned}\right\} \tag{13}$$

Substituting (12) and (13) gives by solving the 9 equations (11) the values of the coefficients a, b, c, A', B', C', C'', β, γ:

$$\left.\begin{aligned} a &= \frac{r_1}{x_1}, \\ b &= \frac{2r - r_1}{2x - x_1}, \\ c &= \frac{r}{x}, \end{aligned}\right\} \tag{14}$$

$$\left.\begin{aligned} \beta &= \alpha_1, \\ \gamma &= \alpha, \end{aligned}\right\} \tag{15}$$

$$\left.\begin{aligned} A' &= -\frac{2e}{z_1}, \\ B' &= -\frac{2e_0}{2r - r_1}, \\ C' &= -\frac{e_0}{r}, \end{aligned}\right\} \tag{16}$$

$$C'' = +\frac{e}{z}, \tag{17}$$

and thus the *integral equations of the rectifier* are

$$i_0 - 2i_1 = A\varepsilon^{-a\theta} - \frac{2e}{z_1}\sin(\theta - \alpha_1), \tag{18}$$

$$i_0 = B\varepsilon^{-b\theta} - \frac{2e_0}{2r - r_1}, \tag{19}$$

and

$$i_0' = C\varepsilon^{-c\theta} - \frac{e_0}{r} + \frac{e}{z}\sin(\theta - \alpha), \tag{20}$$

where a, b, c are given by equations (14), α and α_1 by equations (12) and (13), and A, B, C are integration constants given by the terminal conditions of the problem.

22. These terminal conditions are:

$$\left.\begin{aligned} |i_1|_{\theta=0} &= 0, \\ |i_0|_{\theta=0} &= |i_0'|_{\theta=\pi}, \\ \text{and} \qquad |i_1|_{\theta=\theta_0} &= |i_0|_{\theta=\theta_0} = |i_0'|_{\theta=\theta_0}. \end{aligned}\right\} \tag{21}$$

That is, at $\theta = 0$ the anode current $i_1 = 0$. After half a period, or $\pi = 180°$, the rectified current repeats the same value. At $\theta = \theta_0$, all three currents i_1, i_0, i_0' are identical.

The four equations (21) determine four constants, A, B, C, θ_0. Substituting these constants in equations (18), (19), (20) gives the equations of the rectified current i_0, i_0', and of the anode currents i_1 and $i_2 = i_0 - i_1$, determined by the constants of the system, Z, Z_1, e_0, and by the impressed e.m.f., e.

In the constant-current mercury-arc rectifier system of arc lighting, e, the secondary generated voltage of the constant-current transformer, varies with the load, by the regulation of the transformer, and the rectified current, i_0, i_0', is required to remain constant, or rather its average value.

Let then be given as condition of the problem the average value i of the rectified current, 4 amperes in a magnetite or mercury arc lamp circuit, 5 or 6.6 or 9.6 amperes in a carbon arc lamp circuit.

Assume as fair approximation that the pulsating rectified current i_0, i_0' has its mean value i at the moment, $\theta = 0$. This then gives the additional equation

$$|i_0|_{\theta=0} = i, \tag{22}$$

and from the five equations (21) and (22) the five constants A, B, C, θ_0, e are determined.

Substituting (22), (18), (19), (20) in equations (21) gives

$$\left.\begin{aligned} A &= i - \frac{2e}{z_1} \sin \alpha_1, \\ B &= i + \frac{2e_0}{2r - r_1}, \\ C &= \varepsilon^{c\pi} \left\{ i + \frac{e_0}{r} - \frac{e}{z} \sin \alpha \right\} \end{aligned}\right\} \tag{23}$$

$$-A\varepsilon^{-a\theta_0} - \frac{2e}{z_1}\sin(\alpha_1 - \theta_0) = B\varepsilon^{-b\theta_0} - \frac{2e_0}{2r - r_1}$$

$$= C\varepsilon^{-c\theta_0} - \frac{e_0}{r} - \frac{e}{z}\sin(\alpha - \theta_0). \tag{24}$$

Substituting (23) in (24) gives

$$\frac{2e}{z_1}\left\{\varepsilon^{-a\theta_0}\sin\alpha_1 - \sin(\alpha_1 - \theta_0)\right\} = i\left\{\varepsilon^{-a\theta_0} + \varepsilon^{-b\theta_0}\right\}$$

$$- \frac{2e_0}{2r - r_1}\left\{1 - \varepsilon^{-b\theta_0}\right\} \tag{25}$$

and

$$\frac{e}{z}\left\{\varepsilon^{c(\pi-\theta_0)}\sin\alpha + \sin(\alpha - \theta_0)\right\} = i\left\{\varepsilon^{c(\pi-\theta_0)} - \varepsilon^{-b\theta_0}\right\}$$

$$+ \frac{2e_0}{2r - r_1}\left\{1 - \varepsilon^{-b\theta_0}\right\} + \frac{e_0}{r}\left\{\varepsilon^{c(\pi-\theta_0)} - 1\right\}, \tag{26}$$

and eliminating e from these two equations gives

$$\frac{\varepsilon^{c(\pi-\theta_0)}\sin\alpha + \sin(\alpha - \theta_0)}{\varepsilon^{-a\theta_0}\sin\alpha_1 - \sin(\alpha_1 - \theta_0)}$$

$$= \frac{2z}{z_1}\,\frac{\left\{\varepsilon^{c(\pi-\theta_0)} - \varepsilon^{-b\theta_0}\right\} + \dfrac{2e_0}{i(2r - r_1)}\left\{1 - \varepsilon^{-b\theta_0}\right\} + \dfrac{e_0}{ir}\left\{\varepsilon^{c(\pi-\theta_0)} - 1\right\}}{\left\{\varepsilon^{-a\theta_0} + \varepsilon^{-b\theta_0}\right\} - \dfrac{2e_0}{i(2r - r_1)}\left\{1 - \varepsilon^{-b\theta_0}\right\}} \tag{27}$$

Equation (27) determines angle θ_0, and by successive substitution in (26), (23), e, A, B, C are found.

Equation (27) is transcendental, and therefore has to be solved by approximation, which however is very rapid.

As first approximation, $a\theta_0 = b\theta_0 = c\theta_0 = 0$; $\alpha = \alpha_1 = 90°$ or $\frac{\pi}{2}$ and substituting these values in (27) gives

$$\frac{\varepsilon^{c\pi} + \cos\theta_1}{1 - \cos\theta_1} = \frac{2z}{z_1}\,\frac{\left(\varepsilon^{c\pi} - 1\right)\left(1 + \dfrac{e_0}{ir}\right)}{2}$$

and

$$\cos \theta_1 = \frac{\frac{z}{z_1}\left(\varepsilon^{c\pi} - 1\right)\left(1 + \frac{e_0}{ir}\right) - \varepsilon^{c\pi}}{\frac{z}{z_1}\left(\varepsilon^{c\pi} - 1\right)\left(1 + \frac{e_0}{ir}\right) + 1}. \tag{28}$$

This value of θ_1 substituted in the exponential terms of equation (27) gives a simple trigonometric equation in θ_0, from which follows the second approximation θ_2, and, by interpolation, the final value,

$$\theta_0 = \theta_2 + \frac{(\theta_2 - \theta_1)^2}{\theta_1}. \tag{29}$$

23. For instance, let $e_0 = 950$, $i = 3.8$, the constants of the circuit being $Z_1 = 10 - 185\,j$ and $Z = 50 - 1000\,j$.

Herefrom follows

$$a = 0.054,\ b = 0.050, \text{ and } c = 0.050, \tag{14}$$

$$\alpha_1 = 86.9° \text{ and } \alpha = 87.1°. \tag{15}$$

From equation (28) follows as first approximation, $\theta_1 = 47.8°$; as second approximation, $\theta_2 = 44.2°$.

Hence, by (29),

$$\theta = 44.4°.$$

Substituting a in (26) gives $e = 2100$,
hence, the effective value of transformer secondary voltage,

$$\frac{2\,e}{\sqrt{2}} = 2980 \text{ volts}$$

and, from (23),

$$A = -18.94,\ B = 24.90,\ C = 24.20.$$

Therefore, the equations of the currents are

$$i_0 = 24.90\,\varepsilon^{-0.050\,\theta} - 21.10,$$

$$i_0' = 24.20\,\varepsilon^{-0.050\,\theta} - 19.00 + 2.11 \sin(\theta - 87.1°),$$

$$i_1 = 12.45\,\varepsilon^{-0.050\,\theta} + 9.47\,\varepsilon^{-0.054\,\theta} - 10.58 + 11.35 \sin(\theta - 86.9°),$$

and

$$i_2 = i_0 - i_1.$$

The effective or equivalent alternating secondary current of the transformer, which corresponds to the primary load current, that is, primary current minus exciting current, is

$$i' = i_1 - i_2.$$

From these equations are calculated the numerical values of rectified current i_0, i_0', of anode current i_1, and of alternating current i', and plotted as curves in Fig. 12.

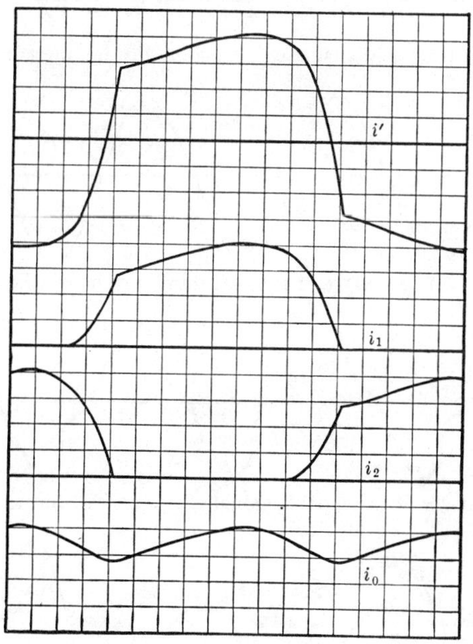

FIG. 12. Current waves of constant-current mercury arc rectifier.

24. As illustrations of the above phenomena are shown in Fig. 66 the performance curves of a small constant-current rectifier, and in Figs. 14 to 15 oscillograms of this rectifier.

Interesting to note is the high frequency oscillation at the termination of the jump of the potential difference *cC* (Fig. 7) which represents the transient term resulting from the electrostatic capacity of the transformer. At the end of the period of overlap of the two rectifying arcs one of the anode currents reaches

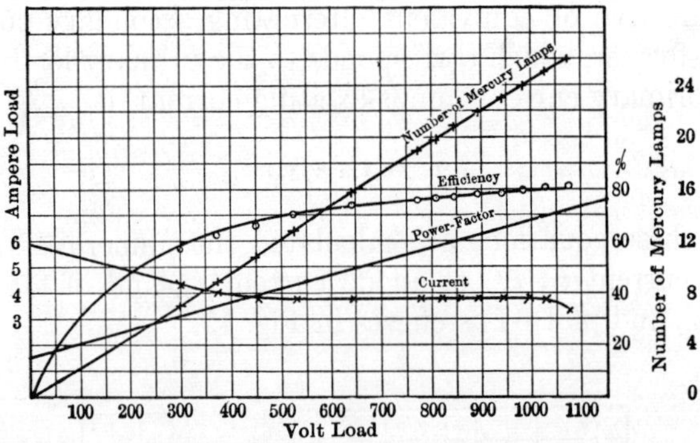

FIG. 13. Results from tests made on a constant-current mercury arc rectifier.

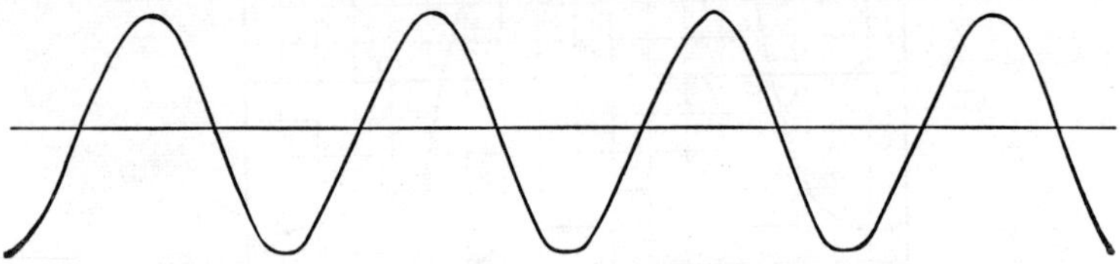

FIG. 14. Supply e.m.f. to constant-current rectifier.

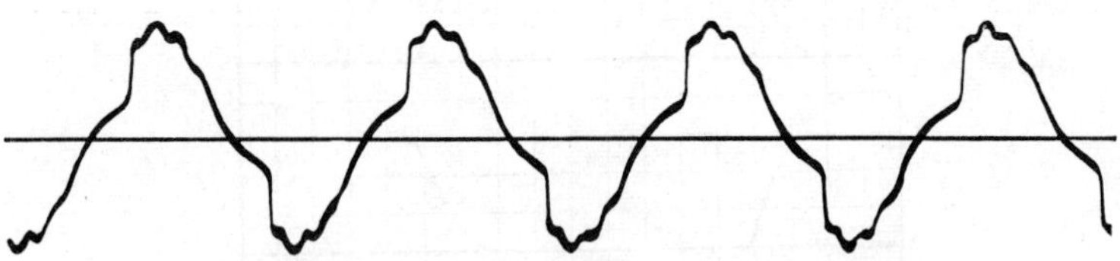

FIG. 15. Secondary terminal e.m.f. of transformer.

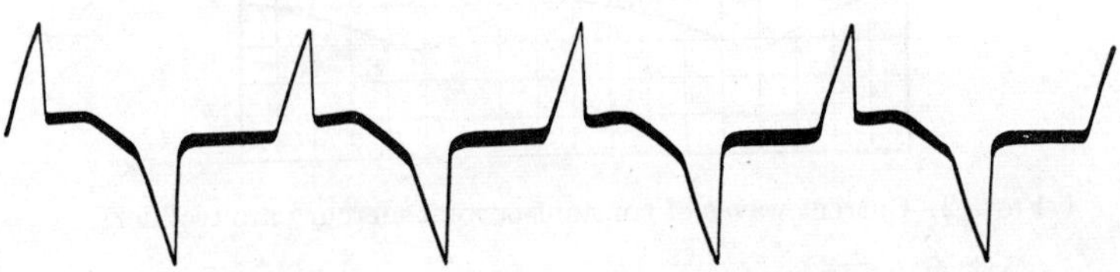

FIG. 16. E.m.f. across a.c. reactive coils.

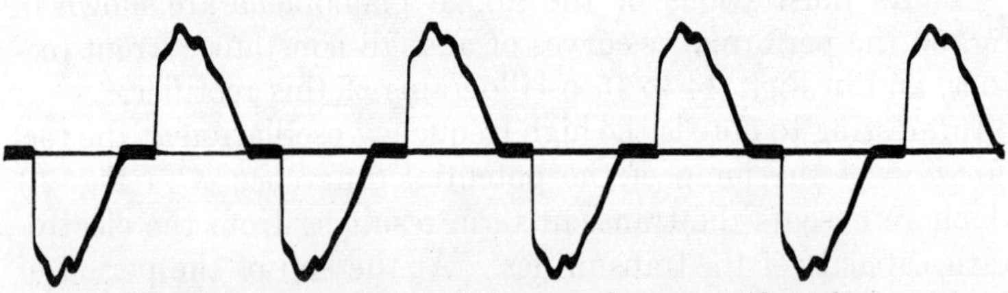

FIG. 17. Alternating e.m.f. impressed upon rectifier tube.

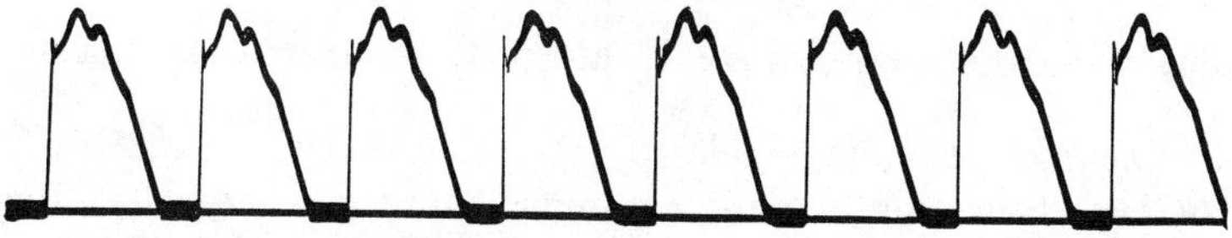

FIG. 18. Unidirectional e.m.f. produced between rectifier neutral and transformer neutral.

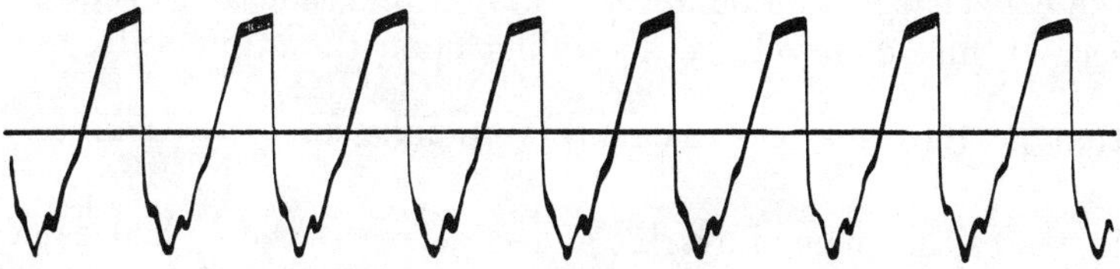

FIG. 19. E.m.f. across d.c. reactive coils.

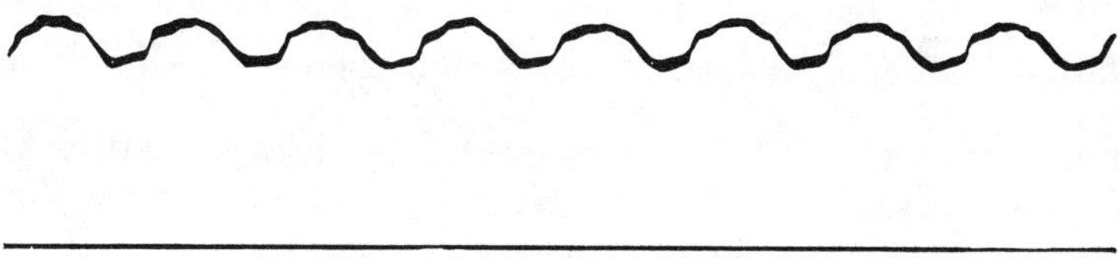

FIG. 20. Rectified e.m.f. supplied to arc circuit.

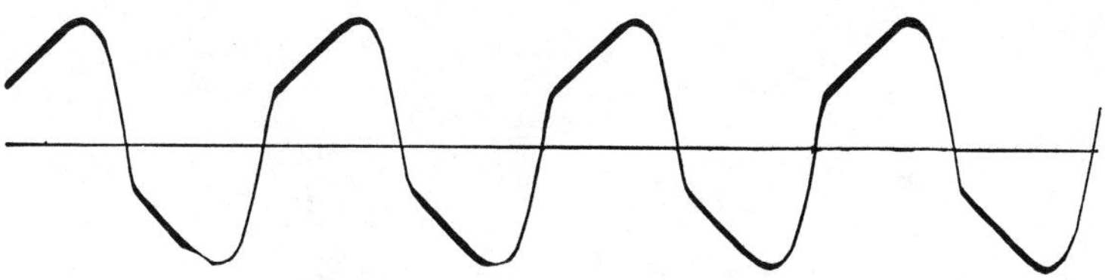

FIG. 21. Primary supply current.

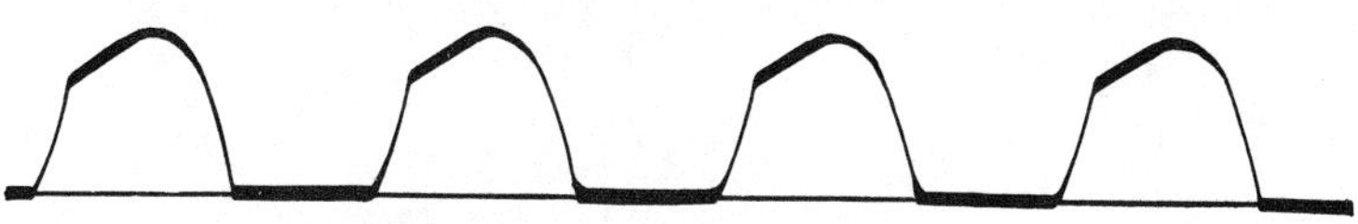

FIG. 22. Current in rectifying arcs.

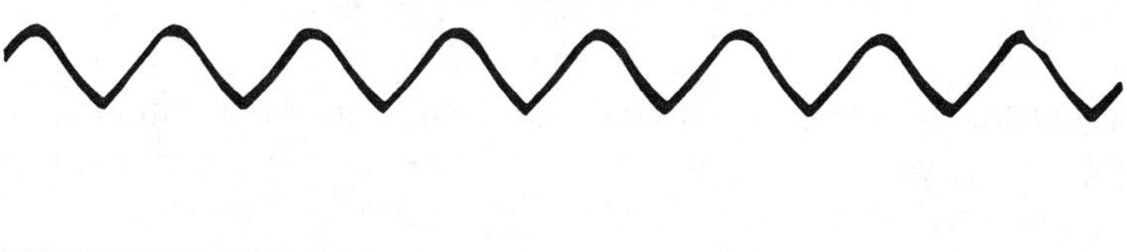

FIG. 23. Rectified current in arc circuit.

zero and stops, and so its $L \frac{di}{dt}$ abruptly changes; that is, a sudden change of voltage takes place in the circuit *aACDc* or *bBCDc*. Since this circuit contains distributed capacity, that of the transformer coil *ACBC* respectively, the line, etc., and inductance, an oscillation results of a frequency depending upon the capacity and inductance, usually a few thousand cycles per second, and of a voltage depending upon the impressed e.m.f.; that is, the $L \frac{di}{dt}$ of the circuit. An increase of inductance L increases the angle of overlap and so decreases the $\frac{di}{dt}$, hence does not greatly affect the amplitude, but decreases the frequency of this oscillation. An increase of $\frac{di}{dt}$ at constant L, as resulting from a decrease of the angle of overlap by delayed starting of the arc, caused by a defective rectifier, however increases the amplitude of this oscillation, and if the electrostatic capacity is high, and therefore the damping out of the oscillation slow, the

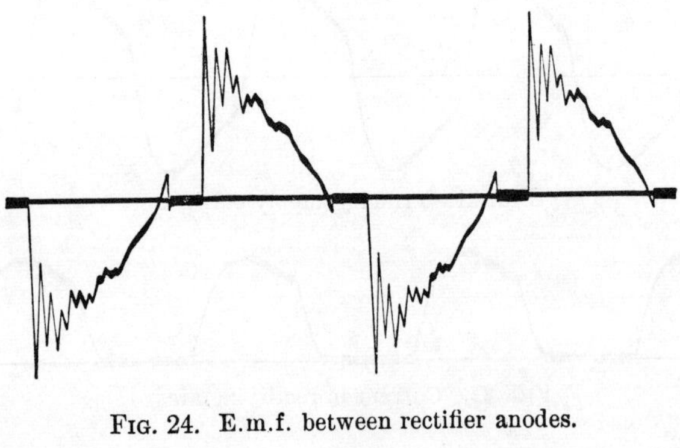

FIG. 24. E.m.f. between rectifier anodes.

oscillation may reach considerable values, as shown in oscillogram, Fig. 24, of the potential difference *ab*. In such cases, if the second half wave of the oscillation reaches below the zero value of the e.m.f. wave *ab*, the rectifying arc is blown out and a disruptive discharge may result.

VI. Equivalent Sine Waves.

25. The curves of voltage and current, in the mercury-arc rectifier system, as calculated in the preceding from the constants of the circuit, consist of successive sections of exponential or of exponential and trigonometric character.

In general, such wave structures, built up of successive sections of different character, are less suited for further calculation. For most purposes, they can be replaced by their equivalent sine waves, that is, sine waves of equal effective value and equal power.

The actual current and e.m.f. waves of the arc rectifier thus may be replaced by their equivalent sine waves, for general calculation, except when investigating the phenomena resulting from the discontinuity in the change of current, as the high frequency oscillation at the end and to a lesser extent at the beginning of the period of overlap of the rectifying arcs, and similar phenomena.

In a constant-current mercury arc rectifier system, of which the exact equations or rather groups of equations of currents and of e.m.fs. were given in the preceding, let i_0 = the mean value of direct current; e_0 = the mean value of direct or rectified voltage; i = the effective value of equivalent sine wave of secondary current of transformer feeding the rectifier; e = the effective value of equivalent sine wave of total e.m.f. generated in the transformer secondary coils, hence, $\frac{e}{2}$ = the effective equivalent sine wave of generated e.m.f. per secondary transformer coil, and θ_0 = the angle of overlap of rectifying arcs.

The secondary generated e.m.f., e, is then represented by a sine wave curve I, Fig. 25 with $e\sqrt{2}$ as maximum value.

Neglecting the impedance voltage of the secondary circuit during the time when only one arc exists and the current changes are very gradual, the terminal voltage between the rectifier anodes, e_1, is given by curve II, Fig. 25, with $e\sqrt{2}$ as maximum value. This curve is identical with e, except during the angle of overlap θ_0, when e_1 is zero. Due to the impedance of the reactive coils in the anode leads, curve II differs slightly from I, but the difference is so small that it can be neglected in deriving the equivalent sine wave, and this impedance considered afterwards as inserted into the equivalent sine-wave circuit.

The rectified voltage, e_2, is then given by curve III, Fig. 25, with a maximum value of $\frac{e}{2}\sqrt{2} = \frac{e}{\sqrt{2}}$ and zero value during the angle of overlap θ_0, or rather a value $= e_a$, the e.m.f. consumed by the rectifying arc (13 to 18 volts).

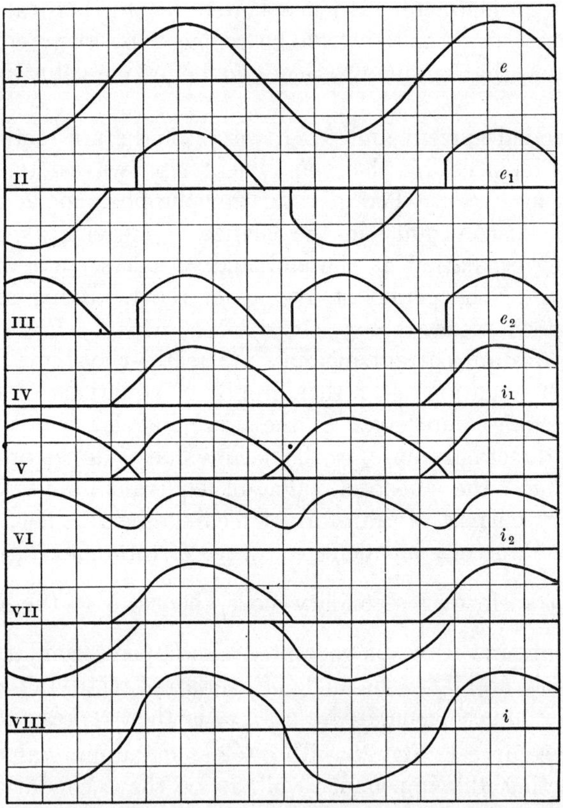

FIG. 25. E.m.f. and current curves in a mercury arc rectifier system.

The direct voltage e_0, when neglecting the effective resistance of the reactive coils, is then the mean value of the rectified voltage, e_2, of curve III, hence is

$$e_0 = \frac{e}{\sqrt{2}} \frac{1}{\pi} \int_{\theta_0}^{\pi} \sin \theta \, d\theta$$

$$= \frac{e\,(1 + \cos\theta_0)}{\sqrt{2}\,\pi};$$

or,

$$e = e_0 \frac{\pi\sqrt{2}}{1 + \cos\theta_0}.$$

If e_a = the mercury arc voltage, r_0 = the effective resistance of reactive coils and i_0 = the direct current, more correctly it is

$$e = (e_0 + e_a + r_0 i_0) \frac{\pi\sqrt{2}}{1 + \cos\theta_0}.$$

The effective alternating voltage between the rectifier anodes is the $\sqrt{\text{mean square of } e_1}$, curve II, hence is

$$e_1 = e\sqrt{2}\sqrt{\frac{1}{\pi}\int_{\theta_0}^{\pi} \sin^2\theta\, d\theta}$$

$$= e\sqrt{2}\sqrt{\frac{1}{\pi}\left[\frac{\theta}{2} - \frac{\sin 2\theta}{4}\right]_{\theta_0}^{\pi}}$$

$$= e\sqrt{2}\sqrt{\frac{\pi - \theta_0}{2\pi} + \frac{\sin 2\theta_0}{4\pi}}$$

$$= e\sqrt{1 - \frac{2\theta_0 - \sin 2\theta_0}{2\pi}},$$

and the drop of voltage in the reactive coils in the anode leads, caused by the overlap of the arcs, thus is

$$e - e_1 = e\left\{1 - \sqrt{1 - \frac{2\theta_0 - \sin 2\theta_0}{2\pi}}\right\}.$$

26. Let i' = the maximum variation of direct current from mean value i_0, hence, $i_2 = i_0 + i'$ = the maximum value of rectified current, and therefore also the maximum value of anode current.

The anode current thus has a maximum value i_2, and each half wave has a duration $\pi + \theta_0$, as shown by curve IV, Fig. 25.

The direct current, i_0, is then given by the superposition or addition of the two anode currents shown in curves V, and is given in curve VI.

The effective value of the equivalent alternating secondary current of the transformer is derived by the subtraction of the two anode currents, or their superposition in reverse direction, as shown by curves VII, and is given by curve VIII.

Each impulse of anode current covers an angle $\pi + \theta_0$, or somewhat more than one half wave.

Denoting, however, each anode wave by π, that is, considering each anode impulse as one half wave (which corresponds to a lower frequency $\left.\frac{\pi}{\pi + \theta_0}\right)$, then, referred to the anode impulse as half wave, the angle of overlap is

$$\theta_1 = \frac{\pi}{\pi + \theta_0} \theta_0.$$

The direct current, i_0, is the mean value of the anode current curves V, VI, and, assuming the latter as equivalent sine waves of maximum value $i_2 = i_0 + i'$, the direct current, i_0, is

$$\begin{aligned} i_0 &= i_2 \frac{1}{\pi - \theta_1} \int_0^\pi \sin \theta' \, d\theta' \\ &= \frac{2\, i_2}{\pi - \theta_1} \\ &= \frac{2\,(\pi + \theta_0)\, i_2}{\pi^2}, \end{aligned}$$

and

$$i_2 = i_0 \frac{\pi - \theta_1}{2};$$

or,

$$i_2 = \frac{\pi^2 i_0}{2\,(\pi + \theta_0)},$$

and the pulsation of the direct current, $i' = i_2 - i_0$, is

$$i' = i_0 \left\{ \frac{\pi^2}{2\,(\pi + \theta_0)} - 1 \right\}.$$

The effective value of the secondary current, as equivalent sine wave in one transformer coil, is the $\sqrt{\text{mean square}}$ of curves VII, VIII, or, assuming this current as existing in both transformer secondary coils in series — actually it alternates, one half wave in one, the other in the other transformer coil — is half this value, or

$$i = \frac{i_2}{2}\sqrt{\frac{1}{\pi - \theta_1}\left\{\int_{\theta_1}^{\pi-\theta_1} \sin^2\theta' d\theta' + \int_0^{\theta_1} [\sin \theta' + \sin (\theta' - \theta_1)]^2 d\theta'\right\}}$$

$$= \frac{i_2}{2}\sqrt{\frac{1}{\pi - \theta_1}\left\{\int_0^{\pi} \sin^2 \theta' \, d\theta' + \int_0^{\theta_1} 2 \sin \theta' \sin (\theta' - \theta_1) \, d\theta'\right\}}$$

$$= \frac{i_2}{2}\sqrt{\frac{1}{\pi - \theta_1}\left\{\frac{\pi}{2} + \Big[\theta' \cos \theta_1 - \sin (2\,\theta' - \theta_1)\Big]_0^{\theta_1}\right\}}$$

$$= \frac{i_2}{2}\sqrt{\frac{1}{\pi - \theta_1}\left\{\frac{\pi}{2} + \theta_1 \cos \theta_1 - \sin \theta_1\right\}};$$

or, substituting

$$i_2 = i_0 \frac{\pi - \theta_1}{2},$$

$$i = \frac{i_0}{4}\sqrt{(\pi - \theta_1)\left\{\frac{\pi}{2} + \theta_1 \cos \theta_1 - \sin \theta_1\right\}}$$

$$= \frac{i_0}{2}\frac{\pi}{2\sqrt{2}}\sqrt{\left(1 - \frac{\theta_1}{\pi}\right)\left(1 + \frac{2\,\theta_1}{\pi}\cos \theta_1 - \frac{2}{\pi}\sin \theta_1\right)};$$

or, substituting

$$\theta_1 = \frac{\pi\theta_0}{\pi + \theta_0},$$

$$i = \frac{i_0}{2}\frac{\pi}{2\sqrt{2}}\sqrt{1 - \frac{\theta_0}{\pi + \theta_0} + \frac{2\,\theta_0\pi}{(\pi + \theta_0)^2}\cos\frac{\theta_0\pi}{\pi + \theta_0} - \frac{2}{\pi + \theta_0}\sin\frac{\theta_0\pi}{\pi + \theta_0}},$$

where $\frac{\pi}{2\sqrt{2}}$ = ratio of effective value to mean value of sine wave.

27. An approximate representation by equivalent sine waves, if e_0 = the mean value of direct terminal voltage, i_0 = the mean value of direct current, is therefore as follows:

The secondary generated e.m.f. of the transformer is

$$e = (e_0 + e_a + r_0 i_0)\frac{\pi\sqrt{2}}{1 + \cos \theta_0};$$

the secondary current of the transformer is

$$= \frac{i_0}{2} \frac{\pi}{2\sqrt{2}} \sqrt{1 - \frac{\theta_0}{\pi + \theta_0} + \frac{2\,\theta_0 \pi}{(\pi + \theta_0)^2} \cos \frac{\theta_0 \pi}{\pi + \theta_0} - \frac{2}{\pi + \theta_0} \sin \frac{\theta_0 \pi}{\pi + \theta_0}};$$

the pulsation of the direct current is

$$i' = i_0 \left\{ \frac{\pi^2}{2\,(\pi + \theta_0)} - 1 \right\};$$

the anode voltage of the rectifier is

$$e_1 = e \sqrt{1 - \frac{2\,\theta_0 - \sin 2\,\theta_0}{2\,\pi}},$$

and herefrom follows the apparent efficiency of rectification, $\frac{e_0 i_0}{ei}$, the power factor, the efficiency, etc.

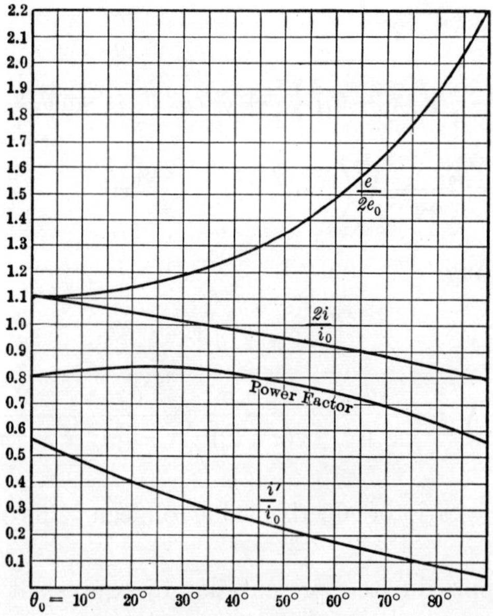

FIG. 26. E.m.f. and current ratio and secondary power factor of constant-current mercury arc rectifier.

From the equivalent sine waves, e and i, of the transformer secondary, and their phase angle, the primary impressed e.m.f. and the primary current of the transformer, and thereby the power factor, the efficiency, and the apparent efficiency of the system, are calculated in the usual manner.

In the secondary circuit, the power factor is below unity essentially due to wave shape distortion, less due to lag of current.

As example are shown, in Fig. 26, with the angle of overlap θ_0 as abscissas, the ratio of voltages, $\frac{e}{2\,e_0}$; the ratio of currents, $\frac{2\,i}{i_0}$; the current pulsation, $\frac{i'}{i_0}$, and the power factor of the secondary circuit.

PART II

TRANSIENTS IN SPACE

CHAPTER I

INTRODUCTION

1. The preceding sections deal with transient phenomena in time, that is, phenomena occurring during the time when a change or transition takes place between one condition of a circuit and another. The time, t, then is the independent variable, electric quantities as current, e.m.f., etc., the dependent variables.

Similar transient phenomena also occur in space, that is, with space, distance, length, etc., as independent variable. Such transient phenomena then connect the conditions of the electric quantities at one point in space with the electric quantities at another point in space, as, for instance, current and potential difference at the generator end of a transmission line with those at the receiving end of the line, or current density at the surface of a solid conductor carrying alternating current, as the rail return of a single-phase railway, with the current density at the center or in general inside of the conductor, or the distribution of alternating magnetism inside of a solid iron, as a lamina of an alternating-current transformer, etc. In such transient phenomena in space, the electric quantities, which appear as functions of space or distance, are not the instantaneous values, as in the preceding chapters, but are alternating currents, e.m.fs., etc., characterized by intensity and phase, that is, they are periodic functions of time, and the analytical method of dealing with such phenomena therefore introduces two independent variables, time t and distance l, that is, the electric quantities are periodic functions of time and transient functions of space.

The introduction of the complex quantities, as representing the alternating wave by a constant algebraic number, eliminates the time t as variable, so that, in the denotation by complex quantities, the transient phenomena in space are functions of one independent variable only, distance l, and thus lead to the same equations as the previously discussed phenomena, with the difference, however, that here, in dealing with space phenomena, the dependent variables, current, e.m.f., etc., are complex quantities, while in the previous discussion they appeared as instantaneous values, that is, real quantities.

Otherwise the method of treatment and the general form of the equations are the same as with transient functions of time.

2. Some of the cases in which transient phenomena in space are of importance in electrical engineering are:

(*a*) Circuits containing distributed capacity and self-inductance, as long-distance energy transmission lines, long-distance telephone circuits, multiple spark-gaps, as used in some forms of high potential lightning arresters (multi-gap arrester), etc.

(*b*) The distribution of alternating current in solid conductors and the increase of effective resistance and decrease of effective inductance resulting therefrom.

(*c*) The distribution of alternating magnetic flux in solid iron, or the screening effect of eddy currents produced in the iron, and the apparent decrease of permeability and increase of power consumption resulting therefrom.

(*d*) The distribution of the electric field of a conductor through space, resulting from the finite velocity of propagation of the electric field, and the variation of self-inductance and mutual inductance and of capacity of a conductor without return, as function of the frequency, in its effect on wireless telegraphy.

(*e*) Conductors conveying very high frequency currents, as lightning discharges, wireless telegraph and telephone currents, etc.

Only the current and voltage distribution in the long distance transmission line can be discussed more fully in the following, and the investigation of the other phenomena only indicated in outline, or the phenomena generally discussed, as lighting conductors.

CHAPTER II

LONG-DISTANCE TRANSMISSION LINE

3. If an electric impulse is sent into a conductor, as a transmission line, this impulse travels along the line at the velocity of light (approximately), or 188,000 miles (3×10^{10} cm.) per second. If the line is open at the other end, the impulse there is reflected and returns at the same velocity. If now at the moment when the impulse arrives at the starting point a second impulse, of opposite direction, is sent into the line, the return of the first impulse adds itself, and so increases the second impulse; the return of this increased second impulse adds itself to the third impulse, and so on; that is, if alternating impulses succeed each other at intervals equal to the time required by an impulse to travel over the line and back, the effects of successive impulses add themselves, and large currents and high e.m.fs. may be produced by small impulses, that is, low impressed alternating e.m.fs., or inversely, when once started, even with zero impressed e.m.f., such alternating currents traverse the lines for some time, gradually decreasing in intensity by the energy consumption in the conductor, and so fading out.

The condition of this phenomenon of electrical resonance thus is that alternating impulses occur at time intervals equal to the time required for the impulse to travel the length of the line and back; that is, the time of one half wave of impressed e.m.f. is the time required by light to travel twice the length of the line, or the time of one complete period is the time light requires to travel four times the length of the line; in other words, the number of periods, or frequency of the impressed alternating e.m.fs., in resonance condition, is the velocity of light divided by four times the length of the line; or, in free oscillation or resonance condition, the length of the line is one quarter wave length.

If then l = length of line, S = speed of light, the frequency of oscillations or natural period of the line is

$$f_0 = \frac{S}{4\,l}, \tag{1}$$

or, with l given in miles, hence S = 188,000 miles per second, it is

$$f_0 = \frac{47{,}000}{l} \text{ cycles.} \tag{2}$$

To get a resonance frequency as low as commercial frequencies, as 25 or 60 cycles, would require l = 1880 miles for f_0 = 25 cycles, and l = 783 miles for f_0 = 60 cycles.

It follows herefrom that many existing transmission lines are such small fractions of a quarter-wave length of the impressed frequency that the change of voltage and current along the line can be assumed as linear, or at least as parabolic; that is, the line capacity can be represented by a condenser in the middle of the line, or by condensers in the middle and at the two ends of the line, the former of four times the capacity of either of the two latter (the first approximation giving linear, the second a parabolic distribution).

For further investigation of these approximations see Dover edition Volume I, Part II.

If, however, the wave of impressed e.m.f. contains appreciable higher harmonics, some of the latter may approach resonance frequency and thus cause trouble. For instance, with a line of 150 miles length, the resonance frequency is f_0 = 313 cycles per second, or between the 5th harmonic and the 7th harmonic, 300 and 420 cycles of a 60-cycle system; fairly close to the 5th harmonic.

The study of such a circuit of distributed capacity thus becomes of importance with reference to the investigation of the effects of higher harmonics of the generator wave.

In long-distance telephony the important frequencies of speech probably range from 100 to 2000 cycles. For these frequencies the wave length varies from $\frac{S}{l}$ = 1880 miles down to 94 miles, and a telephone line of 1000 miles length would thus contain from about one-half to 11 complete waves of the impressed frequency. For long-distance telephony the phenomena occurring in the line thus can be investigated only by considering the complete equation of distributed capacity and inductance as so-called "wave transmission" and the phenomena thus essentially differ from those in a short energy transmission line.

4. Therefore in very long circuits, as in lines conveying alternating currents of high value at high potential over extremely long distances, by overhead conductors or underground cables, or with very feeble currents at extremely high frequency, such as telephone currents, the consideration of the *line resistance*, which consumes e.m.fs. in phase with the current, and of the *line reactance*, which consumes e.m.fs. in quadrature with the current, is not sufficient for the explanation of the phenomena taking place in the line, but several other factors have to be taken into account.

In long lines, especially at high potentials, the *electrostatic capacity* of the line is sufficient to consume noticeable currents. The charging current of the line condenser is proportional to the difference of potential and is one-fourth period ahead of the e.m.f. Hence, it either increases or decreases the main current, according to the relative phase of the main current and the e.m.f.

As a consequence the current changes in intensity, as well as in phase, in the line from point to point; and the e.m.fs. consumed by the resistance and inductance, therefore, also change in phase and intensity from point to point, being dependent upon the current.

Since no insulator has an infinite resistance, and since at high potentials not only leakage over surfaces but even direct *escape of electricity* into the air takes place by "brush discharge," or "corona," we have to recognize the existence of a current approximately proportional and in phase with the e.m.f. of the line. This current represents consumption of power, and is therefore analogous to the e.m.f. consumed by resistance, while the capacity current and the e.m.f. of inductance are wattless or reactive.

Furthermore, the alternating current passing over the line produces in all neighboring conductors secondary currents, which react upon the primary current and thereby introduce e.m.fs. of *mutual inductance* into the primary circuit. Mutual inductance is neither in phase nor in quadrature with the current, and can therefore be resolved into a *power component* of mutual inductance in phase with the current, which acts as an increase of resistance, and into a *reactive component* in quadrature with the current, which appears as a self-inductance.

This mutual inductance is not always negligible, as, for instance, its disturbing influence in telephone circuits shows.

The alternating potential of the line induces, by *electrostatic influence*, electric charges in neighboring conductors outside of the circuit, which retain corresponding opposite charges on the line wires. This electrostatic influence requires the expenditure of a current proportional to the e.m.f. and consisting of a *power component* in phase with the e.m.f. and a *reactive component* in quadrature thereto.

The alternating electromagnetic field of force set up by the line current produces in some materials a loss of power by *magnetic hysteresis*, or an expenditure of e.m.f. in phase with the current, which acts as an increase of resistance. This electromagnetic hysteresis loss may take place in the conductor proper if iron wires are used, and may then be very serious at high frequencies such as those of telephone currents.

The effect of *eddy currents* has already been referred to under "mutual inductance," of which it is a power component.

The alternating electrostatic field of force expends power in dielectrics by what is called *dielectric hysteresis*. In concentric cables, where the electrostatic gradient in the dielectric is comparatively large, the dielectric hysteresis may at high potentials consume considerable amounts of power. The dielectric hysteresis appears in the circuit as consumption of a current whose component in phase with the e.m.f. is the *dielectric power current*, which may be considered as the power component of the charging current.

Besides this there is the apparent increase of ohmic resistance due to *unequal distribution of current*, which, however, is usually not large enough to be noticeable at low frequencies.

Also, especially at very high frequency, energy is *radiated* into space, due to the finite velocity of the electric field, and can be represented by power components of current and of voltage respectively.

5. This gives, as the most general case and per unit length of line,

E.m.fs. consumed in phase with the current, I, and $= rI$, representing consumption of power, and due to *resistance,* and its apparent increase by unequal current distribution; to the power component of *mutual inductance:* to *secondary currents;* to the power component of *self-inductance:* to *electromagnetic hysteresis;* and to *electromagnetic radiation.*

E.m.fs. consumed in quadrature with the current, I, and $= xI$, reactive, and due to *self-inductance* and *mutual inductance.*

Currents consumed in phase with the e.m.f., E, and $= gE$, representing consumption of power, and due: to *leakage* through the insulating material, brush discharge or corona; to the power component of *electrostatic influence;* to the power component of *capacity,* or *dielectric hysteresis,* and to *electrostatic radiation.*

Currents consumed in quadrature with the e.m.f., E, and $= bE$, being reactive, and due to *capacity* and *electrostatic influence.*

Hence we get four constants per unit length of line, namely: Effective resistance, r; effective reactance, x; effective conductance, g, and effective susceptance, $b = -b_c$ (b_c being the absolute value of susceptance). These constants represent the coefficients per unit length of line of the following: e.m.f. consumed in phase with the current; e.m.f. consumed in quadrature with the current; current consumed in phase with the e.m.f., and current consumed in quadrature with the e.m.f.

6. This line we may assume now as supplying energy to a *receiver circuit of any description,* and determine the current and e.m.f. at any point of the circuit.

That is, an e.m.f. and current (differing in phase by any desired angle) may be given at the terminals of the receiving circuit. To be determined are the e.m.f. and current at any point of the line, for instance, at the generator terminals; or the impedance, $Z_1 = r_1 + jx_1$, or admittance, $Y_1 = g_1 - jb_1$, of the receiver circuit, and e.m.f., E_0, at generator terminals are given; the current and e.m.f. at any point of circuit to be determined, etc.

7. Counting now the distance, l, from a point 0 of the line which has the e.m.f.

$$\dot{E}_1 = e_1 - je_1'$$

and the current

$$\dot{I}_1 = i_1 - ji_1',$$

and counting l positive in the direction of rising power and negative in the direction of decreasing power, at any point l, in the line differential dl the leakage current is

$$\dot{E}g\, dl$$

and the capacity current is

$$j\dot{E}b\, dl;$$

hence, the total current consumed by the line differential dl is

$$\begin{aligned} d\dot{I} &= \dot{E}\,(g + jb)\, dl \\ &= \dot{E}Y\, dl, \end{aligned}$$

or

$$\frac{d\dot{I}}{dl} = Y\dot{E}. \tag{1}$$

In the line differential dl the e.m.f. consumed by resistance is

$$\dot{I}r\, dl,$$

the e.m.f. consumed by inductance is

$$j\dot{I}x\, dl;$$

hence, the total e.m.f. consumed by the line differential dl is

$$\begin{aligned} d\dot{E} &= \dot{I}\,(r + jx)\, dl \\ &= \dot{I}Z\, dl, \end{aligned}$$

or

$$\frac{d\dot{E}}{dl} = Z\dot{I}. \tag{2}$$

These *fundamental differential equations* (1) and (2) are symmetrical with respect to $\dot{I}$ and $\dot{E}$.

Differentiating these equations (1) and (2) gives

$$\left.\begin{aligned} \frac{d^2\dot{I}}{dl^2} &= Y\frac{d\dot{E}}{dl} \\ \text{and}\qquad \frac{d^2\dot{E}}{dl^2} &= Z\frac{d\dot{I}}{dl}, \end{aligned}\right\} \tag{3}$$

and substituting (1) and (2) in (3) gives the *differential equations of $\dot{E}$ and $\dot{I}$*, thus:

$$\frac{d^2\dot{E}}{dl^2} = YZ\dot{E} \tag{4}$$

and

$$\frac{d^2\dot{I}}{dl^2} = YZ\dot{I}. \tag{5}$$

These differential equations are identical in form, and consequently I and E are functions differing by their integration constants or by their limiting conditions only.

These equations are of the form

$$\frac{d^2\dot{U}}{dl^2} = ZY\dot{U}$$

and are integrated by

$$\dot{U} = A\varepsilon^{Vl},$$

where ε is the basis of the natural logarithms, $= 2.718283$.

Choosing equation (5), which is integrated by

$$\dot{I} = A\varepsilon^{Vl}, \tag{6}$$

and differentiating (6) twice gives

$$\frac{d^2\dot{I}}{dl^2} = \dot{V}^2\dot{A}\varepsilon^{Vl},$$

and substituting (6) in (5), the factor $\dot{A}\varepsilon^{Vl}$ cancels, and we have

$$V^2 = ZY,$$

or

$$V = \sqrt{ZY}, \tag{7}$$

hence, the general integral,

$$\dot{I} = \dot{A}_1\varepsilon^{+Vl} - \dot{A}_2\varepsilon^{-Vl}. \tag{8}$$

By equation (1),

$$\dot{E} = \frac{1}{Y}\frac{d\dot{I}}{dl},$$

and substituting herein equation (8) gives

$$\dot{E} = \frac{V}{Y}\left\{\dot{A}_1\varepsilon^{+Vl} + \dot{A}_2\varepsilon^{-Vl}\right\}, \tag{9}$$

or, substituting (7),

$$\dot{E} = \sqrt{\frac{Z}{Y}}\left\{ \dot{A}_1\varepsilon^{+Vl} + \dot{A}_2\varepsilon^{-Vl} \right\}. \tag{10}$$

The integration constants A_1 and A_2 in (8), (9), (10), in general, are complex quantities. The coefficient of the exponent, V, as square root of the product of two complex quantities, also is a complex quantity, therefore may be written

$$V = \alpha + j\beta, \tag{11}$$

and substituting for V, Z and Y gives

$$(\alpha + j\beta)^2 = (r + jx)(g + jb),$$

or

$$(\alpha^2 - \beta^2) + 2\,j\alpha\beta = (rg - xb) + j\,(rb + gx),$$

and this resolves into the two separate equations

$$\left.\begin{aligned} \alpha^2 - \beta^2 &= rg - xb \\ 2\,\alpha\beta &= rb + gx, \end{aligned}\right\} \tag{12}$$

since, when two complex quantities are equal, their real terms as well as their imaginary terms must be equal.

Equations (12) squared and added give

$$\begin{aligned} (\alpha^2 + \beta^2)^2 &= (rg - xb)^2 + (rb + xg)^2 \\ &= (r^2 + x^2)(g^2 + b^2) \\ &= z^2y^2; \end{aligned}$$

hence,

$$\alpha^2 + \beta^2 = zy, \tag{13}$$

and from (12) and (13),

$$\left.\begin{aligned} \alpha &= \sqrt{\tfrac{1}{2}(zy + rg - xb)} \\ \text{and}\qquad \beta &= \sqrt{\tfrac{1}{2}(zy - rg + xb)}. \end{aligned}\right\} \tag{14}$$

Equations (8) and (10) now assume the form

$$\left.\begin{aligned} \dot{I} &= \dot{A}_1\varepsilon^{+(\alpha+j\beta)l} - \dot{A}_2\varepsilon^{-(\alpha+j\beta)l} \\ \text{and}\qquad \dot{E} &= \sqrt{\frac{Z}{Y}}\left\{ A_1\varepsilon^{+(\alpha+j\beta)l} + A_2\varepsilon^{-(\alpha+j\beta)l} \right\}. \end{aligned}\right\} \tag{15}$$

Substituting for the exponential function with an imaginary exponent the trigonometric expression

$$\varepsilon^{\pm j\beta l} = \cos \beta l \pm j \sin \beta l, \tag{16}$$

equations (15) assume the form

$$\left.\begin{aligned} \dot{I} &= \dot{A}_1\varepsilon^{+al}(\cos \beta l + j \sin \beta l) - \dot{A}_2\varepsilon^{-al}(\cos \beta l - j \sin \beta l) \\ &\text{and} \\ \dot{E} &= \sqrt{\frac{Z}{Y}}\left\{ \dot{A}_1\varepsilon^{+al}(\cos \beta l + j \sin \beta l) + \dot{A}_2\varepsilon^{-al}(\cos \beta l - j \sin \beta l) \right\}, \end{aligned}\right\} \tag{17}$$

where $\dot{A}_1$ and $\dot{A}_2$ are the constants of integration.

The distribution of current $\dot{I}$ and voltage $\dot{E}$ along the circuit, therefore, is represented by the sum of two products of exponential and trigonometric functions of the distance l. Of these terms, the one, with factor $\dot{A}\varepsilon^{+al}$, increases with increasing distance l, that is, increases towards the generator, while the other, with factor $\dot{A}_2\varepsilon^{-al}$, decreases towards the generator and thus increases with increasing distance from the generator. The phase angle of the former decreases, that of the latter increases towards the generator, and the first term thus can be called the main wave, the second term the reflected wave.

At the point $l = 0$, by equations (17) we have

$$\dot{I}_0 = \dot{A}_1 - \dot{A}_2,$$

$$\dot{E}_0 = \sqrt{\frac{Z}{Y}}\left\{ \dot{A}_1 + \dot{A}_2 \right\},$$

and the ratio

$$\frac{\dot{A}_2}{\dot{A}_1} = m\,(\cos \tau - j \sin \tau),$$

where τ may be called the angle of reflection, and m the ratio of amplitudes of reflected and main wave at the reflection point.

8. The general integral equations of current and voltage distribution (17) can be written in numerous different forms.

Substituting $-\dot{A}_2$ instead of $+\dot{A}_2$, the sign between the terms reverses, and the current appears as the sum, the voltage as difference of main and reflected wave.

Rearranging (17) gives

$$\left.\begin{aligned} \dot{I} &= (\dot{A}_1\varepsilon^{+al} - \dot{A}_2\varepsilon^{-al})\cos\beta l + j\,(\dot{A}_1\varepsilon^{+al} + \dot{A}_2\varepsilon^{-al})\sin\beta l \\ &\text{and} \\ \dot{E} &= \sqrt{\frac{Z}{Y}}\left\{(\dot{A}_1\varepsilon^{+al} + \dot{A}_2\varepsilon^{-al})\cos\beta l + j\,(\dot{A}_1\varepsilon^{+al} - \dot{A}_2\varepsilon^{-al})\sin\beta l\right\}. \end{aligned}\right\} \quad (18)$$

Substituting (7) gives

$$\sqrt{\frac{Z}{Y}} = \frac{Z}{V} = \frac{V}{Y}, \quad (19)$$

and substituting

$$\left.\begin{aligned} \frac{\dot{A}_1}{V} &= \dot{B}_1 \\ \text{and} \quad \frac{\dot{A}_2}{V} &= \dot{B}_2, \end{aligned}\right\}$$

or

$$\left.\begin{aligned} \frac{\dot{A}_1}{Y} &= \dot{C}_1 \\ \text{and} \quad \frac{\dot{A}_2}{Y} &= \dot{C}_2, \end{aligned}\right\}$$

changes equations (17) to the forms,

$$\left.\begin{aligned} \dot{I} &= V\left\{\dot{B}_1\varepsilon^{+al}(\cos\beta l + j\sin\beta l) - \dot{B}_2\varepsilon^{-al}(\cos\beta l - j\sin\beta l)\right\} \\ &\text{and} \\ \dot{E} &= Z\left\{\dot{B}_1\varepsilon^{+al}(\cos\beta l + j\sin\beta l) + \dot{B}_2\varepsilon^{-al}(\cos\beta l - j\sin\beta l),\right\} \end{aligned}\right\} \quad (20)$$

or

$$\left.\begin{aligned} \dot{I} &= Y\left\{\dot{C}_1\varepsilon^{+al}(\cos\beta l + j\sin\beta l) - \dot{C}_2\varepsilon^{-al}(\cos\beta l - j\sin\beta l)\right\} \\ &\text{and} \\ \dot{E} &= V\left\{\dot{C}_1\varepsilon^{+al}(\cos\beta l + j\sin\beta l) + \dot{C}_2\varepsilon^{-al}(\cos\beta l - j\sin\beta l)\right\} \end{aligned}\right\} \quad (21)$$

Substituting in (17)

$$\frac{\dot{A}_1}{\sqrt{Y}} = \dot{D}_1 \text{ and } \frac{\dot{A}_2}{\sqrt{Y}} = \dot{D}_2$$

gives

$$\left.\begin{aligned} \dot{I} &= \sqrt{Y}\left\{\dot{D}_1\varepsilon^{+al}(\cos\beta l + j\sin\beta l) - \dot{D}_2\varepsilon^{-al}(\cos\beta l - j\sin\beta l)\right\} \\ &\text{and} \\ \dot{E} &= \sqrt{Z}\left\{\dot{D}_1\varepsilon^{+al}(\cos\beta l + j\sin\beta l) + \dot{D}_2\varepsilon^{-al}(\cos\beta l - j\sin\beta l)\right\}. \end{aligned}\right\} \quad (22)$$

Reversing the sign of l, that is, counting the distance in the opposite direction, or positive for decreasing power, from the generator towards the receiving circuit, and not, as in equations (17) to (22), from the receiving circuit towards the generator, exchanges the position of the two terms; that is, the first term, or the main wave, decreases with increasing distance, and lags; the second term, or the reflected wave, increases with the distance, and leads.

Equations (17) thus assume the form

$$\left.\begin{aligned} \dot{I} &= \dot{A}_1\varepsilon^{-al}(\cos\beta l - j\sin\beta l) - \dot{A}_2\varepsilon^{+al}(\cos\beta l + j\sin\beta l) \\ &\text{and} \\ \dot{E} &= \sqrt{\frac{Z}{Y}}\left\{\dot{A}_1\varepsilon^{-al}(\cos\beta l - j\sin\beta l) + \dot{A}_2\varepsilon^{+al}(\cos\beta l + j\sin\beta l)\right\}, \end{aligned}\right\} \quad (23)$$

and correspondingly equations (18) to (22) modify.

9. The two integration constants contained in equations (17) to (23) require two conditions for their determination, such as current and voltage at one point of the circuit, as at the generator or at the receiving end; or current at one point, voltage at the other; or voltage at one point, as at the generator, and ratio of voltage and current at the other end, as the impedance of the receiving circuit.

Let the current and voltage (in intensity as well as phase, that is, as complex quantities) be given at one point of the circuit, and counting the distance l from this point, the terminal conditions are

$$\left.\begin{aligned} l &= 0, \\ \dot{I} &= \dot{I}_0 = i_0 - ji_0', \\ \text{and} \quad \dot{E} &= \dot{E}_0 = e_0 - je_0'. \end{aligned}\right\} \quad (24)$$

Substituting (24) in (17) gives

$$\left.\begin{aligned} \dot{I}_0 &= \dot{A}_1 - \dot{A}_2 \\ \text{and}\quad \dot{E}_0 &= \sqrt{\frac{Z}{Y}}\left(\dot{A}_1 + \dot{A}_2\right); \end{aligned}\right\}$$

hence,

$$\left.\begin{aligned} \dot{A}_1 &= \frac{1}{2}\left\{\dot{I}_0 + \dot{E}_0\sqrt{\frac{Y}{Z}}\right\} \\ \text{and}\quad \dot{A}_2 &= -\frac{1}{2}\left\{\dot{I}_0 - \dot{E}_0\sqrt{\frac{Y}{Z}}\right\}, \end{aligned}\right\}$$

and substituted in (17) gives

$$\left.\begin{aligned} \dot{I} = \frac{1}{2}\Bigg\{&\left(\dot{I}_0 + \dot{E}_0\sqrt{\frac{Y}{Z}}\right)\varepsilon^{+al}(\cos\beta l + j\sin\beta l) \\ &+\left(\dot{I}_0 - \dot{E}_0\sqrt{\frac{Y}{Z}}\right)\varepsilon^{-al}(\cos\beta l - j\sin\beta l)\Bigg\} \\ \text{and}\quad & \\ \dot{E} = \frac{1}{2}\Bigg\{&\left(\dot{E}_0 + \dot{I}_0\sqrt{\frac{Z}{Y}}\right)\varepsilon^{+al}(\cos\beta l + j\sin\beta l) \\ &+\left(\dot{E}_0 - \dot{I}_0\sqrt{\frac{Z}{Y}}\right)\varepsilon^{-al}(\cos\beta l - j\sin\beta l)\Bigg\}. \end{aligned}\right\}. \quad (25)$$

If then $\dot{I}_0$ and $\dot{E}_0$ are the current and voltage respectively at the receiving end or load end of a circuit of length l_0, equations (25) represent current and voltage at any point of the circuit, from the receiving end $l = 0$ to the generator end $l = l_0$.

If $\dot{I}_0$ and $\dot{E}_0$ are the current and voltage at the generator terminals, since in equations (17) l is counted towards rising power, in the present case the receiving end of the line is represented by $l = -l_0$; that is, the negative values of l represent the distance from the generator end, along the line. In this case it is more convenient to reverse the sign of l, that is, use equations (22) and the distribution of current and voltage at distance l from the generator terminals. $\dot{I}$, $\dot{E}$ are then given by

$$\left.\begin{aligned} \dot{I} &= \frac{1}{2}\left\{\left(\dot{I}_0 + \dot{E}_0\sqrt{\frac{Y}{Z}}\right)\varepsilon^{-al}(\cos\beta l - j\sin\beta l)\right. \\ &\qquad \left. + \left(\dot{I}_0 - \dot{E}_0\sqrt{\frac{Y}{Z}}\right)\varepsilon^{+al}(\cos\beta l + j\sin\beta l)\right\} \\ \text{and} & \\ \dot{E} &= \frac{1}{2}\left\{\left(\dot{E}_0 + \dot{I}_0\sqrt{\frac{Z}{Y}}\right)\varepsilon^{-al}(\cos\beta l - j\sin\beta l)\right. \\ &\qquad \left. + \left(\dot{E}_0 - \dot{I}_0\sqrt{\frac{Z}{Y}}\right)\varepsilon^{+al}(\cos\beta l + j\sin\beta l)\right\}. \end{aligned}\right\} \quad (26)$$

10. Assume that the character of the load, that is, the impedance, $\frac{\dot{E}_1}{\dot{I}_1} = Z_1 = r_1 + jx_1$, or admittance, $\frac{\dot{I}_1}{\dot{E}_1} = Y_1 = \frac{1}{Z_1} = g_1 - jb_1$, of the receiving circuit and the voltage $\dot{E}_0$ at the generator end of the circuit be given.

Let l_0 = length of circuit, and counting distance l from the generator end, for $l = 0$ we have

$$\dot{E} = \dot{E}_0;$$

this substituted in equation (23) gives

$$\dot{E}_0 = \sqrt{\frac{Z}{Y}}(\dot{A}_1 + \dot{A}_2). \quad (27)$$

However, for $l = l_0$,

$$\frac{\dot{E}}{\dot{I}} = Z_1;$$

substituting (23) herein gives

$$Z_1 = \sqrt{\frac{Z}{Y}}\frac{\dot{A}_1\varepsilon^{-al_0}(\cos\beta l_0 - j\sin\beta l_0) + \dot{A}_2\varepsilon^{+al_0}(\cos\beta l_0 + j\sin\beta l_0)}{\dot{A}_1\varepsilon^{-al_0}(\cos\beta l_0 - j\sin\beta l_0) - \dot{A}_2\varepsilon^{+al_0}(\cos\beta l_0 + j\sin\beta l_0)};$$

hence, substituting (19) and expanding,

$$\frac{\dot{A}_1\varepsilon^{-al_0}(\cos\beta l_0 - j\sin\beta l_0)}{\dot{A}_2\varepsilon^{+al_0}(\cos\beta l_0 + j\sin\beta l_0)} = \frac{VZ_1 + Z}{VZ_1 - Z},$$

or $$\dot{A}_2 = \dot{A}_1\frac{VZ_1 - Z}{VZ_1 + Z}\varepsilon^{-2al_0}(\cos 2\beta l_0 - j\sin 2\beta l_0),$$

and denoting the complex factor by

$$\dot{C} = \frac{VZ_1 - Z}{VZ_1 + Z}\varepsilon^{-2\,al_0}(\cos 2\,\beta l_0 - j \sin \beta l_0), \tag{28}$$

which may be called the reflection constant, we have

$$\dot{A}_2 = \dot{C}\dot{A}_1,$$

and by (27),

$$\left.\begin{aligned} \dot{A}_1 &= \frac{\dot{E}_0}{1+\dot{C}}\sqrt{\frac{Y}{Z}} \\ \text{and} \qquad \dot{A}_2 &= \frac{\dot{E}_0\,\dot{C}}{1+\dot{C}}\sqrt{\frac{Y}{Z}}; \end{aligned}\right\} \tag{29}$$

hence, substituted in (17),

$$\left.\begin{aligned} \dot{I} &= \frac{\dot{E}_0}{1+\dot{C}}\sqrt{\frac{Y}{Z}}\left\{\varepsilon^{-al}(\cos\beta l - j\sin\beta l) - \dot{C}\varepsilon^{+al}(\cos\beta l + j\sin\beta l)\right\} \\ &\text{and} \\ \dot{E} &= \frac{\dot{E}_0}{1+\dot{C}}\left\{\varepsilon^{-al}(\cos\beta l - j\sin\beta l) + \dot{C}\varepsilon^{+al}(\cos\beta l + j\sin\beta l)\right\}. \end{aligned}\right\} \tag{30}$$

11. As an example, consider the problem of delivering, in a three-phase system, 200 amperes per phase, at 90 per cent power factor lag at 60,000 volts per phase (or between line and neutral) and 60 cycles, at the end of a transmission line 200 miles in length, consisting of two separate circuits in multiple, each consisting of number 00 B. and S. wire with 6 feet distance between the conductors.

Number 00 B. and S. wire has a resistance of 0.42 ohms per mile, and at 6 feet distance from the return conductor an inductance of 2.4 mh. and capacity of 0.015 mf. per mile.

The two circuits in multiple give, at 60 cycles, the following line constants per mile: $r = 0.21$ ohm, $L = 1.2 \times 10^{-3}$ henry, and $C = 0.03 \times 10^{-6}$ farad; hence,

$$\begin{aligned} x &= 2\,\pi f L = 0.45, \\ Z &= 0.21 + 0.45\,j, \\ z &= 0.50, \end{aligned}$$

and, neglecting the conductance ($g = 0$),

$$\begin{aligned} b &= 2\,\pi f C = 11 \times 10^{-6}, \\ Y &= 11 \times 10^{-6}\,j, \\ y &= 11 \times 10^{-6}, \end{aligned}$$

and
$$\left.\begin{aligned}\alpha &= 0.524 \times 10^{-3},\\ \beta &= 2.285 \times 10^{-3},\end{aligned}\right\} \qquad (31)$$
$$V = (0.524 + 2.285\,j)\,10^{-3},$$
$$\sqrt{\frac{Y}{Z}} = \frac{V}{Z} = (4.53 + 0.9\,j)\,10^{-3}$$
and
$$\sqrt{\frac{Z}{Y}} = \frac{V}{Y} = (0.208 - 0.047\,j)\,10^{+3}.$$

Counting the distance l from the receiving end, and choosing the receiving voltage as zero vector, we have

$$l = 0,$$
$$\dot{E} = \dot{E}_0 = e_0 = 60{,}000 \text{ volts},$$

and the current of 200 amperes at 90 per cent power factor,

$$\dot{I} = \dot{I}_0 = i_0 + ji_0' = 180 - 87\,j,$$

and substituting these values in equations (25) gives

$$\left.\begin{aligned}\dot{I} &= (226 - 14.4\,j)\,\varepsilon^{+al}(\cos\beta l + j\sin\beta l) - (46 + 72.6\,j)\,\varepsilon^{-al}\\ &\qquad(\cos\beta l - j\sin\beta l), \text{ in amperes,}\\ &\text{and}\\ \dot{E} &= (46.7 - 13.3\,j)\,\varepsilon^{+al}(\cos\beta l + j\sin\beta l) + (13.3 + 13.3\,j)\,\varepsilon^{-al}\\ &\qquad(\cos\beta l - j\sin\beta l), \text{ in kilovolts,}\end{aligned}\right\} \qquad (32)$$

where α and β are given by above equations (31).

From equations (32) the following results are obtained.

Receiving end of line,	$l = 0$		
$\dot{I} = 180 - 87\,j$	$i = 200$ amp.	$\tan\theta_1 = 0.483$	$\theta_1 = 26°$
$\dot{E} = 60 \times 10^3$	$e = 60{,}000$ volts		$\theta_2 = 0$
		power factor,	0.90 lag
Middle of line,	$l = 100$		
$\dot{I} = 177 - 18\,j$	$i = 178$ amp.	$\tan\theta_1 = +0.102$	$\theta_1 = 6°$
$\dot{E} = (66.2 + 6.9\,j)\,10^3$	$e = 66{,}400$ volts	$\tan\theta_2 = -0.104$	$\theta_2 = -6°$
			$\theta_1 - \theta_2 = \theta = 12°$
		power factor, $\cos\theta$	$= 0.979$ lag.
Generator end of line,	$l = 200$		
$\dot{I} = 165.7 + 56\,j$	$i = 175$ amp.	$\tan\theta_1 = -0.338$	$\theta_1 = -19°$
$\dot{E} = (69 + 15\,j)\,10^3$	$e = 70{,}700$ volts	$\tan\theta_2 = -0.218$	$\theta_2 = -12°$
			$\theta_1 - \theta_2 = \theta = -7°$
		power factor, $\cos\theta$	$= 0.993$ lead.

As seen, the current decreases from the receiving end to the middle of the line, but from there to the generator remains practically constant. The voltage increases more in the receiving half of the line than in the generator half. The power-factor is practically unity from the middle of the line to the generator.

12. It is interesting to compare with above values the values derived by neglecting the distributed character of resistance, inductance, and capacity.

From above constants per mile it follows, for the total line of 200 miles length, $r_0 = 42$ ohms, $x_0 = 90$ ohms, and $b_0 = 2.2 \times 10^{-3}$ mho; hence,

$$Z_0 = 42 + 90\,j$$
$$\text{and } Y_0 = 2.2\,j\,10^{-3}.$$

(1) Neglecting the line capacity altogether, with $\underset{\cdot}{I}_0$ and $\underset{\cdot}{E}_0$ at the receiver terminals, at the generator terminals we have

$$\underset{\cdot}{I}_1 = \underset{\cdot}{I}_0$$

and

$$\underset{\cdot}{E}_1 = \underset{\cdot}{E}_0 + Z_0 \underset{\cdot}{I}_0;$$

hence,

$\underset{\cdot}{I}_1 = 180 - 87\,j$	$i_1 = 200$ amp.	$\tan\theta_1 = 0.483$	$\theta_1 = +26°$
$\underset{\cdot}{E}_1 = (75.4 + 12.6\,j)\,10^3$	$e_1 = 76{,}400$ volts	$\tan\theta_2 = -0.167$	$\theta_2 = -9°$
			$\theta_1 - \theta_2 = \theta = +34°$
			power factor, $\cos\theta = 0.83$ lag.

These values are extremely inaccurate, voltage and current at generator too high and power factor too low.

(2) Representing the line capacity by a condenser at the generator end, that is, adding the condenser current at the generator end,

$$\underset{\cdot}{I}_1 = \underset{\cdot}{I}_0 + Y_0 \underset{\cdot}{E}_1$$

and

$$\underset{\cdot}{E}_1 = \underset{\cdot}{E}_0 + Z_0 \underset{\cdot}{I}_0;$$

hence,

$\underset{\cdot}{I}_1 = 152 + 89\,j$	$i_1 = 176$ amp.	$\tan\theta_1 = -0.585$	$\theta_1 = -30°$
$\underset{\cdot}{E}_1 = (75.4 + 12.6\,j)10^3$	$e_1 = 76{,}400$ volts	$\tan\theta_2 = -0.167$	$\theta_2 = -9°$
			$\theta_1 - \theta_2 = \theta = -21°$
			power factor, $\cos\theta = 0.93$ lead.

As seen, the current is approximately correct, but the voltage is far too high and the power factor is still low, but now leading.

(3) Representing the line capacity by a condenser at the receiving end, that is, adding the condenser current at the load,

$$\dot{I}_1 = \dot{I}_0 + Y_0\dot{E}_0$$

and

$$\dot{E}_1 = \dot{E}_0 + Z_0\dot{I}_1;$$

hence,

$\dot{I}_1 = 180 + 45j$ $i_1 = 186$ amp. $\tan\theta_1 = -.250$ $\theta_1 = -14°$

$\dot{E}_1 = (63.5 + 18.1j)\,10^3$ $e_1 = 66{,}000$ volts $\tan\theta_2 = -.285$ $\theta_2 = -16°$

$\theta_1 - \theta_2 = \theta = +2°$

power factor, $\cos\theta = 1.00$

In this case the voltage e_1 is altogether too low, the current somewhat high, but the power factor fairly correct.

(4) Taking the average of the values of (2) and of (3) gives

$\dot{I}_1 = 166 + 67j$ $i_1 = 179$ amp. $\tan\theta_1 = -0.403$ $\theta_1 = -22°$

$\dot{E}_1 = (69.4 + 15.3j)\,10^3$ $e_1 = 71{,}100$ volts $\tan\theta_2 = -0.220$ $\theta_2 = -12°$

$\theta_1 - \theta_2 = \theta = -10°$

power-factor, $\cos\theta = 0.985$ lead.

As seen by comparing these average values with the exact result as derived above, these values are not very different, but constitute a fair approximation in the present case. Such a close coincidence of this approximation with the exact result can, however, not be counted upon in all instances.

13. In the equations (17) to (23) the length

$$l_w = \frac{2\pi}{\beta} \tag{33}$$

is a complete wave length, which means that in the distance $\frac{2\pi}{\beta}$ the phases of the components of current and of e.m.f. repeat, and that in half this distance they are just opposite.

Hence, the remarkable condition exists that in a very long line at different points the currents are simultaneously in opposite directions and the e.m.fs. are opposite.

The difference of space phase τ between current I and e.m.f. E at any point l of the line is determined by the equation

$$m\,(\cos \tau - j \sin \tau) = \frac{\dot{E}}{\dot{I}}, \tag{34}$$

where m is a constant.

Hence, τ varies from point to point, oscillating around a medium position, τ_∞, which it approaches at infinity.

This difference of phase, τ_∞, towards which current and e.m.f. tend at infinity, is determined by the expression

$$m\,(\cos \tau_\infty - j \sin \tau_\infty) = \left[\frac{\dot{E}}{\dot{I}}\right]_{l=\infty},$$

or, substituting for $\dot{E}$ and $\dot{I}$ their values from equations (23), and since $\varepsilon^{-al} = 0$, and $\dot{A}_1 \varepsilon^{al} (\cos \beta l + j \sin \beta l)$ cancels,

$$m\,(\cos \tau_\infty + j \sin \tau_\infty) = \sqrt{\frac{Z}{Y}} = \frac{V}{Y} = \frac{\alpha + j\beta}{g + jb}$$

$$= \frac{(\alpha g + \beta b) - j\,(\alpha b - \beta g)}{b^2 + g^2};$$

hence,

$$\tan \tau_\infty = \frac{+\,\alpha b - \beta g}{\alpha g + \beta b}. \tag{35}$$

14. This angle, $\tau_\infty = 0$; that is, current and e.m.f. come more and more in phase with each other when

$$\alpha b - \beta g = 0; \text{ that is,}$$

$$\alpha \div \beta = g \div b, \text{ or}$$

$$\frac{\alpha^2 - \beta^2}{2\,\alpha\beta} = \frac{g^2 - b^2}{2\,gb};$$

substituting (12) gives

$$\frac{gr - bx}{gx + br} = \frac{g^2 - b^2}{2\,gb};$$

hence, expanding,

$$r \div x = g \div b; \tag{36}$$

that is, *the ratio of resistance to inductance equals the ratio of leakage to capacity.*

This angle, $\tau_\infty = 45°$; that is, current and e.m.f. differ by one-eighth period if $+ \alpha b - \beta g = \alpha g + \beta b$, or

$$\frac{\alpha}{\beta} = \frac{b + g}{b - g},$$

which gives $$rg + xb = 0, \tag{37}$$

which means that two of the four line-constants, either g and x or g and b, must be zero.

The case where $g = 0 = x$, that is, a line having only resistance and distributed capacity but no self-inductance, is approximately realized in concentric or multiple-conductor cables, and in these the space-phase angle tends towards 45 degrees lead for infinite length.

15. As an example are shown the characteristic curves of a transmission line of the relative constants,

$r : x : g : b = 8 : 32 : 1.25 \times 10^{-4} : 25 \times 10^{-4}$ and $e = 25{,}000$, $i = 200$ at the receiving circuit, for the conditions

(*a*) Non-inductive load in the receiving circuit, Fig. 80.

(*b*) Wattless receiving circuit of 90 time-degrees lag, Fig. 81.

(*c*) Wattless receiving circuit of 90 time-degrees lead, Fig. 82.

These curves are determined graphically by constructing the topographic circuit characteristics in polar coördinates as explained in Dover edition Volume I, Part II, and deriving corresponding values of current, potential difference, and phase angle therefrom.

As seen from these diagrams, for wattless receiving circuit, current and e.m.f. oscillate in intensity inversely to each other, with an amplitude of oscillation gradually decreasing when passing from the receiving circuit towards the generator, while the space-phase angle between current and e.m.f. oscillates between lag and lead with decreasing amplitude. Approximately maxima and minima of current coincide with minima and maxima of e.m.f. and zero phase angles.

For such graphical constructions, polar coördinate paper and two angles α and δ are desirable, the angle α being the angle between current and change of e.m.f., $\tan \alpha = \frac{x}{r} = 4$, and the

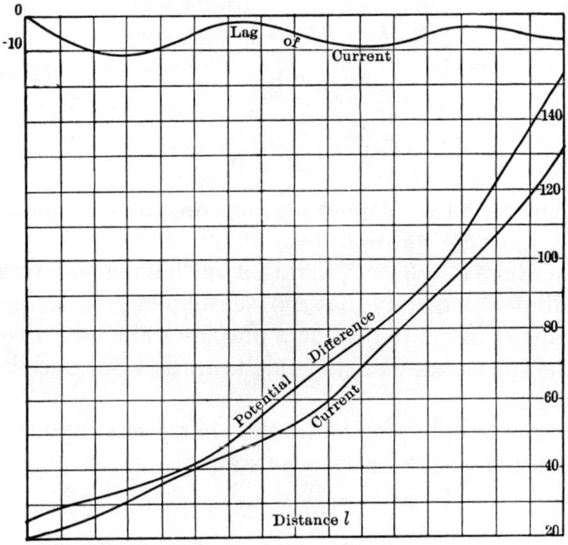

FIG. 1. Current, e.m.f. and space-phase angle between current and e.m.f. in a transmission line. Non-inductive load.

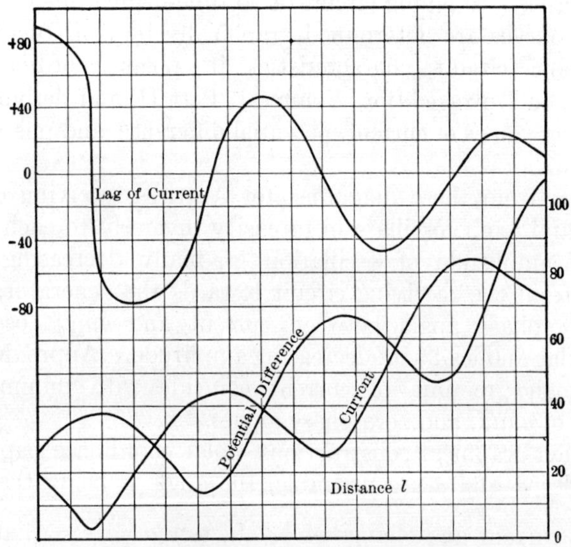

FIG. 2. Current, e.m.f. and space-phase angle between current and e.m.f. in a transmission line. Inductive load.

angle δ the angle between e.m.f. and change of current, $\tan \delta = \frac{b}{g} = 20$ in above instance.

With non-inductive load, Fig. 1, these oscillations of intensity have almost disappeared, and only traces of them are noticeable in the fluctuations of the space-phase angle and the relative values of current and e.m.f. along the line.

Towards the generator end of the line, that is, towards rising power, the curves can be extended indefinitely, approaching more and more the conditions of non-inductive circuit. Towards decreasing power, however, all curves ultimately reach the conditions of a wattless receiving circuit, as Figs. 2 and 3, at the point where the total energy input into the line has been consumed therein, and at this point the two curves for lead and for lag join each other as shown in Fig. 4, the one being a prolongation of the other, and the power in the line reverses. Thus in Fig. 4 energy flows from both sides of the line towards the point of zero power marked by 0, where the current and e.m.f. are in quadrature with each other, the current being leading with regard to the power from the left and lagging with regard to the power from the right side of the diagram.

16. It is of interest to investigate some special cases of such circuits of distributed constants.

(A) *Open circuit at the end of the line.*

Assuming a constant alternating e.m.f. $\dot{E}_1$ impressed upon a circuit at one end while the other end of the circuit is open.

Counting the distance l from the open end of the line, and denoting the length of the line by l_0, for $l = 0$,

$$\dot{I} = \dot{I}_0 = 0,$$

and for $l = l_0$,

$$\dot{E} = \dot{E}_1;$$

hence, substituting in equations (17),

$$0 = \dot{A}_1 - \dot{A}_2,$$

$$\dot{E}_1 = \sqrt{\frac{Z}{Y}} \left\{ \dot{A}_1 \varepsilon^{+al_0}(\cos \beta l_0 + j \sin \beta l_0) + \dot{A}_2 \varepsilon^{-al_0}(\cos \beta l_0 - j \sin \beta l_0) \right\};$$

hence,

$$\dot{A}_2 = \dot{A}_1 = \dot{A}$$

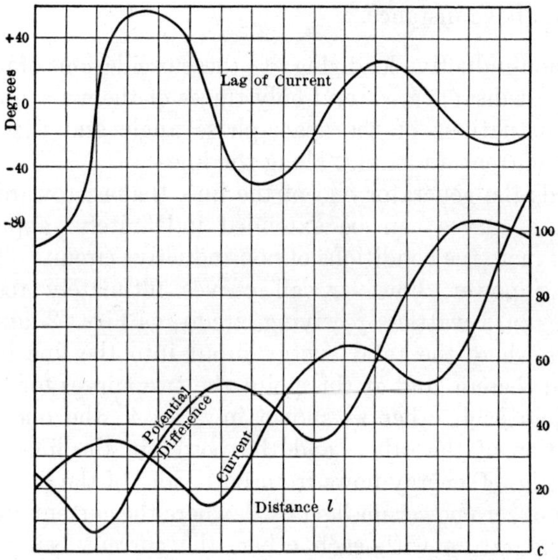

FIG. 3. Current, e.m.f. and space-phase angle between current and e.m.f. in a transmission line. Anti-inductive load.

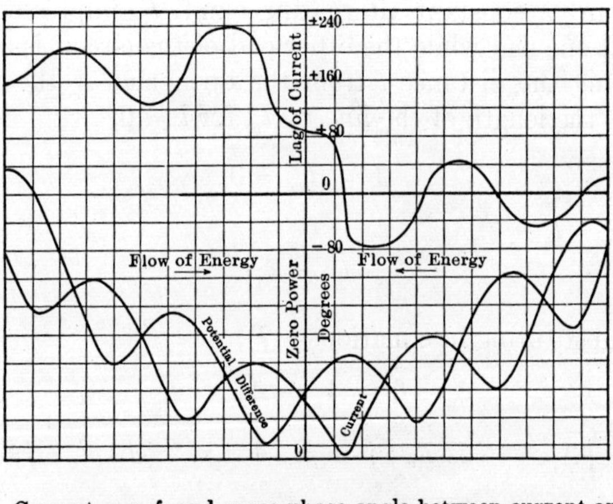

FIG. 4. Current, e.m.f. and space-phase angle between current and e.m.f. in a transmission line.

and

$$\dot{A} = \frac{\dot{E}_1\sqrt{\frac{Y}{Z}}}{\varepsilon^{+al_0}(\cos\beta l_0 + j\sin\beta l_0) + \varepsilon^{-al_0}(\cos\beta l_0 - j\sin\beta l_0)}$$

$$= \frac{\dot{E}_1\sqrt{\frac{Y}{Z}}}{(\varepsilon^{+al_0} + \varepsilon^{-al_0})\cos\beta l_0 + j(\varepsilon^{+al_0} - \varepsilon^{-al_0})\sin\beta l_0};$$

hence, substituting in (17),

$$\left.\begin{aligned} \dot{I} &= \dot{E}_1\sqrt{\frac{Y}{Z}}\,\frac{(\varepsilon^{+al} - \varepsilon^{-al})\cos\beta l + j(\varepsilon^{+al} + \varepsilon^{-al})\sin\beta l}{(\varepsilon^{+al_0} + \varepsilon^{-al_0})\cos\beta l_0 + j(\varepsilon^{+al_0} - \varepsilon^{-al_0})\sin\beta l_0} \\ &\text{and} \\ \dot{E} &= \dot{E}_1\,\frac{(\varepsilon^{+al} + \varepsilon^{-al})\cos\beta l + j(\varepsilon^{+al} - \varepsilon^{-al})\sin\beta l}{(\varepsilon^{+al_0} + \varepsilon^{-al_0})\cos\beta l_0 + j(\varepsilon^{+al_0} - \varepsilon^{-al_0})\sin\beta l_0}. \end{aligned}\right\} \quad (38)$$

At $l = 0$, or the open end of the line, by equations (38),

$$\dot{I}_0 = 0$$

and

$$\dot{E}_0 = \frac{2\dot{E}_1}{(\varepsilon^{+al_0} + \varepsilon^{-al_0})\cos\beta l_0 + j(\varepsilon^{+al_0} - \varepsilon^{-al_0})\sin\beta l_0}. \quad (39)$$

The absolute values of I and E follow from equations (38) and (39):

$$I = E_1\sqrt{\frac{y}{z}}\sqrt{\frac{(\varepsilon^{+al} - \varepsilon^{-al})^2\cos^2\beta l + (\varepsilon^{+al} + \varepsilon^{-al})^2\sin^2\beta l}{(\varepsilon^{+al_0} + \varepsilon^{-al_0})^2\cos^2\beta l_0 + (\varepsilon^{+al_0} - \varepsilon^{-al_0})^2\sin^2\beta l_0}},$$

which expanded gives

$$\left.\begin{aligned} I &= E_1\sqrt{\frac{y}{z}\,\frac{\varepsilon^{+2al} + \varepsilon^{-2al} - 2\cos 2\beta l}{\varepsilon^{+2al_0} + \varepsilon^{-2al_0} + 2\cos 2\beta l_0}} \\ &\text{and} \\ E &= E_1\sqrt{\frac{\varepsilon^{+2al} + \varepsilon^{-2al} + 2\cos 2\beta l}{\varepsilon^{+2al_0} + \varepsilon^{-2al_0} + 2\cos\beta l_0}} \end{aligned}\right\} \quad (40)$$

and

$$E_0 = \frac{2E_1}{\sqrt{\varepsilon^{+2al_0} + \varepsilon^{-2al_0} + 2\cos 2\beta l_0}}. \quad (41)$$

As function of l, the e.m.f. E or the current I is a maximum or minimum for

$$\frac{d}{dl}\left(\varepsilon^{+2al} + \varepsilon^{-2al} \pm 2 \cos 2 \beta l\right) = 0;$$

hence,

$$\alpha \left(\varepsilon^{+2al} - \varepsilon^{-2al}\right) = \pm 2 \beta \sin 2 \beta l. \tag{42}$$

For $l = 0$, and since α is a small quantity, the left side of (42) also is small, and for values of $\sin 2 \beta l$ approximating zero, that is, in the neighborhood of $l = \frac{n\pi}{2\beta}$, or where βl is a multiple of a quadrant, equation (42) becomes zero. At $\beta l \cong 2 n \frac{\pi}{2}$, or the even quadrants, E is a maximum, I a minimum, at $\beta l \cong (2n - 1)\frac{\pi}{2}$, or the odd quadrants, E is a minimum, I a maximum.

The even quadrants, therefore, are nodes of current and wave crests of e.m.f., and the odd quadrants are nodes of e.m.f. and crests of current.

A maximum voltage point, or wave crest, occurs at the open end of the line at $l = 0$, and is given by equation (41). As function of the length l_0 of the line this is a maximum for

$$\frac{d}{dl_0}\left(\varepsilon^{+2al_0} + \varepsilon^{-2al_0} + 2 \cos 2 \beta l_0\right) = 0,$$

or

$$\alpha \left(\varepsilon^{+2al_0} - \varepsilon^{-2al_0}\right) = 2 \beta \sin 2 \beta l_0,$$

or approximately at

$$\left.\begin{aligned} \beta l_0 &= \frac{\pi}{2} \\ l_0 &= n \frac{\pi}{2\beta}; \end{aligned}\right\} \tag{43}$$

that is, when the line is a quarter wave length or an odd multiple thereof.

Substituting in (41), $\beta l_0 = \frac{\pi}{2}$ gives (44)

$$E_0 = \frac{2\,E_1}{\sqrt{\varepsilon^{+2\,al_0} + \varepsilon^{-2\,al_0} - 2}}$$

$$= \frac{2\,E_1}{\varepsilon^{+al_0} - \varepsilon^{-al_0}}. \tag{45}$$

Since

$$\varepsilon^{\pm\,al_0} = 1 \pm \alpha l_0 + \tfrac{1}{2}\,\alpha^2 l_0^{\,2} \pm \frac{1}{2 \times 3}\,\alpha^3 l_0^{\,3} + \pm \ldots,$$

for small values of αl_0 we have

$$\varepsilon^{+al_0} - \varepsilon^{-al_0} = 2\,\alpha l_0$$

and

$$E_0 = \frac{E_1}{\alpha l_0}, \tag{46}$$

which is the maximum voltage that can occur at the open end of a line with voltage E_1 impressed upon it at the other end.

Since, approximately,

$$\beta = \sqrt{xb}$$

$$= 2\,\pi f \sqrt{LC},$$

by (44) we have

$$f = \frac{1}{4\,l_0 \sqrt{LC}}, \tag{47}$$

the frequency which at the length of line l_0 produces maximum voltage at the open end.

For the constants in the example discussed in paragraph 11 we have $l_0 = 200$ miles, $r = 0.21$ ohm, $L = 1.2 \times 10^{-3}$ henry, $C = 0.03 \times 10^{-6}$ farad, $g = 0$, $f = 208$ cycles per sec., $x = 1.57$ ohms, $z = 1.58$ ohms, $b = 39 \times 10^{-6}$ mho, $\alpha = 0.53 \times 10^{-3}$, and $E_0 = 9.3\,E_1$.

(B) *Line grounded at the end.*

17. Let the circuit be grounded or connected to the return conductor at one end, $l = 0$, and supplied by a constant impressed e.m.f. E_1 at the other end, $l = l_0$.

Then for $l = 0$,

$$E = E_0 = 0$$

and for $l = l_0$,

$$E = E_1;$$

hence, substituting in (17),

$$0 = A_1 + A_2$$

and

$$E_1 = \sqrt{\frac{Z}{Y}}\left\{A_1 \varepsilon^{+al_0}(\cos \beta l_0 + j \sin \beta l_0) + A_2{}^{-al_0}(\cos \beta l_0 - j \sin \beta l_0)\right\};$$

hence,

$$A_1 = - A_2 = A,$$

and

$$A = \frac{E_1 \sqrt{\frac{Y}{Z}}}{(\varepsilon^{+al_0} - \varepsilon^{-al_0}) \cos \beta l_0 + j\,(\varepsilon^{+al_0} + \varepsilon^{-al_0}) \sin \beta l_0},$$

and, substituting in (17),

$$\left.\begin{aligned} I &= E_1 \sqrt{\frac{Y}{Z}} \frac{(\varepsilon^{+al} + \varepsilon^{-al}) \cos \beta l + j\,(\varepsilon^{+al} - \varepsilon^{-al}) \sin \beta l}{(\varepsilon^{+al_0} - \varepsilon^{-al_0}) \cos \beta l_0 + j\,(\varepsilon^{+al_0} + \varepsilon^{-al_0}) \sin \beta l_0} \\ E &= E_1 \frac{(\varepsilon^{+al} - \varepsilon^{-al}) \cos \beta l + j\,(\varepsilon^{+al} + \varepsilon^{-al}) \sin \beta l}{(\varepsilon^{+al_0} - \varepsilon^{-al_0}) \cos \beta l_0 + j\,(\varepsilon^{+al_0} + \varepsilon^{-al}) \sin \beta l_0}. \end{aligned}\right\} \quad (48)$$

At the grounded end, $l = 0$,

$$I_0 = \frac{2\,E_1 \sqrt{\frac{Y}{Z}}}{(\varepsilon^{+al_0} - \varepsilon^{-al_0}) \cos \beta l_0 + j\,(\varepsilon^{+al_0} + \varepsilon^{-al_0}) \sin \beta l_0}. \quad (49)$$

Substituting (49) in (48) gives

$$\left.\begin{aligned} I &= \tfrac{1}{2} I_0 \{(\varepsilon^{+al} + \varepsilon^{-al}) \cos \beta l + j\,(\varepsilon^{+al} - \varepsilon^{-al}) \sin \beta l\} \\ &\text{and} \\ E &= \tfrac{1}{2} I_0 \sqrt{\frac{Z}{Y}}\left\{(\varepsilon^{+al} - \varepsilon^{-al}) \cos \beta l + j\,(\varepsilon^{+al} + \varepsilon^{-al}) \sin \beta l.\right\} \end{aligned}\right\} \quad (50)$$

In this case nodes of voltage and crests of current appear at $l = 0$ and at the even quadrants, $\beta l = 2n\frac{\pi}{2}$, and nodes of current and crests of voltage appear at the odd quadrants, $\beta l = (2n-1)\frac{\pi}{2}$.

(C) *Infinitely long conductor.*

18. If an e.m.f. $\dot{E}_0$ is impressed upon an infinitely long conductor, that is, a conductor of such length that of the power input no appreciable part reaches the end, we have, for $l = 0$,

$$\dot{E} = \dot{E}_0$$

and for $l = \infty$,

$$\dot{E} = 0 \text{ and } \dot{I} = 0;$$

hence, substituting in (23) gives

$$\dot{A}_2 = 0$$

and

$$\dot{E}_0 = \sqrt{\frac{Z}{Y}}\,\dot{A}_1;$$

hence,

$$\left.\begin{aligned} \dot{I} &= \dot{E}_0\sqrt{\frac{Y}{Z}}\,\varepsilon^{-al}(\cos\beta l - j\sin\beta l) \\ \text{and} \quad \dot{E} &= \dot{E}_0\varepsilon^{-al}(\cos\beta l - j\sin\beta l). \end{aligned}\right\} \qquad (51)$$

From (51) it follows that

$$\frac{\dot{E}}{\dot{I}} = \sqrt{\frac{Z}{Y}};$$

that is, an infinitely long conductor acts like an impedance,

$$Z_1 = \sqrt{\frac{Z}{Y}} = r_1 + jx_1,$$

and the current at every point of the conductor thus has the same space-phase angle to the voltage,

$$\tan\alpha_1 = \frac{x_1}{r_1}.$$

The equivalent impedance of the infinite conductor is

$$Z_1 = \sqrt{\frac{Z}{Y}} = \frac{V}{Y} = \frac{\alpha + j\beta}{g + jb}$$

$$= \frac{\alpha g + \beta b}{y^2} + j\frac{\beta g - \alpha b}{y^2} \tag{52}$$

and the space-phase angle is

$$\tan \alpha_1 = \frac{\beta g - \alpha b}{\alpha g + \beta b} \cdot \tag{53}$$

If $g = 0$ and $x = 0$, we have

$$\alpha = \beta = \sqrt{\frac{br}{2}}$$

and

$$\tan \alpha_1 = 1,$$

or

$$\alpha_1 = 45°;$$

that is, current and e.m.f. differ by one-eighth period.

This is approximately the case in cables, in which the dielectric losses and the inductance are small.

An infinitely long conductor therefore shows the wave propagation in its simplest and most perspicuous form, since the reflected wave is absent.

(D) *Generator feeding into a closed circuit.*

19. Let $l = 0$ be the center of the circuit; then

$$E_l = -E_{-l} \text{ and } I_l = I_{-l};$$

hence, $$E = 0 \text{ at } l = 0,$$

and the equations are the same as those of a line grounded at the end $l = 0$, which have been discussed under (B).

(E) *Line of quarter wave length.*

20. Interesting is the case of a line of quarter wave length.

Let the length l_0 of the line be one quarter wave of the impressed e.m.f.

$$\beta l_0 = \frac{\pi}{2} \cdot \tag{54}$$

To illustrate the general character of the phenomena, we may as first approximation neglect the energy losses in the circuit, that is, assume the resistance r and the conductance g as negligible compared with x and b,

$$r = 0 = g.$$

These values substituted in (14) give

$$\alpha = 0 \text{ and } \beta = \sqrt{xb}. \tag{55}$$

Counting the distance l from the end of the line l_0 we have for $l = 0$,

and

$$\left.\begin{aligned} \dot{E}_0 &= e_0 - je_0' \\ \dot{I}_0 &= i_0 - ji_0', \end{aligned}\right\} \tag{56}$$

and at the beginning of the line for $l = l_0$,

and

$$\left.\begin{aligned} \dot{E}_1 &= e_1 - je_1' \\ \dot{I}_1 &= i_1 - ji_1', \end{aligned}\right\} \tag{57}$$

and by (54) and (55),

$$l_0 = \frac{\pi}{2\sqrt{xb}}\cdot \tag{58}$$

Substituting (56), (57), and (54) in (17) gives

$$\dot{I}_0 = \dot{A}_1 - \dot{A}_2$$

and

$$\dot{E}_0 = \sqrt{\frac{Z}{Y}}(\dot{A}_1 + \dot{A}_2) = \sqrt{\frac{x}{b}}(\dot{A}_1 + \dot{A}_2),$$

or

$$\dot{I}_1 = j(\dot{A}_1 + \dot{A}_2)$$

and

$$\dot{E}_1 = -j\sqrt{\frac{Z}{Y}}(\dot{A}_1 - \dot{A}_2) = +j\sqrt{\frac{x}{b}}(\dot{A}_1 - \dot{A}_2);$$

hence, eliminating A_1 and A_2 gives the relations between the electric quantities at the generator end of the quarter-wave line, $\dot{E}_1$, $\dot{I}_1$, and at the receiving end, $\dot{E}_0$, $\dot{I}_0$:

$$E_0 = -j\sqrt{\frac{x}{b}}\,I_1$$

and (59)

$$I_0 = -j\sqrt{\frac{b}{x_0}}\,E_1,$$

and the absolute values are

$$E_0 = \sqrt{\frac{x}{b}}\,I_1$$

and (60)

$$I_0 = \sqrt{\frac{b}{x}}\,E_1;$$

which means that if the supply voltage E_1 is constant, the output current I_0 is constant and lags 90 space-degrees behind the input voltage; if the supply current I_1 is constant, the output voltage E_0 is constant, and lags 90 space-degrees, and inversely. A quarter-wave line of negligible losses thus converts from constant potential to constant current, or from constant current to constant voltage. (Constant-potential constant-current transformation.)

Multiplying (60) gives

$$E_0 I_0 = I_1 E_1,$$

or

$$E_0 = \frac{E_1 I_1}{I_0};$$

hence, if $I_0 = 0$, that is, the line is open at the end, $E_0 = \infty$, and with a finite voltage supply to a line of quarter-wave length, an infinite (extremely high) voltage is produced at the other end.

Such a circuit thus may be used to produce very high voltages.

Since $x_0 = l_0 x$ = total reactance and $b_0 = l_0 b$ = total susceptance of the circuit, by (58) we have

$$x_0 b_0 = \frac{\pi^2}{4}, \tag{61}$$

or the condition of quarter-wave length.

Substituting $x_0 = 2\,\pi f L_0$ and $b_0 = 2\,\pi f C_0$, we have

$$L_0 C_0 = \frac{1}{16\,f^2}, \tag{62}$$

or

$$f = \frac{1}{4\,\sqrt{L_0 C_0}}, \tag{63}$$

the condition of quarter-wave transmission.

21. If the resistance, r, and the conductance, g, of a quarter-wave circuit are not negligible, substituting (56), (54) and (57) in (17) we have, for $l = 0$,

$$\left.\begin{aligned} I_0 &= A_1 - A_2 \\ \text{and} \qquad E_0 &= \sqrt{\frac{Z}{Y}}\left(A_1 + A_2\right), \end{aligned}\right\} \tag{64}$$

and for $l = l_0$,

$$\left.\begin{aligned} I_1 &= j\left\{A_1\varepsilon^{+al_0} + A_2\varepsilon^{-al_0}\right\} \\ \text{and} \qquad E_1 &= j\sqrt{\frac{Z}{Y}}\left\{A_1\varepsilon^{+al_0} - A_2\varepsilon^{-a}\right\}. \end{aligned}\right\} \tag{65}$$

From (64) it follows that

$$\left.\begin{aligned} A_1 &= \frac{1}{2}\left(I_0 + \sqrt{\frac{Y}{Z}}E_0\right) \\ \text{and} \qquad A_2 &= -\frac{1}{2}\left(I_0 - \sqrt{\frac{Y}{Z}}E_0\right), \end{aligned}\right\} \tag{66}$$

and substituting in (65) and rearranging we have

$$\left.\begin{aligned} I_1 &= j\left\{E_0\sqrt{\frac{Y}{Z}}\,\frac{\varepsilon^{+al_0} + \varepsilon^{-al_0}}{2} + I_0\,\frac{\varepsilon^{+al_0} - \varepsilon^{-al_0}}{2}\right\} \\ \text{and} \qquad E_1 &= j\left\{I_0\sqrt{\frac{Z}{Y}}\,\frac{\varepsilon^{+al_0} + \varepsilon^{-al_0}}{2} + E_0\,\frac{\varepsilon^{+al_0} - \varepsilon^{-al_0}}{2}\right\}, \end{aligned}\right\} \tag{67}$$

or,

$$\left.\begin{aligned}\dot{I}_0 &= -j\left\{\dot{E}_1\sqrt{\frac{Y}{Z}}\frac{\varepsilon^{+al_0}+\varepsilon^{-al_0}}{\varepsilon^{+2al_0}+\varepsilon^{-2al_0}}-\dot{I}_1\frac{\varepsilon^{+al_0}-\varepsilon^{-al_0}}{\varepsilon^{+2al_0}+\varepsilon^{-2al_0}}\right\}\\ &\text{and}\\ \dot{E}_0 &= -j\left\{\dot{I}_1\sqrt{\frac{Z}{Y}}\frac{\varepsilon^{+al_0}+\varepsilon^{-al_0}}{\varepsilon^{+2al}+\varepsilon^{-2al_0}}-\dot{E}_1\frac{\varepsilon^{+al_0}-\varepsilon^{-al_0}}{\varepsilon^{+2al_0}+\varepsilon^{-2al_0}}\right\},\end{aligned}\right\} \tag{68}$$

or, analogous to equation (59),

$$\left.\begin{aligned}\dot{I}_0 &= -\sqrt{\frac{Y}{Z}}\left\{\frac{2j\dot{E}_1}{\varepsilon^{+al_0}+\varepsilon^{-al_0}}+\frac{\dot{E}_0(\varepsilon^{+al_0}-\varepsilon^{-al_0})}{\varepsilon^{+al_0}+\varepsilon^{-al_0}}\right\}\\ &\text{and}\\ \dot{E}_0 &= -\sqrt{\frac{Z}{Y}}\left\{\frac{2j\dot{I}_1}{\varepsilon^{+al_0}+\varepsilon^{-al_0}}+\frac{\dot{I}_0(\varepsilon^{+al_0}-\varepsilon^{-al_0})}{\varepsilon^{+al_0}+\varepsilon^{-al_0}}\right\}.\end{aligned}\right\} \tag{69}$$

In these equations the second term is usually small, due to the factor $(\varepsilon^{+al_0} - \varepsilon^{-al_0})$, and the first term represents constant potential-constant current transformation.

22. In a quarter-wave line, at constant impressed e.m.f. $\dot{E}_1$, the current output $\dot{I}_0$ is approximately constant and lagging 90 degrees behind $\dot{E}_1$; it falls off slightly, however, with increasing load, that is, increasing $\dot{I}_1$, due to the second term in equation (68); the voltage at the end of the line, $\dot{E}_0$, at constant impressed voltage, is approximately proportional to the load, but does not reach infinity at open circuit, but a finite, though high, limiting value.

Inversely, at constant current input the voltage output is approximately constant and the output current proportional to the load.

The deviation from constancy, at constant $\dot{E}_1$, of $\dot{I}_0$, or at constant $\dot{I}_1$, of $\dot{E}_0$, therefore, is due to the second term, with factor $(\varepsilon^{+al_0} - \varepsilon^{-al_0})$.

Substituting (54),

$$al_0 = \frac{\alpha}{\beta}\frac{\pi}{2},$$

hence, al_0 is usually a very small quantity, and $\varepsilon^{\pm al_0} = \varepsilon^{\pm\frac{\alpha}{\beta}\frac{\pi}{2}}$ thus can be represented by the first terms of the series:

$$\varepsilon^{\pm\frac{\alpha}{\beta}\frac{\pi}{2}} = 1 \pm \frac{\alpha}{\beta}\frac{\pi}{2} + \frac{1}{2}\left(\frac{\alpha}{\beta}\frac{\pi}{2}\right)^2 \pm \frac{1}{2\times 3}\left(\frac{\alpha}{\beta}\frac{\pi}{2}\right)^3 + \pm \cdots$$

$$= 1 \pm \frac{\alpha}{\beta}\frac{\pi}{2};$$

hence,

$$\frac{\varepsilon^{+al_0} + \varepsilon^{-al_0}}{2} = 1,$$

and

$$\frac{\varepsilon^{+al_0} - \varepsilon^{-al_0}}{2} = \frac{\alpha}{\beta}\frac{\pi}{2},$$

and, by (69),

$$\left.\begin{aligned} \dot{I}_0 &= -j\sqrt{\frac{Y}{Z}}\dot{E}_1 - \frac{\alpha}{\beta}\frac{\pi}{2}\sqrt{\frac{Y}{Z}}\dot{E}_0 \\ \text{and} \quad \dot{E}_0 &= -j\sqrt{\frac{Z}{Y}}\dot{I}_1 - \frac{\alpha}{\beta}\frac{\pi}{2}\sqrt{\frac{Z}{Y}}\dot{I}_0. \end{aligned}\right\} \tag{70}$$

If r and g are small compared with x and b,

$$\beta = \sqrt{\tfrac{1}{2}(zy - rg + xb)} = \sqrt{xb}$$

and

$$\alpha = \sqrt{\tfrac{1}{2}(zy + rg - xb)};$$

substituting, by the binomial theorem,

$$zy = \sqrt{(r^2 + x^2)(g^2 + b^2)} = xb\left\{\left[1 + \left(\frac{r}{x}\right)^2\right]\left[1 + \left(\frac{g}{b}\right)^2\right]\right\}^{\frac{1}{2}}$$

$$= xb\left\{1 + \frac{1}{2}\left(\frac{r}{x}\right)^2 + \frac{1}{2}\left(\frac{g}{b}\right)^2 + \ldots\right\}$$

gives

$$\alpha = \sqrt{\frac{1}{2}\left\{\frac{1}{2}\left(\frac{r}{x}\right)^2 + \frac{1}{2}\left(\frac{g}{b}\right)^2 + \frac{rg}{bx}\right\}bx}$$

$$= \frac{\sqrt{bx}}{2}\left(\frac{r}{x} + \frac{g}{b}\right),$$

and

$$\frac{\alpha}{\beta} = \frac{1}{2}\left(\frac{r}{x} + \frac{g}{b}\right).$$

The quantity

$$\frac{\alpha}{\beta} = \frac{1}{2}\left(\frac{r}{x} + \frac{g}{b}\right) = u \tag{71}$$

may be called the *time constant of the circuit.*

The equations of *quarter-wave transmission* thus are

$$\left.\begin{aligned} \dot{I}_0 &= -\sqrt{\frac{Y}{Z}}\left\{j\dot{E}_1 + \frac{u\pi}{2}\dot{E}_0\right\} \\ \text{and} \quad \dot{E}_0 &= -\sqrt{\frac{Z}{Y}}\left\{j\dot{I}_1 + \frac{u\pi}{2}\dot{I}_0\right\}, \end{aligned}\right\} \tag{72}$$

and the maximum voltage $\dot{E}_0$, at the open end of the circuit, at constant impressed e.m.f. $\dot{E}_1$, is

$$\dot{I}_0 = 0$$

and

$$\dot{E}_0 = -\frac{2j\dot{E}_1}{u\pi}, \tag{73}$$

and the current input is

$$\begin{aligned} \dot{I}_1 &= j\sqrt{\frac{Y}{Z}}E_0 \\ &= \frac{2\dot{E}_1}{u\pi}\sqrt{\frac{Y}{Z}}, \end{aligned} \tag{74}$$

where, approximately,

$$\sqrt{\frac{Z}{Y}} = \sqrt{\frac{L}{C}}. \tag{75}$$

23. Consider as an example a high potential coil of a transformer with one of its terminals connected to a source of high potential, for testing its insulation to ground, while the other terminal is open.

Assume the following constants per unit length of circuit: $r = 0.1$ ohm, $L = 0.02$ h., $C = 0.01 \times 10^{-6}$ farad, and $g = 0$; then, with a length of circuit $l_0 = 100$, the quarter-wave frequency is, by (47),

$$f = \frac{1}{4\,l_0\sqrt{LC}} = 177 \text{ cycles per sec.,}$$

or very close to the third harmonic of a 60-cycle impressed voltage.

If, therefore, the testing frequency is low, 59 cycles, the circuit is a quarter wave of the third harmonic.

Assuming an impressed e.m.f. of 50,000 volts and 59 cycles, containing a third harmonic of 10 per cent, or $E_1 = 5000$ volts at 177 cycles, for this harmonic, we have $x = 22.2$ ohms and $b = 11.1 \times 10^{-6}$ ohm; hence, $u = 0.00225$
and

$$E_0 = \frac{2}{u\pi} E_1 = 283\, E_1;$$

therefore at $E_1 = 5000$ volts, $E_0 = 1{,}415{,}000$ volts;
that is, infinity, as far as insulation strength is concerned.

Quarter-wave circuits thus may be used, and are used, to produce extremely high voltages, and if a sufficiently high frequency is used — 100,000 cycles and more, as in wireless telegraphy, etc. — the length of the circuit is moderate.

This method of producing high voltages has the disadvantage that it does not give constant potential, but the high voltage is due to the tendency of the circuit to regulate for constant current, which means infinite voltage at infinite resistance or open circuit, but as soon as current is taken off the high potential point the voltage falls. The great advantage of the quarter-wave method of producing high voltage is its simplicity and ease of insulation; as the voltage gradually builds up along the circuit, the high voltage point or end of circuit may be any distance away from the power supply, and thus can easily be made safe.

24. As a quarter-wave circuit converts from constant potential to constant current, it is not possible, with constant voltage impressed upon a circuit of approximately a quarter-wave length, to get constant voltage at the other or receiving end of the circuit. Long before the circuit approaches quarter-wave length, and as soon as it becomes an appreciable part of a quarter wave, this tendency to constant current regulation makes itself felt by great variations of voltage with changes of load at the receiving end of the circuit, constant voltage being impressed upon the generator end; that is, with increasing length of transmission lines the voltage regulation at the receiving end becomes seriously impaired hereby, even if the line resistance is moderate, and the operation of apparatus which require approximate constancy of voltage but do not operate on constant current—as synchronous apparatus—becomes more difficult.

Hence, at the end of very long transmission lines the voltage regulation becomes poor, and synchronous machines tend to instability and have to be provided with powerful steadying devices, giving induction motor features, and with a line approaching quarter-wave length, voltage regulation at the receiving end ceases, unless very powerful voltage controlling devices are used, such as large synchronous condensers, that is, synchronous machines establishing a fixed voltage and controlling the line by automatically drawing leading or lagging currents, in correspondence with the line conditions.

In this case of a line of approximately quarter-wave length, the constant potential-constant current transformation may be used to produce constant or approximately constant voltage at the load, by supplying constant current to the line; that is, the transmission line is made a quarter-wave length by modifying its constants, or choosing the proper frequency, the generators are designed to regulate for constant current and thus give a voltage varying with the load, and are connected in series (with constant current generators series connection is stable, parallel connections unstable) and feed constant current, at variable voltage, into the quarter-wave line. At the receiving end of the line, constant voltage then exists with varying load, or rather a voltage, which slightly falls off with the load, due to the power loss in the line. To maintain constant receiver voltage at all loads, then, would require a slight increase of generator current with increase of load, that is, increase of generator voltage, which can be produced by compounding regulated by the voltage.

In such a quarter-wave transmission the voltage at the receiving end then remains constant, while the current output from the line increases from nothing at no load. At the generator end the current remains approximately constant, increasing from no load to full load by the amount required to take care of the line loss, while the voltage at the generators increases from nearly nothing at no load, with increasing load, approximately proportional thereto.

25. There is, however, a serious limitation imposed upon quarter-wave transmission by considerations of voltage; to use the transmission line economically the voltage throughout it should not differ much, since the insulation of the line depends on the maximum, the efficiency of transmission, however, on the average voltage, and a line in which the voltage at the two ends is very different is uneconomical.

To use line copper and line insulation economically, in a quarter-wave transmission, the voltages at the two ends should be approximately equal at maximum load. These voltages are related to each other and to the current by the line constants, by equations (72).

By these equations (neglecting the term with u), reduced to absolute values, we have approximately

$$i_0 = \sqrt{\frac{y}{z}}\, e_1$$

and

$$e_0 = \sqrt{\frac{z}{y}}\, i_1,$$

and if $e_1 = e_0$,

$$i_0 = \sqrt{\frac{y}{z}}\, e_0;$$

hence, the power is

$$p_0 = e_0 i_0 = \sqrt{\frac{y}{z}}\, e_0^2,$$

or

$$e_0^2 = p_0 \sqrt{\frac{z}{y}}; \tag{76}$$

hence, the voltage e_0 required to transmit the power p_0 without great potential differences in the line depends on the power p_0 and the line constants, and inversely.

26. As an example of a quarter-wave transmission may be considered the transmission of 60,000 kilowatts over a distance of 700 miles, for the supply of a general three-phase distribution system, of 95 per cent power factor, lag.

The design of the transmission line is based on a compromise between different and conflicting requirements: economy in first cost requires the highest possible voltage and smallest conductor section, or high power loss in the line; economy of operation requires high voltage and large conductor section, or low power loss; reliability of operation of the line requires lowest permissible voltage and therefore large conductor section or high power loss; reliability of operation of the receiving system requires good voltage regulation and thus low line resistance, etc., etc.

Assume that the maximum effective voltage between the line conductors is limited to 120,000, and that there are two separate pole lines, each carrying three wires of 500,000 circular mils cross section, placed 6 feet between wires, and provided with a grounded neutral.

If there were no energy losses in the line and no increase of capacity due to insulators, etc., the speed of propagation would be the velocity of light, $S = 188{,}000$ miles per second, and the quarter-wave frequency of a line of $l_0 = 700$ miles would be

$$f = \frac{S}{4\,l_0} = 67 \text{ cycles per sec.};$$

hence, fairly close to the standard frequency of 60 cycles.

The loss of power in the line, and thus the increase of inductance by the magnetic field inside of the conductor (which would not exist in a conductor of perfect conductivity or zero resistance loss), the increase of capacity by insulators, poles, etc., lowers the frequency below that corresponding to the velocity of light and brings it nearer to 60 cycles.

In a line as above assumed the constants per mile of double conductor are: $r = 0.055$ ohm; $L = 0.001$ henry, and $C = 0.032 \times 10^{-6}$ farad, and, neglecting the conductance, $g = 0$, the quarter-wave frequency is

$$f = \frac{1}{4\,l_0\sqrt{LC}} = 63 \text{ cycles per sec.}$$

Either then the frequency of 63 cycles per second, or slightly above standard, may be chosen, or the line inductance or line capacity increased, to bring the frequency down to 60 cycles.

Assuming the inductance increased to $L = 0.0011$ henry gives $f = 60$ cycles per second, and the line constants then are $l_0 = 700$ miles; $f = 60$ cycles per second; $r = 0.055$ ohm; $L = 0.0011$ henry; $C = 0.032 \times 10^{-6}$ farad, and $g = 0$; hence, $x = 0.415$ ohm; $z = 0.42$ ohm; $Z = 0.055 + 0.415\,j$ ohm;

$b = 12.1 \times 10^{-6}$ mho; $y = 12.1 \times 10^{-6}$ mho, and $Y = j\ 12.1 \times 10^{-6}$ mho, and

$$\sqrt{\frac{Z}{Y}} = 186 + 12\,j,$$

$$\sqrt{\frac{z}{y}} = 186,$$

$$u = \frac{1}{2}\left(\frac{r}{x} + \frac{g}{b}\right) = 0.066,$$

$$\beta = 2.247 \times 10^{-3}$$

$$\alpha = u\beta = 0.148 \times 10^{-3}.$$

At 60,000 kilowatts total input, or 20,000 kilowatts per line, and 120,000 volts between lines, or $\frac{120{,}000}{\sqrt{3}} = 69{,}000$ volts per line, and about 95 per cent power factor, the current input at full load is 306 amp. per line (of two conductors in multiple).

To get at full load $p = 20 \times 10^6$ watts, approximately the same voltage at both ends of the line, by equation (76), we must have

$$e^2 = p\sqrt{\frac{z}{y}},$$

or

$$e = 61{,}000 \text{ volts}.$$

Assuming therefore at the receiving end the voltage of 110,000 between the lines, or, 63,500 volts per line, and choosing the output current as zero vector, and counting the distance from the receiving end towards the generator, we have for $l = 0$,

$$\dot{I} = \dot{I}_0 = i_0,$$

and the voltage, at 95 per cent power factor, or $\sqrt{1 - 0.95^2} = 0.312$ inductance factor, is

$$\dot{E} = \dot{E}_0 = e_0\,(0.95 + 0.312\,j)$$
$$= 60{,}300 + 19{,}800\,j.$$

Substituting these values in equations (72) gives

$$i_0 = \frac{\{-\,j\dot{E}_1 - 0.104\,(60{,}300 + 19{,}800\,j)\}}{186 - 12\,j},$$

$$60{,}300 + 19{,}800\,j = -\,(186 - 12\,j)\,\{j\dot{I}_1 + 0.104\,i_0\};$$

hence, $-j\dot{E}_1 = (186 - 12j)\,i_0 + (6250 + 2060j)$

and
$$-j\dot{I}_1 = \frac{60{,}300 - 19{,}800j}{186 - 12j} + 0.104\,i_0$$
$$= 317 + 128j + 0.104\,i_0,$$

and the absolute values are

$$e_1 = \sqrt{(186\,i_0 + 6250)^2 + (12\,i_0 - 2060)^2}$$

and $$i_1 = \sqrt{(317 + 0.104\,i_0)^2 + 128^2};$$

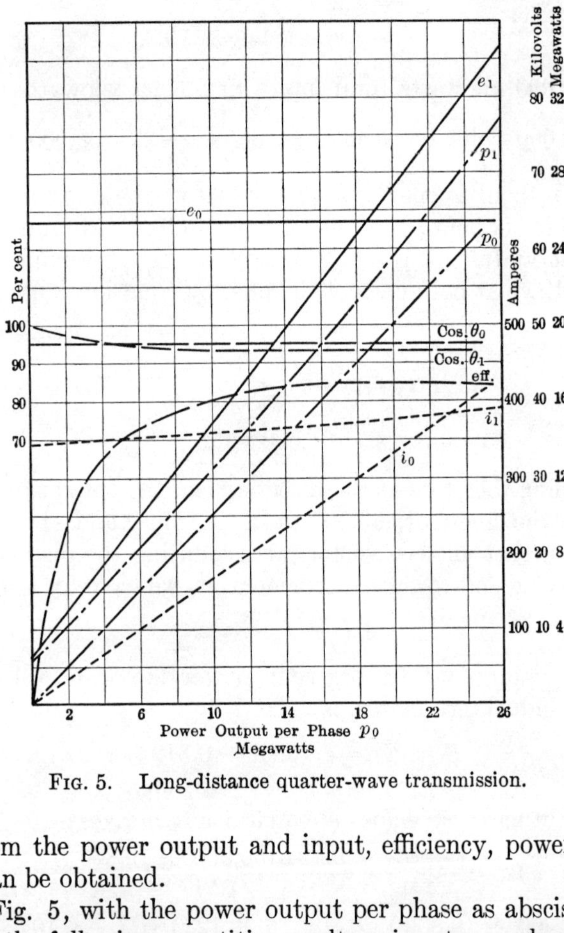

FIG. 5. Long-distance quarter-wave transmission.

herefrom the power output and input, efficiency, power factor, etc., can be obtained.

In Fig. 5, with the power output per phase as abscissas, are shown the following quantities: voltage input e_1 and output e_0, in drawn lines; amperes input i_1 and output i_0, in dotted lines; power input p_1 and output p_0, in dash-dotted lines, and efficiency and power factor in dashed lines.

As seen, the power factor at the generator is above 93 per cent leading, and the efficiency reaches nearly 85 per cent.

At full load input of 20,000 kilowatts per phase, and 95 per cent power factor, lagging, of the output, the generator voltage is 58,500, or still 8 per cent below the output voltage of 63,500. The generator voltage equals the output voltage at 10 per cent overload, and exceeds it by 14 per cent at 25 per cent overload.

To maintain constant voltage at the output side of the line, the generator current has to be increased from 342 amperes at no load to 370 amperes at full load, or by 8.2 per cent, and inversely, at constant-current input, the output voltage would drop off, from no load to full load, by about 8 per cent. This, with a line of 15 per cent resistance drop, is a far closer voltage regulation than can be produced by constant potential supply, except by the use of synchronous machines for phase control.

CHAPTER III

THE NATURAL PERIOD OF THE TRANSMISSION LINE

27. An interesting application of the equations of the long distance transmission line given in the preceding chapter can be made to the determination of the *natural period of a transmission line;* that is, the frequency at which such a line discharges an accumulated charge of atmospheric electricity (lightning), or oscillates because of a sudden change of load, as a break of circuit, or in general a change of circuit conditions, as closing the circuit, etc.

The discharge of a condenser through a circuit containing self-inductance and resistance is oscillating (provided the resistance does not exceed a certain critical value depending upon the capacity and the self-inductance); that is, the discharge current alternates with constantly decreasing intensity. The frequency of this oscillating discharge depends upon the capacity C and the self-inductance L of the circuit, and to a much lesser extent upon the resistance, so that, if the resistance of the circuit is not excessive, the frequency of oscillation can, by neglecting the resistance, be expressed with fair, or even close, approximation by the formula

$$f = \frac{1}{2\pi\sqrt{CL}}.$$

An electric transmission line represents a circuit having capacity as well as self-inductance; and thus when charged to a certain potential, for instance, by atmospheric electricity, as by induction from a thunder-cloud passing over or near the line, the transmission line discharges by an oscillating current.

Such a transmission line differs, however, from an ordinary condenser in that with the former the capacity and the self-inductance are distributed along the circuit.

In determining the frequency of the oscillating discharge of such a transmission line, a sufficiently close approximation is obtained by neglecting the resistance of the line, which, at the relatively high frequency of oscillating discharges, is small compared with the reactance. This assumption means that the dying out of the discharge current through the influence of the resistance of the circuit is neglected, and the current assumed as an alternating current of approximately the same frequency and the same intensity as the initial waves of the oscillating discharge current. By this means the problem is essentially simplified.

28. Let l_0 = total length of a transmission line; l = the distance from the beginning of the line; r = resistance per unit length; x = reactance per unit length = $2\,\pi f L$, where L = inductance per unit length; g = conductance from line to return (leakage and discharge into the air) per unit length; b = capacity susceptance per unit length = $2\,\pi f C$, where C = capacity per unit length.

Neglecting the line resistance and line conductance,

$$r = 0 \text{ and } g = 0,$$

the line constants α and β, by equations (14), Chapter II, then assume the form

$$\alpha = 0 \text{ and } \beta = \sqrt{xb}, \qquad (1)$$

and the line equations (17) of Chapter II become

$$\dot{I} = (\dot{A}_1 - \dot{A}_2)\cos\beta l + j\,(\dot{A}_1 + \dot{A}_2)\sin\beta l$$

and

$$\dot{E} = \sqrt{\frac{Z}{Y}}\left\{(\dot{A}_1 + \dot{A}_2)\cos\beta l + j\,(\dot{A}_1 - \dot{A}_2)\sin\beta l\right\},$$

or writing

$$\dot{A}_1 - \dot{A}_2 = \dot{C}_1 \text{ and } \dot{A}_1 + \dot{A}_2 = \dot{C}_2,$$

and substituting

$$\sqrt{\frac{Z}{Y}} = \sqrt{\frac{x}{b}} = \sqrt{\frac{L}{C}} \qquad (2)$$

we have

$$\dot{I} = \dot{C}_1 \cos\beta l + j\dot{C}_2 \sin\beta l$$

and

$$\dot{E} = \sqrt{\frac{L}{C}}\left\{C_2 \cos\beta l + jC_1 \sin\beta l\right\}. \qquad (3)$$

A free oscillation of a circuit implies that energy is neither supplied to the circuit nor abstracted from it. This means that at both ends of the circuit, $l = 0$ and $l = l_0$, the power equals zero.

If this is the case, the following conditions may exist:

(1) The current is zero at one end, the voltage zero at the other end.

(2) Either the current is zero at both ends or the voltage is zero at both ends.

(3) The circuit has no end but is closed upon itself.

(4) The current is in quadrature with the voltage. This case does not represent a free oscillation, since the frequency depends also on the connected circuit, but rather represents a line supplying a wattless or reactive load.

In free oscillation the circuit thus must be either open or grounded at its ends or closed upon itself.

(1) *Circuit open at one end, grounded at other end.*

29. Assuming the circuit grounded at $l = 0$, open at $l = l_0$, we have for $l = 0$,

$$\dot{E} = \dot{E}_0 = 0,$$

and for $l = l_0$,

$$\dot{I} = \dot{I}_1 = 0;$$

hence, substituting in equations (3), at $l = 0$,

$$\dot{C}_2 = 0;$$

hence,

and

$$\left.\begin{aligned} \dot{I} &= \dot{C}_1 \cos \beta l \\ \dot{E} &= j\sqrt{\frac{L}{C}}\, C_1 \sin \beta l, \end{aligned}\right\} \tag{4}$$

and at $l = l_0$,

$$C_1 \cos \beta l_0 = 0,$$

and since C_1 cannot be zero without the oscillation disappearing altogether,

$$\cos \beta l_0 = 0; \tag{5}$$

hence,

$$\beta l_0 = (2n - 1)\frac{\pi}{2}, \tag{6}$$

where $n = 1, 2, 3 \ldots$ or any integer and

$$\beta l = (2\,n - 1)\frac{\pi}{2\,l_0}\,l. \tag{7}$$

Substituting (1) in (6) gives

$$\beta = \sqrt{xb} = \frac{(2\,n - 1)\,\pi}{2\,l_0}, \tag{8}$$

or substituting for x and b, $x = 2\,\pi f L$ and $b = 2\,\pi f C$, gives

$$2\,\pi f\sqrt{LC} = \frac{(2\,n - 1)\,\pi}{2\,l_0};$$

or

$$f = \frac{(2\,n - 1)}{4\,l_0\sqrt{LC}} \tag{9}$$

is the *frequency of oscillation* of the circuit.

The lowest frequency or *fundamental frequency of oscillation* is, for $n = 1$,

$$f_1 = \frac{1}{4\,l_0\sqrt{LC}}, \tag{10}$$

and besides this fundamental frequency, all its *odd multiples* or higher harmonics may exist in the oscillation

$$f = (2\,n - 1)\,f_1. \tag{11}$$

Writing $L_0 = l_0 L$ = total inductance, and $C_0 = l_0 C$ = total capacity of the circuit, equation (9) assumes the form

$$f_1 = \frac{1}{4\sqrt{L_0 C_0}}. \tag{12}$$

The fundamental frequency of oscillation of a transmission line open at one end and grounded at the other, and having a total inductance L_0 and a total capacity C_0, is, neglecting energy losses,

$$f_1 = \frac{1}{4\sqrt{L_0 C_0}},$$

while the frequency of oscillation of a localized inductance L_0 and localized capacity C_0, that is, the frequency of discharge of a condenser C_0 through an inductance L_0, is

$$f = \frac{1}{2\pi\sqrt{L_0 C_0}}. \tag{13}$$

The difference is due to the distributed character of L_0 and C_0 in the transmission line and the resultant phase displacement between the elements of the line, which causes the inductance and capacity of the line elements, in their effect on the frequency, not to add but to combine to a resultant, which is the projection of the elements of a quadrant, on the diameter, or $\frac{2}{\pi}$ times the sum, just as, for instance, the resultant m.m.f. of a distributed armature winding of n turns of i amperes is not ni but $\frac{2}{\pi}\,ni$.

Hence, the effective inductance of a transmission line in free oscillation is

$$L_0' = \frac{2}{\pi} l_0 L$$

and the effective capacity is (14)

$$C_0' = \frac{2}{\pi} l_0 C,$$

and using the effective values L_0' and C_0', the fundamental frequency, equation (11), then appears in the form

$$f_1 = \frac{1}{2\pi\sqrt{L_0' C_0'}}; \tag{15}$$

that is, the same value as found for the condenser discharge.

In comparing with localized inductances and capacities, the distributed capacity and inductance, in free oscillation, thus are represented by their effective values (13) and (14).

30. Substituting in equations (4),

$$\dot{C}_1 = c_1 + jc_2, \tag{16}$$

gives

$$\dot{I} = (c_1 + jc_2)\cos\beta l$$

and (17)

$$\dot{E} = \sqrt{\frac{L}{C}}\,(c_2 - jc_1)\sin\beta l.$$

By the definition of the complex quantity as vector representation of an alternating wave the cosine component of the wave is represented by the real, the sine component by the imaginary term; that is, a wave of the form $c_1 \cos 2\pi ft + c_2 \sin 2\pi ft$ is represented by $c_1 + jc_2$, and inversely, the equations (17), in their analytic expression, are

$$\left.\begin{aligned} i &= (c_1 \cos 2\pi ft + c_2 \sin 2\pi ft) \cos \beta l \\ \text{and} \quad e &= \sqrt{\frac{L}{C}} (c_2 \cos 2\pi ft - c_1 \sin 2\pi ft) \sin \beta l. \end{aligned}\right\} \quad (18)$$

Substituting (7) and (11) in (18), and writing

$$\theta = 2\pi f_1 t \text{ and } \tau = \frac{\pi l}{2 l_0} \quad (19)$$

gives

$$\left.\begin{aligned} i &= \{c_1 \cos (2n-1)\theta + c_2 \sin (2n-1)\theta\} \cos (2n-1)\tau \\ &= c \cos (2n-1)(\theta - \gamma) \cos (2n-1)\tau \\ \text{and} & \\ e &= \sqrt{\frac{L}{C}} \left\{ c_2 \cos (2n-1)\theta - c_1 \sin (2n-1)\theta \right\} \sin (2n-1)\tau \\ &= -\sqrt{\frac{L}{C}} c \sin (2n-1)(\theta - \gamma) \sin (2n-1)\tau, \end{aligned}\right\} \quad (20)$$

where

$$\tan (2n-1)\gamma = \frac{c_2}{c_1} \quad \text{and} \quad c = \sqrt{c_1^2 + c_2^2}. \quad (21)$$

In the denotation (19), θ represents the *time angle*, with the complete cycle of the fundamental frequency of oscillation as one revolution or 360 degrees, and τ represents the *distance angle*, with the length of the line as a quadrant or 90 degrees. That is, distances are represented by angles, and the whole line is a quarter wave of the fundamental frequency of oscillation. This form of free oscillation may be called *quarter-wave oscillation*.

The fundamental or lowest discharge wave or oscillation of the circuit then is

$$\left.\begin{aligned} i_1 &= c \cos (\theta - \gamma_1) \cos \tau \\ \text{and} \quad e_1 &= -\sqrt{\frac{L}{C}} c \sin (\theta - \gamma_1) \sin \tau. \end{aligned}\right\} \quad (22)$$

With this wave the voltage is a maximum at the open end of the line, $l = l_0$, and gradually decreases to zero at the other end or beginning of the line, $l = 0$.

The current is zero at the open end of the line, and gradually increases to a maximum at $l = 0$, or the grounded end of the line.

Thus the relative intensities of current and potential along the line are as represented by Fig. 6, where the current is shown as I, the voltage as E.

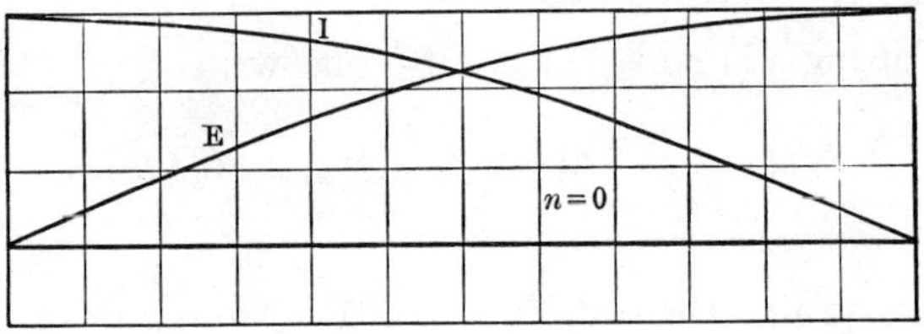

FIG. 6. Discharge of current and e.m.f. along a transmission line open at one end. Fundamental discharge frequency.

The next higher discharge frequency, for $n = 2$, gives

$$\left.\begin{aligned} i_3 &= c_3 \cos 3\,(\theta - \gamma_3) \cos 3\,\tau \\ \text{and} \qquad e_3 &= -\,c_3 \sqrt{\frac{L}{C}} \sin 3\,(\theta - \gamma_3) \sin 3\,\tau. \end{aligned}\right\} \tag{23}$$

Here the voltage is again a maximum at the open end of the line, $l = l_0$, or $\tau = \frac{\pi}{2} = 90°$, and gradually decreases, but reaches zero at two-thirds of the line, $l = \frac{2\,l_0}{3}$, or $\tau = \frac{\pi}{3} = 60°$, then increases again in the opposite direction, reaches a second but opposite maximum at one-third of the line, $l = \frac{l_0}{3}$, or $\tau = \frac{\pi}{6} = 30°$, and decreases to zero at the beginning of the line. There is thus a node of voltage at a point situated at a distance of two-thirds of the length of the line.

The current is zero at the end of the line, $l = l_0$, rises to a maximum at a distance of two-thirds of the length of the line, decreases to zero at a distance of one-third of the length of the line, and rises again to a second but opposite maximum at the beginning of the line, $l = 0$. The current thus has a node at a point situated at a distance of one-third of the length of the line.

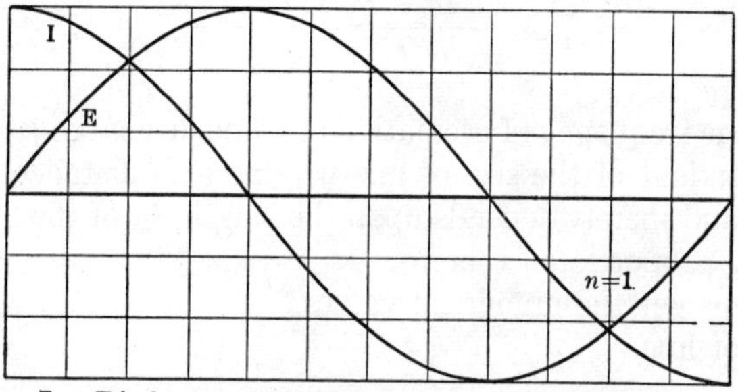

FIG. 7. Discharge of current and e.m.f. along a transmission line open at one end.

The discharge waves, $n = 2$, are shown in Fig. 7, those with $n = 3$, with two nodal points, in Fig. 8.

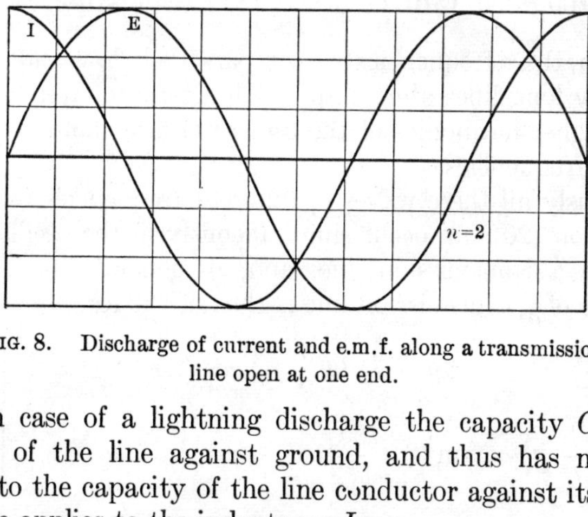

FIG. 8. Discharge of current and e.m.f. along a transmission line open at one end.

31. In case of a lightning discharge the capacity C_0 is the capacity of the line against ground, and thus has no direct relation to the capacity of the line conductor against its return. The same applies to the inductance L_0.

If d = diameter of line conductor, l_h = height of conductor above ground, and l_0 = length of conductor, the capacity is

$$C_0 = \frac{1.11 \times 10^{-6} l_0}{2 \log \frac{4 l_h}{d}}, \text{in mf.}$$

the self-inductance is

$$L_0 = 2 \times 10^{-6} l_0 \log \frac{4 l_h}{d}, \text{ in mh.} \qquad (24)$$

The fundamental frequency of oscillation, by substituting (24) in (10), is

$$f_1 = \frac{1}{4\sqrt{C_0 L_0}} = \frac{7.5 \times 10^9}{l_0}; \tag{25}$$

that is, the frequency of oscillation of a line discharging to ground is independent of the size of line wire and its distance from the ground, and merely depends upon the length, l_0, of the line, being inversely proportional thereto.

We thus get the numerical values,

Length of line

$$= \begin{cases} 10 \quad 20 \quad 30 \quad 40 \quad 50 \quad 60 \quad 80 \quad 100 \text{ miles} \\ 1.6 \quad 3.2 \quad 4.8 \quad 6.4 \quad 8 \quad 9.6 \quad 12.8 \quad 16 \times 10^6 \text{ cm.} \end{cases}$$

hence frequency,

$f_1 =$ 4700 2350 1570 1175 940 783 587 470 cycles per sec.

As seen, these frequencies are comparatively low, and especially with very long lines almost approach alternator frequencies.

The higher harmonics of the oscillation are the odd multiples of these frequencies.

Obviously all these waves of different frequencies represented in equation (20) can occur simultaneously in the oscillating discharge of a transmission line, and, in general, the oscillating discharge of a transmission line is thus of the form

$$\left.\begin{aligned} i &= \sum_1^\infty {}^n c_n \cos (2n-1)(\theta - \gamma_n) \cos (2n-1)\tau, \\ e &= -\sqrt{\frac{L}{C}} \sum_1^\infty {}^n c_n \sin (2n-1)(\theta - \gamma_n) \sin (2n-1)\tau. \end{aligned}\right\} \tag{26}$$

A simple harmonic oscillation as a line discharge would require a sinoidal distribution of potential on the transmission line at the instant of discharge, which is not probable, so that probably all lightning discharges of transmission lines or oscillations produced by sudden changes of circuit conditions are complex waves of many harmonics, which in their relative magnitude depend upon the initial charge and its distribution — that is, in the case of the lightning discharge, upon the atmospheric electrostatic field of force.

The fundamental frequency of the oscillating discharge of a transmission line is relatively low, and of not much higher magnitude than frequencies in commercial use in alternating-current circuits. Obviously, the more nearly sinoidal the distribution of potential before the discharge, the more the low harmonics predominate, while a very unequal distribution of potential, that is a very rapid change along the line, causes the higher harmonics to predominate.

32. As an example the discharge of a transmission line may be investigated, the line having the following constants per mile: $r = 0.21$ ohm; $L = 1.2 \times 10^{-3}$ henry; $C = 0.03 \times 10^{-6}$ farad, and of the length $l_0 = 200$; hence, by equations (10), (19), $f_1 = 208$ cycles per sec.; $\theta = 1315\ t$, and $\tau = 0.00785\ l$, when charged to a uniform voltage of $e_0 = 60{,}000$ volts but with no current in the line before the discharge, and the line then grounded at one end, $l = 0$, while open at the other end, $l = l_0$.

Then, for $t = 0$ or $\theta = 0$, $i = 0$ for all values of τ except $\tau = 0$; hence, by (26),

$$\cos (2n - 1) \gamma_n = 0,$$

and thus

$$(2n - 1) \gamma_n = \frac{\pi}{2} \tag{27}$$

and

$$\cos (2n - 1)(\theta - \gamma_n) = \sin (2n - 1) \theta,$$
$$\sin (2n - 1)(\theta - \gamma_n) = -\cos (2n - 1) \theta;$$

hence,

$$\left.\begin{aligned} i &= \sum_1^\infty {}^n c_n \sin (2n - 1) \theta \cos (2n - 1) \tau \\ \text{and} \quad & \\ e &= \sqrt{\frac{L}{C}} \sum_1^\infty {}^n c_n \cos (2n - 1) \theta \sin (2n - 1) \tau. \end{aligned}\right\} \tag{28}$$

Also for $t = 0$, or $\theta = 0$, $e = e_0$ for all values of τ except $\tau = 0$; hence, by (28),

$$e_0 = \sqrt{\frac{L}{C}} \sum_1^\infty {}^n c_n \sin (2n - 1) \tau. \tag{29}$$

From equation (29), the coefficients c_n are determined in the usual manner of evaluating a Fourier series, that is, by multiplying with $\sin(2m-1)\tau$ (or $\cos(2m-1)\tau$) and integrating:

$$\int_0^{\pi} e_0 \sin(2m-1)\tau\, d\tau =$$

$$\sqrt{\frac{L}{C}} \sum_1^{\infty}{}_n c_n \int_0^{\pi} \sin(2n-1)\tau \sin(2m-1)\tau\, d\tau.$$

Since

$$\int_0^{\pi} \sin(2n-1)\tau \sin(2m-1)\tau\, d\tau$$

$$= \int_0^{\pi} \frac{\cos 2(n-m)\tau - \cos 2(n+m-1)\tau}{2}\, d\tau,$$

which is zero for $n \neq m$, while for $m = n$ the term

$$\int_0^{\pi} \frac{\cos 2(n-m)\tau}{2}\, d\tau = \int_0^{\pi} \frac{d\tau}{2} = \frac{\pi}{2}$$

and

$$\int_0^{\pi} e_0 \sin(2n-1)\tau\, d\tau = -e_0\left[\frac{\cos(2n-1)\tau}{2n-1}\right]_0^{\pi} = +\frac{2e_0}{2n-1},$$

we have

$$\frac{2e_0}{2n-1} = c_n \frac{\pi}{2}\sqrt{\frac{L}{C}}$$

and

$$c_n = \frac{4e_0}{(2n-1)\pi}\sqrt{\frac{C}{L}}; \tag{30}$$

hence,

$$i = \frac{4}{\pi} e_0 \sqrt{\frac{C}{L}} \sum_1^{\infty}{}_n \frac{\sin(2n-1)\theta \cos(2n-1)\tau}{2n-1}$$

$$= \frac{4}{\pi} e_0 \sqrt{\frac{C}{L}} \left\{\sin\theta\cos\tau + \frac{\sin 3\theta\cos 3\tau}{3} + \frac{\sin 5\theta \cos 5\tau}{5} + \cdots\right\} \tag{31}$$

$$= 382\left\{\sin\theta\cos\tau + \frac{\sin 3\theta\cos 3\tau}{3} + \frac{\sin 5\theta\cos 5\tau}{5} + \cdots\right\},$$

in amperes,

and

$$e = \frac{4}{\pi} e_0 \sum_1^\infty {}_n \frac{\cos (2n - 1) \theta \sin (2n - 1) \tau}{2n - 1}$$

$$= \frac{4}{\pi} e_0 \left\{ \cos \theta \sin \tau + \frac{\cos 3 \theta \sin 3 \tau}{3} + \frac{\cos 5 \theta \sin 5 \tau}{5} + \cdots \right\} \quad (32)$$

$$= 76{,}400 \left\{ \cos \theta \sin \tau + \frac{\cos 3 \theta \sin 3 \tau}{3} + \frac{\cos 5 \theta \sin 5 \tau}{5} + \cdots \right\},$$

in volts.

33. As further example, assume now that this line is short-circuited at one end, $l = 0$, while supplied with 25-cycle alternating power at the other end, $l = l_0$, and that the generator voltage drops, by the short circuit, to 30,000, and then the line cuts off from the generating system at about the maximum value of the short-circuit current, that is, at the moment of zero value of the impressed e.m.f.

At a frequency of $f_0 = 25$ cycles, the reactance per unit length of line or per mile is

$$x = 2 \pi f_0 L = 0.188 \text{ ohm}$$

and the impedance is

$$z = \sqrt{r^2 + x^2} = 0.283 \text{ ohm},$$

or, for the total line,

$$z_0 = l_0 z = 56.6 \text{ ohms};$$

hence, the approximate short-circuit current

$$i = \frac{e}{z_0} = \frac{30{,}000}{56.6} = 530 \text{ amp.},$$

and its maximum value is

$$i_0 = 530 \times \sqrt{2} = 750 \text{ amp.}$$

Therefore, in equations (26), at time $t = 0$, or $\theta = 0$, $e = 0$ for all values of τ except $\tau = \frac{\pi}{2}$; hence,

$$\sin (2n - 1) \gamma_n = 0,$$

or,

$$\gamma_n = 0,$$

and thus

$$\left.\begin{aligned} i &= \sum_1^{\infty}{}^n c_n \cos (2n-1)\,\theta \cos (2n-1)\,\tau \\ \text{and} \\ e &= -\sqrt{\frac{L}{C}} \sum_1^{\infty}{}^n c_n \sin (2n-1)\,\theta \sin (2n-1)\,\tau. \end{aligned}\right\} \tag{33}$$

However, at $t = 0$, or $\theta = 0$, for all values of τ except $\tau = \frac{\pi}{2}$, $i = i_0$;
hence, substituting in (33),

$$i_0 = \sum_1^{\infty}{}^n c_n \cos (2n-1)\,\tau. \tag{34}$$

From equation (34), the coefficients c_n are determined in the same manner as in the preceding example, by multiplying with $\cos (2n-1)\,\tau$ and integrating, as

$$c_n = -(-1)^n \frac{4\,i_0}{(2n-1)\,\pi}; \tag{35}$$

hence,

$$\begin{aligned} i &= -\frac{4\,i_0}{\pi} \sum_1^{\infty}{}^n (-1)^n \frac{\cos (2n-1)\,\theta \cos (2n-1)\,\tau}{2n-1} \\ &= \frac{4\,i_0}{\pi} \left\{ \cos\theta \cos\tau - \frac{\cos 3\,\theta \cos 3\,\tau}{3} + \frac{\cos 5\,\theta \cos 5\,\tau}{5} - + \cdots \right\} \\ &= 956 \left\{ \cos\theta \cos\tau - \frac{\cos 3\,\theta \cos 3\,\tau}{3} + \frac{\cos 5\,\theta \cos 5\,\tau}{5} - + \cdots \right\}, \end{aligned} \tag{36}$$

in amperes,

and

$$\begin{aligned} e &= -\frac{4\,i_0}{\pi} \sqrt{\frac{L}{C}} \sum_1^{\infty}{}^n (-1)^n \frac{\sin (2n-1)\,\theta \sin (2n-1)\,\tau}{2n-1} \\ &= \frac{4\,i_0}{\pi} \sqrt{\frac{L}{C}} \left\{ \sin\theta \sin\tau - \frac{\sin 3\,\theta \sin 3\,\tau}{3} + \frac{\sin 5\,\theta \sin 5\,\tau}{5} - + \cdots \right\} \\ &= 191{,}200 \left\{ \sin\theta \sin\tau - \frac{\sin 3\,\theta \sin 3\,\tau}{3} + \frac{\sin 5\,\theta \sin 5\,\tau}{5} - + \cdots \right\}, \end{aligned} \tag{37}$$

in volts.

The maximum voltage is reached at time $\theta = \frac{\pi}{2}$, and is

$$e = \frac{4\, i_0}{\pi}\sqrt{\frac{L}{C}}\left\{\sin\tau + \frac{\sin 3\,\tau}{3} + \frac{\sin 5\,\tau}{5} + \cdots\right\},$$

and since the series

$$\sin\tau + \frac{\sin 3\,\tau}{3} + \frac{\sin 5\,\tau}{5} + \cdots = \frac{\pi}{2},$$

the maximum voltage is

$$e = i_0\sqrt{\frac{L}{C}} = 300{,}000 \text{ volts.}$$

As seen, very high voltages may be produced by the interruption of the short-circuit current.

(2*a*) *Circuit grounded at both ends.*

34. The method of investigation is the same as in paragraph 29; the terminal conditions are, for $l = 0$,

$$\dot{E} = 0,$$

and for $l = l_0$,

$$\dot{E} = 0.$$

Substituting $l = 0$ into equations (3) gives

$$C_2 = 0;$$

hence,

$$\left.\begin{aligned} \dot{I} &= C_1 \cos\beta l, \\ \dot{E} &= jC_1\sqrt{\frac{L}{C}}\sin\beta l. \end{aligned}\right\} \qquad (38)$$

Substituting $l = l_0$ in (38) gives

$$\dot{E}_1 = 0 = jC_1\sqrt{\frac{L}{C}}\sin\beta l_0;$$

hence,

$$\sin\beta l_0 = 0, \text{ or } \beta l = n\pi, \qquad (39)$$

and, in the same manner as in (1),

$$\beta l = n\frac{\pi}{l_0}l = n\tau; \tag{40}$$

that is, the length of the line, l_0, represents one half wave, or $\tau = \pi$, or a multiple thereof.

$$f = \frac{n}{2\,l_0\sqrt{LC}} = \frac{n}{2\sqrt{L_0C_0}}, \tag{41}$$

and the *fundamental frequency of oscillation* is

$$f_1 = \frac{1}{2\,l_0\sqrt{LC}}, \tag{42}$$

and

$$f = nf_1; \tag{43}$$

that is, the line can oscillate at a fundamental frequency f_1, for which the length, l_0, of the line is a half wave, and at all multiples or higher harmonics thereof, the *even* ones as well as the *odd* ones.

This kind of oscillation may be called a *half-wave oscillation.*

35. Unlike the quarter-wave oscillation, which contains only the odd higher harmonics of the fundamental wave, the half-wave oscillation also contains the even harmonics of the fundamental frequency of oscillation.

Substituting $\dot{C}_1 = c_1 + jc_2$ into (38) gives

$$\left.\begin{aligned} \dot{I} &= (c_1 + jc_2)\cos\beta l \\ \text{and} \qquad \dot{E} &= (c_2 - jc_1)\sqrt{\frac{L}{C}}\sin\beta l, \end{aligned}\right\} \tag{44}$$

and replacing the complex imaginary by the analytic expression, that is, the real term by $\cos 2\,\pi ft$, the imaginary term by $\sin 2\,\pi ft$, gives

$$\left.\begin{aligned} i &= \{c_1\cos 2\,\pi ft + c_2\sin 2\,\pi ft\}\cos\beta l \\ \text{and} \qquad e &= \sqrt{\frac{L}{C}}\{c_2\cos 2\,\pi ft - c_1\sin 2\,\pi ft\}\sin\beta l, \end{aligned}\right\}$$

and substituting

we have
$$\left.\begin{aligned} 2\,\pi f_1 t &= \theta, \\ 2\,\pi f t &= n\theta; \end{aligned}\right\} \qquad (45)$$

then (44) gives, by (40):

$$i = (c_1 \cos n\theta + c_2 \sin n\theta) \cos n\tau$$

and

$$e = \sqrt{\frac{L}{C}}\,(c_2 \cos n\theta - c_1 \sin n\theta) \sin n\tau; \qquad (46)$$

or writing

$$c_1 = c \cos n\gamma$$

and

$$c_2 = c \sin n\gamma \qquad (47)$$

gives

$$i = c \cos n\,(\theta - \gamma) \cos n\tau$$

and

$$e = - c\,\sqrt{\frac{L}{C}}\,\sin n\,(\theta - \gamma) \sin n\tau, \qquad (48)$$

and herefrom the *general equations of this half-wave oscillation* are

$$i = \sum_1^{\infty}{}^n\, c_n \cos n\,(\theta - \gamma_n) \cos n\tau$$

and

$$e = -\sqrt{\frac{L}{C}}\sum_1^{\infty}{}^n\, c_n \sin n\,(\theta - \gamma_n) \sin n\tau \qquad (49)$$

(2*b*) *Circuit open at both ends.*

36. For $l = 0$ we have

$$\dot{I} = 0;$$

hence,

$$\dot{C}_1 = 0$$

and

$$\dot{I} = j C_2 \sin \beta l$$

and

$$\dot{E} = \sqrt{\frac{L}{C}}\,C_2 \cos \beta l, \qquad (50)$$

while for $l = l_0$, $\dot{I} = 0$;

hence,

$$\sin \beta l_0 = 0, \text{ or } \beta l_0 = n\pi; \tag{51}$$

that is, the circuit performs a *half-wave oscillation* of *fundamental frequency*,

$$f_1 = \frac{1}{2\, l_0 \sqrt{LC}} \tag{52}$$

and all its higher harmonics, the even ones as well as the odd ones have a frequency

$$f = nf_1, \tag{53}$$

and the final equations are

$$\left.\begin{aligned} i &= -\sum_1^\infty n\, c_n \sin n\,(\theta - \gamma) \sin n\tau \\ \text{and} \qquad & \\ e &= \sqrt{\frac{L}{C}} \sum_1^\infty n\, c_n \cos n\,(\theta - \gamma) \cos n\tau, \end{aligned}\right\} \tag{54}$$

where $\theta = 2\,\pi f_1 t$ and $\tau = \frac{\pi}{l_0} l.$ (55)

(3) *Circuit closed upon itself.*

37. If a circuit of length l_0 is closed upon itself, then the free oscillation of such a circuit is characterized by the condition that current and voltage at $l = l_0$ are the same as at $l = 0$, since $l = l_0$ and $l = 0$ are the same point of the circuit.

Substituting this condition in equations (3) gives

$$\left.\begin{aligned} \dot{I} = \dot{C}_1 &= \dot{C}_1 \cos \beta l_0 + j C_2 \sin \beta l_0 \\ \text{and} \qquad & \\ \sqrt{\frac{C}{L}}\, \dot{E} = \dot{C}_2 &= \dot{C}_2 \cos \beta l_0 + j C_1 \sin \beta l_0; \end{aligned}\right\} \tag{56}$$

herefrom follows

$$\left.\begin{aligned} C_1 (1 - \cos \beta l_0) &= + j C_2 \sin \beta l_0, \\ C_2 (1 - \cos \beta l_0) &= + j C_1 \sin \beta l_0, \end{aligned}\right\} \tag{57}$$

hence, $(1 - \cos \beta l_0)^2 = - \sin^2 \beta l_0$

or

$$\cos \beta l_0 = 1; \tag{58}$$

hence,

$$\beta l_0 = 2\,n\pi; \tag{59}$$

that is, the circuit must be a complete wave or a multiple thereof.

The free oscillation of a circuit which is closed upon itself is a *full-wave oscillation*, containing a *fundamental wave* of frequency

$$f_1 = \frac{1}{l\sqrt{LC}}, \tag{60}$$

and all the higher harmonics thereof, the *even* ones as well as the *odd* ones,

$$f = nf_1. \tag{61}$$

Substituting in (3),

$$\left.\begin{aligned} \dot{C}_1 &= c_1' + jc_1'' \\ \text{and} \qquad \dot{C}_2 &= c_2' + jc_2'' \end{aligned}\right\}$$

gives

$$\left.\begin{aligned} \dot{I} &= (c_1' + jc_1'')\cos\beta l + (c_2'' - jc_2')\sin\beta l \\ \text{and} \qquad \dot{E} &= \sqrt{\frac{L}{C}}\,\{(c_2' + jc_2'')\cos\beta l + (c_1'' - jc_1')\sin\beta l\}. \end{aligned}\right\} \tag{62}$$

Substituting the analytic expression,

$$c_1' + jc_1'' = c_1'\cos 2\,\pi ft + c_1''\sin 2\,\pi ft, \text{ etc.,}$$

also

$$\left.\begin{aligned} 2\,\pi f_1 t &= \theta, \\ 2\,\pi ft &= n\theta, \end{aligned}\right\} \tag{63}$$

$$\left.\begin{aligned} \text{and} \qquad \beta l &= \frac{2\,\pi n}{l_0}\,l = n\tau, \\ \text{where} \qquad \tau &= \frac{2\,\pi}{l_0}\,l, \end{aligned}\right\} \tag{64}$$

that is, the length of the circuit, $l = l_0$, is represented by the angle $\tau = 2\,\pi$, or a complete cycle, this gives

$$\begin{aligned} I &= (c_1' \cos n\theta + c_1'' \sin n\theta) \cos n\tau \\ &\quad + (c_2'' \cos n\theta - c_2' \sin n\theta) \sin n\tau \end{aligned}$$

and

$$\begin{aligned} E &= \sqrt{\frac{L}{C}} \{ (c_2' \cos n\theta + c_2'' \sin n\theta) \cos n\theta \\ &\quad + (c_1'' \cos n\theta - c_1' \sin n\theta) \sin n\tau \}, \end{aligned} \tag{65}$$

or writing

$$\begin{aligned} c_1' &= a \cos n\gamma \\ c_1'' &= a \sin n\gamma \\ c_2' &= b \cos n\chi \\ c_2'' &= b \sin n\chi \end{aligned}$$

gives

$$i = a \cos n (\theta - \gamma) \cos n\tau - b \sin n (\theta - \chi) \sin n\tau$$

and

$$e = \sqrt{\frac{L}{C}} \{ b \cos n (\theta - \chi) \cos n\tau - a \sin n (\theta - \gamma) \sin n\tau \}. \tag{66}$$

Thus in its most general form the full-wave oscillation gives the equations

$$\begin{aligned} i &= \sum_1^\infty {}^n \{ a_n \cos n (\theta - \gamma_n) \cos n\tau - b_n \sin n (\theta - \chi_n) \sin n\tau \} \\ e &= \sqrt{\frac{L}{C}} \sum_1^\infty {}^n \{ b_n \cos n (\theta - \chi_n) \cos n\tau - a_n \sin n (\theta - \gamma_n) \sin n\tau \}, \end{aligned} \tag{67}$$

where

$$\begin{aligned} \theta &= 2 \pi f_1 t, \\ f_1 &= \frac{1}{l_0 \sqrt{LC}}, \\ \tau &= \frac{2 \pi}{l_0} l, \end{aligned} \tag{68}$$

and a_n, γ_n and b_n, χ_n are groups of four integration constants.

38. With a short circuit at the end of a transmission line, the drop of potential along the line varies fairly gradually and uniformly, and the instantaneous rupture of a short circuit — as by a short-circuiting arc blowing itself out explosively —

causes an oscillation in which the lower frequencies predominate, that is, a low-frequency high-power surge. A spark discharge from the line, a sudden high voltage charge entering the line locally, as directly by a lightning stroke, or indirectly by induction during a lightning discharge elsewhere, gives a distribution of potential which momentarily is very non-uniform, changes very abruptly along the line, and thus gives rise mainly to very high harmonics, but as a rule does not contain to any appreciable extent the lower frequencies; that is, it causes a high-frequency oscillation, more or less local in extent, and while of high voltage, of rather limited power, and therefore less destructive than a low-frequency surge.

At the frequencies of the high-frequency oscillation neither capacity nor inductance of the transmission line is perfectly constant: the inductance varies with the frequency, by the increasing screening effect or unequal current distribution in the conductor; the capacity increases by brush discharge over the insulator surface, by the increase of the effective conductor diameter due to corona effect, etc. The frequencies of the very high harmonics are therefore not definite but to some extent variable, and since they are close to each other they overlap; that is, at very high frequencies the transmission line has no definite frequency of oscillation, but can oscillate with any frequency.

A long-distance transmission line has a definite natural period of oscillation, of a relatively low fundamental frequency and its overtones, but can also oscillate with any frequency whatever, provided that this frequency is very high.

This is analogous to waves formed in a body of water of regular shape: large standing waves have a definite wave length, depending upon the dimensions of the body of water, but very short waves, ripples in the water, can have any wave length, and do not depend on the size of the body of water.

A further investigation of oscillations in conductors with distributed capacity, inductance, and resistance requires, however, the consideration of the resistance, and so leads to the investigation of phenomena transient in space as well as in time, which are discussed in Part III of this volume.

39. In the equations discussed in the preceding, of the free oscillations of a circuit containing uniformly distributed resistance, inductance, capacity, and conductance, the energy losses in the circuit have been neglected, and voltage and current therefore appear alternating instead of oscillating. That is, these equations represent only the initial or maximum values of the phenomenon, but to represent it completely an exponential function of time enters as factor, which, as will be seen in Section IV, is of the form

$$\varepsilon^{-ut}, \tag{69}$$

where $u = \frac{1}{2}\left(\frac{r}{L} + \frac{g}{C}\right)$ may be called the "time constant" of the circuit.

While quarter-wave oscillations occasionally occur, and are of serious importance, the occurrence of half-wave oscillations and especially of full-wave oscillations of the character discussed before, that is, of a uniform circuit, is less frequent.

When in a circuit, as a transmission line, a disturbance or oscillation occurs while this circuit is connected to other circuits — as the generating system and the receiving apparatus — as is usually the case, the disturbance generally penetrates into the circuits connected to the circuit in which the disturbance originated, that is, the entire system oscillates, and this oscillation usually is a full-wave oscillation; that is, the oscillation of a circuit closed upon itself; occasionally a half-wave oscillation. For instance, if in a transmission system comprising generators, step-up transformers, high-potential lines, step-down transformers, and load, a short circuit occurs in the line, the circuit comprising the load, the step-down transformers, and the lines from the step-down transformers to the short circuit is left closed upon itself without power supply, and its stored energy is, therefore, dissipated as a full-wave oscillation. Or, if in this system an excessive load, as the dropping out of step of a synchronous converter, causes the circuit to open at the generating station, the dissipation of the stored energy — in this case that of the excessive current in the system — occurs as a full-wave oscillation, if the line cuts off from the generating station on the low-tension side of the step-up transformers, and the oscillating circuit comprises the high-tension coils of the step-up transformers, the transmission line, step-down transformers, and load. If the line disconnects from the generating system on the high-potential side of the step-up transformers, the oscillation is a half-wave oscillation, with the two ends of the oscillating circuit open.

Such oscillating circuits, however, — representing the most frequent and most important case of high-potential disturbances in transmission systems,—cannot be represented by the preceding equations since they are not circuits of uniformly distributed constants but compound circuits comprising several sections of different constants, and therefore of different ratios of energy consumption and energy storage, $\frac{r}{L}$ and $\frac{g}{C}$. During the free oscillation of such circuits an energy transfer takes place between the different sections of the circuit, and energy flows from those sections in which the energy consumption is small compared with the energy storage, as transformer coils and highly inductive loads, to those sections in which the energy consumption is large compared with the energy storage, as the more non-inductive parts of the system. This introduces into the equations exponential functions of the distance as well as the time, and requires a study of the phenomenon as one transient in distance as well as in time. The investigation of the oscillation of a compound circuit, comprising sections of different constants, is treated in Part III of this volume.

CHAPTER IV

DISTRIBUTED CAPACITY OF HIGH-POTENTIAL TRANSFORMERS

40. In the high-potential coils of transformers designed for very high voltages phenomena resulting from distributed capacity occur.

In transformers for very high voltages — 100,000 volts and more, or even considerably less in small transformers — the high-potential coil contains a large number of turns, a great length of conductor, and therefore its electrostatic capacity is appreciable, and such a coil thus represents a circuit of distributed resistance, inductance, and capacity somewhat similar to a transmission line.

The same applies to reactive coils, etc., wound for very high voltages, and even in smaller reactive coils at very high frequency.

This capacity effect is more marked in smaller transformers, where the size of the iron core and therewith the voltage per turn is less, and therefore the number of turns greater than in very large transformers, and at the same time the exciting current and the full-load current are less; that is, the charging current of the conductor more comparable with the load current of the transformer or reactive coil.

It is, however, much more serious in large transformers, since in such the resistance is smaller compared to the inductance and capacity, and therefore the damping of any high frequency oscillation less, the possibility of the formation of sustained and cumulative oscillations greater.

However, even in large transformers and at moderately high voltages, capacity effects occur in transformers, if the frequency is sufficiently high, as is the case with the currents produced in overhead lines by lightning discharges, or by arcing grounds resulting from spark discharges between conductor and ground, or in starting or disconnecting the transformer. With such frequencies, of many thousand cycles, the internal capacity of the transformer becomes very marked in its effect on the distribution of voltage and current, and may produce dangerous high-voltage points in the transformer.

The distributed capacity of the transformer, however, is different from that of a transmission line.

In a transmission line the distributed capacity is shunted capacity, that is, can be represented diagrammatically by condensers shunted across the circuit from line to line, or, what amounts to the same thing, from line to ground and from ground to return line, as shown diagrammatically in Fig. 9.

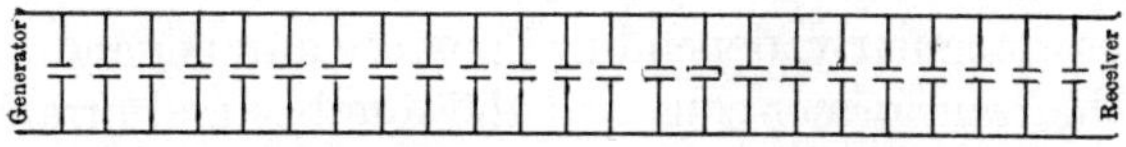

FIG. 9. Distributed capacity of a transmission line.

The high-potential coil of the transformer also contains shunted capacity, or capacity from the conductor to ground, and so each coil element consumes a charging current proportional to its potential difference against ground. Assuming the circuit as insulated, and the middle of the transformer coil at ground potential, the charge consumed by unit length of the coil increases from zero at the center to a maximum at the ends. If one terminal of the circuit is grounded, the charge consumed by the coil increases from zero at the grounded terminal to a maximum at the ungrounded terminal.

In addition thereto, however, the transformer coil also contains a capacity between successive turns and between successive layers. Starting from one point of the conductor, after a certain

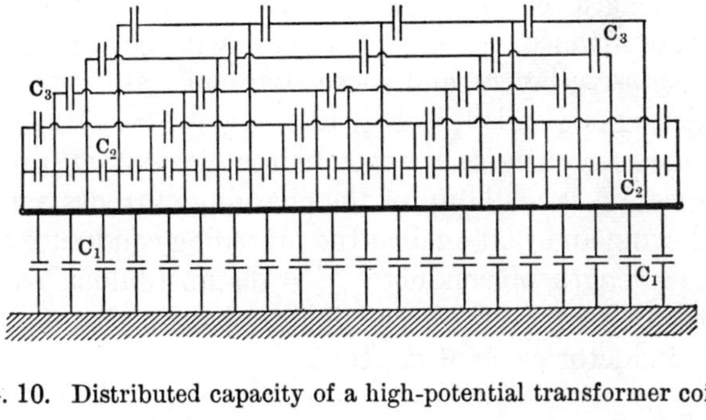

FIG. 10. Distributed capacity of a high-potential transformer coil.

length, the length of one turn, the conductor reapproaches the first point in the next adjacent turn. It again approaches the first point at a different and greater distance in the next adjacent layer.

first point at a different and greater distance in the next adjacent layer.

A transformer high-potential coil can be represented diagrammatically as a conductor, Fig. 10. C_1 represents the capacity against ground, C_2 represents the capacity between adjacent turns, and C_3 the capacity between adjacent layers of the coil.

The capacities C_2 and C_3 are not uniformly distributed but more or less irregularly, depending upon the number and arrangement of the transformer coils and the number and arrangement of turns in the coil. As approximation, however, the capacities C_2 and C_3 can be assumed as uniformly distributed capacity between successive conductor elements. If l = length of conductor, they may be assumed as a capacity between l and $l + dl$, or as a capacity across the conductor element dl.

This approximation is permissible in investigating the general effect of the distributed capacity, but omits the effect of the irregular distribution of C_2 and C_3, which leads to local oscillations of higher frequencies, extending over sections of the circuit, such as individual transformer coils, and may cause destructive voltage across individual transformer coils, without the appearance of excessive voltages across the main terminals of the transformer.

41. Let then, in the high-potential coil of a high-voltage transformer, e = the e.m.f. generated per unit length of conductor, as, for instance, per turn; $Z = r + jx$ = the impedance per unit length; $Y = g + jb$ = the capacity admittance against ground per unit length of conductor, and $Y' = pY$ = the capacity effective admittance representing the capacity between successive turns, successive layers, and successive coils, as represented by the condensers C_2 and C_3 in Fig. 10.

The charging current of a conductor element dl, due to the admittance Y', is made up of the charging currents against the next following and that against the preceding conductor element.

Let l_0 = length of conductor; l = distance along conductor; $\dot{E}$ = potential at point l, or conductor element dl, and $\dot{I}$ = current in conductor element dl; then

$$d\dot{E} = \frac{d\dot{E}}{dl}\, dl = \text{the potential difference between successive}$$

conductor elements or turns.

$Y' \frac{d\dot{E}}{dl} dl$ = the charging current between one conductor element and the next conductor element or turn.

$- Y' \frac{d(\dot{E} - d\dot{E})}{dl} dl$ = the charging current between one conductor element and the preceding conductor element or turn, hence,

$Y' \frac{d^2\dot{E}}{dl^2} dl$ = the charging current of one conductor element due to capacity between adjacent conductors or turns.

If now the distance l is counted from the point of the conductor, which is at ground potential, $Y\dot{E}dl$ = the charging current of one conductor element against ground, and

$$d\dot{I} = \left\{ Y\dot{E} + Y' \frac{d^2\dot{E}}{dl^2} \right\} dl = Y \left\{ \dot{E} + p \frac{d^2\dot{E}}{dl^2} \right\} dl$$

is the total current consumed by a conductor element.

However, the e.m.f. consumed by impedance equals the e.m.f. consumed per conductor element; thus

$$d\dot{E}_z = Z\dot{I}\, dl.$$

This gives the two differential equations:

$$\frac{d\dot{I}}{dl} = Y \left\{ \dot{E} + p \frac{d^2\dot{E}}{dl^2} \right\} \tag{1}$$

and

$$e - \frac{d\dot{E}}{dl} = Z\dot{I}. \tag{2}$$

Differentiating (2) and substituting in (1) gives

$$\frac{d^2\dot{E}}{dl^2} = ZY \left\{ \dot{E} + p \frac{d^2\dot{E}}{dl^2} \right\};$$

transposing,

$$\frac{d^2\dot{E}}{dl^2} = \frac{-\dot{E}}{p - \frac{1}{ZY}}, \tag{3}$$

or

$$\frac{d^2\dot{E}}{dl^2} = -a^2\dot{E}, \tag{4}$$

where
$$a^2 = \frac{1}{p - \frac{1}{ZY}}. \tag{5}$$

If $\frac{1}{ZY}$ is small compared with p, we have, approximately,
$$a^2 = \frac{1}{p} \tag{6}$$

and
$$\dot{E} = \dot{A} \cos al + \dot{B} \sin al, \tag{7}$$

and since, for $l = 0$, $\dot{E} = 0$, if the distance l is counted from the point of zero potential, we have
$$\dot{E} = \dot{B} \sin al, \tag{8}$$

and the current is given by equation (2) as
$$\dot{I} = \frac{1}{Z}\left\{e - \frac{d\dot{E}}{dl}\right\}; \tag{9}$$

substituting (8) in (9) gives
$$\dot{I} = \frac{1}{Z}\left\{e - a\dot{B} \cos al\right\}. \tag{10}$$

If now $\dot{I}_1$ = the current at the transformer terminals, $l = l_0$, we have, from (10),
$$Z\dot{I}_1 = e - a\dot{B} \cos al_0$$

and
$$\dot{B} = \frac{e - Z\dot{I}_1}{a \cos al_0}; \tag{11}$$

substituting in (8) and (10),
$$\left.\begin{aligned} \dot{E} &= (e - Z\dot{I}_1)\frac{\sin al}{a \cos al_0} \\ \text{and} \quad \dot{I} &= \frac{1}{Z}\left\{(e - Z\dot{I}_1)\frac{\cos al}{\cos al_0} - e\right\}; \end{aligned}\right\} \tag{12}$$

for $\dot{I}_1 = 0$, or open circuit of the transformer, this gives
$$\left.\begin{aligned} \dot{E} &= e\frac{\sin al}{a \cos al_0} \\ \text{and} \quad \dot{I} &= \frac{e}{Z}\left(\frac{\cos al}{\cos al_0} - 1\right). \end{aligned}\right\} \tag{13}$$

The e.m.f., E, thus is a maximum at the terminals,

$$E_1 = \frac{e}{a} \tan al_0,$$

the current a maximum at the zero point of potential, $l = 0$, where

$$I_0 = \frac{e}{Z}\left(\frac{1}{\cos al_0} - 1\right). \tag{14}$$

42. Of all industrial circuits containing distributed capacity and inductance, the high potential coils of large high voltage power transformers probably have the lowest attenuation constant of oscillations, that is the lowest ratio of r to $\sqrt{LC}$, and high frequency oscillations occurring in such circuits thus die out at a slower rate, hence are more dangerous than in most other industrial circuits. Nearest to them in this respect are the armature circuits of large high voltage generators, and similar considerations apply to them.

As the result hereof, the possibility of the formation of continual and cumulative oscillations, in case of the presence of a source of high frequency power, as an arc or a spark discharge in the system, is greater in high potential transformer coils than in most other circuits. Regarding such cumulative oscillations and their cause and origin, see the chapters on "Instability of Electric Circuits," in Dover edition Volume I, Part III.

The frequency of oscillation of the high potential circuit of large high voltage power transformers usually is of the magnitude of 10,000 to 30,000 cycles; the frequency of oscillation of individual transformer coils of this circuit is usually of the magnitude of 30,000 to 100,000 cycles. There then are the danger frequencies of large high voltage transformers.

CHAPTER V

DISTRIBUTED SERIES CAPACITY

43. The capacity of a transmission line, cable, or high-potential transformer coil is shunted capacity, that is, capacity from conductor to ground, or from conductor to return conductor, or shunting across a section of the conductor, as from turn to turn or layer to layer of a transformer coil.

In some circuits, in addition to this shunted capacity, distributed series capacity also exists, that is, the circuit is broken at frequent and regular intervals by gaps filled with a dielectric or insulator, as air, and the two faces of the conductor ends thus constitute a condenser in series with the circuit. Where the elements of the circuit are short enough so as to be represented, approximately, as conductor differentials, the circuit constitutes a circuit with distributed series capacity.

An illustration of such a circuit is afforded by the so-called "multi-gap lightning arrester," as shown diagrammatically in Fig. 11, which consists of a large number of metal cylinders p, q . . . , with small spark gaps between the cylinders, connected between line L and ground G. This arrangement, Fig. 11, can be represented diagrammatically by Fig. 12. Each cylinder has a capacity C_0 against ground, a capacity C against the adjacent cylinder, a resistance r,— usually very small,— and an inductance L.

The series of insulator discs of a high voltage suspension—or strain—insulator also forms such a circuit.

If such a series of n equal capacities or spark gaps is connected across a constant supply voltage e_0, each gap has a voltage $e = \frac{e_0}{n}$. If, however, the supply voltage is alternating, the voltage does not divide uniformly between the gaps, but the potential difference is the greater, that is, the potential gradient steeper the nearer the gap is to the line L, and this distribution of potential becomes the more non-uniform the higher the frequency; that is, the greater the charging current of the capacity of the cylinder against the ground. The charging currents against ground, of all the cylinders from q to the ground G, Figs. 11 and 12, must pass the gap between the adjacent cylinders p and q; that is, the charging current of the condenser represented by two adjacent

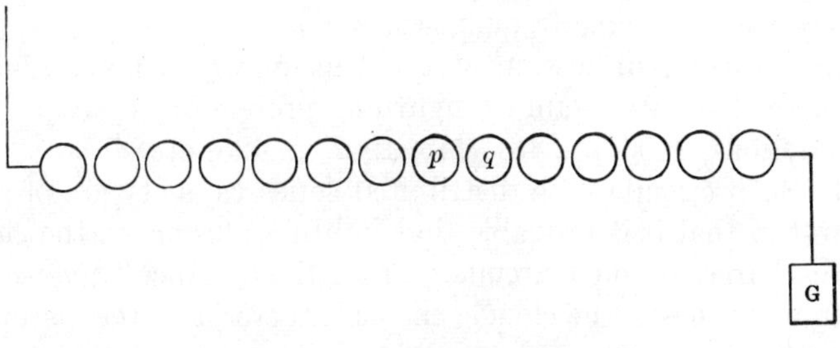

FIG. 11. Multi-gap lightning arrester.

cylinders p and q is the sum of all the charging currents from q to G; and as the potential difference between the two cylinders p and q is proportional to the charging current of the condenser

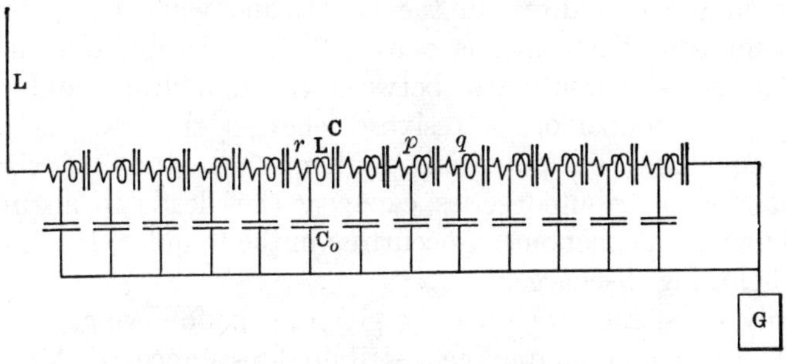

FIG. 12. Equivalent circuit of a multi-gap lightning arrester.

formed by these two cylinders, C, this potential difference increases towards L, being, at each point proportional to the vector sum of all the charging currents, against ground, of all the cylinders between this point and ground.

The higher the frequency, the more non-uniform is the potential gradient along the circuit and the lower is the total supply voltage required to bring the maximum potential gradient, near the line L, above the disruptive voltage, that is, to initiate the discharge. Thus such a multigap structure is discriminating regarding frequency; that is, the discharge voltage with increasing frequency, does not remain constant, but decreases with increase of frequency, when the frequency becomes sufficiently high to give appreciable charging currents. Hence high frequency oscillations discharge over such a structure at lower voltage than machine frequencies.

For a further discussion of the feature which makes such a multigap structure useful for lightning protection, see A. I. E. E. Transactions, 1906, pp. 431, 448, 1907, p. 425, etc.

44. Such circuits with distributed series capacity are of great interest in that it is probable that lightning flashes in the clouds are discharges in such circuits. From the distance traversed by lightning flashes in the clouds, their character, and the disruptive strength of air, it appears certain that no potential difference can exist in the clouds of such magnitude as to cause a disruptive discharge across a mile or more of space. It is probable that as the result of condensation of moisture, and the lack of uniformity of such condensation, due to the gusty nature of air currents, a non-uniform distribution of potential is produced between the rain drops in the cloud; and when the potential gradient somewhere in space exceeds the disruptive value, an oscillatory discharge starts between the rain drops, and gradually, in a number of successive discharges, traverses the cloud and equalizes the potential gradient. A study of circuits containing distributed series capacity thus leads to an understanding of the phenomena occurring in the thunder cloud during the lightning discharge.*

Only a general outline can be given in the following.

45. In a circuit containing distributed resistance, conductance, inductance, shunt, and series capacity, as the multigap lightning arrester, Fig. 11, represented electrically as a circuit in Fig. 12, let r = the effective resistance per unit length of circuit, or per circuit element, that is, per arrester cylinder; g = the shunt conductance per unit length, representing leakage, brush discharge, electrical radiation, etc.; L = the inductance per unit length of circuit; C = the series capacity per unit length of circuit, or circuit element, that is, capacity between adjacent arrester cylinders, and C_0 = the shunt capacity per unit length of circuit, or circuit element, that is, capacity between arrester cylinder and

* See paper, "Lightning and Lightning Protection," N.E.L.A., 1907. Reprinted and enlarged in "General Lectures on Electrical Engineering," by Author.

ground. If then f = the frequency of impressed e.m.f., the series impedance per unit length of circuit is

$$Z = r + j\,(x - x_c); \tag{1}$$

the shunt admittance per unit length of circuit is

$$Y = g + jb, \tag{2}$$

where

$$\left.\begin{aligned} x &= 2\,\pi f L, \\ x_c &= \frac{1}{2\,\pi f C}, \\ b &= 2\,\pi f C_0; \end{aligned}\right\} \tag{3}$$

or the absolute values are

$$\left.\begin{aligned} z &= \sqrt{r^2 + (x - x_c)^2} \\ \text{and} \qquad y &= \sqrt{g^2 + b^2}. \end{aligned}\right\} \tag{4}$$

If the distance along the circuit from line L towards ground G is denoted by l, the potential difference between point l and ground by $\dot{E}$, and the current at point l by $\dot{I}$, the differential equations of the circuit are *

$$\frac{d\dot{E}}{dl} = Z\dot{I} \tag{5}$$

and

$$\frac{d\dot{I}}{dl} = Y\dot{E}. \tag{6}$$

Differentiating (5) and substituting (6) therein gives

$$\frac{d^2\dot{E}}{dl^2} = YZ\dot{E}. \tag{7}$$

Equation (7) is integrated by

$$\dot{E} = A_1\varepsilon^{-al} + A_2\varepsilon^{+al}, \tag{8}$$

where

$$a = \sqrt{YZ} = \alpha + j\beta, \tag{9}$$

$$\left.\begin{aligned} \alpha &= \sqrt{\tfrac{1}{2}\{yz + gr - b\,(x - x_c)\}} \\ \text{and} \qquad \beta &= \sqrt{\tfrac{1}{2}\{yz - gr + b\,(x - x_c)\}}. \end{aligned}\right\} \tag{10}$$

* Part II, Chapter II, paragraph 7, p. 65.

Substituting (10) in (8) and eliminating the imaginary exponents by the substitution of trigonometric functions,

$$E = A_1 \varepsilon^{-al} (\cos \beta l - j \sin \beta l) + A_2 \varepsilon^{+al} (\cos \beta l + j \sin \beta l). \quad (11)$$

46. However, if n = the total length of circuit from line L to ground G, or total number of arrester cylinders between line and ground, for $l = n$,

$$E = 0, \quad (12)$$

and for $l = 0$,

$$E = e_0 = \text{the impressed e.m.f.} \quad (13)$$

Substituting (12) and (13) into (11) gives

$$0 = A_1 \varepsilon^{-an} (\cos \beta n - j \sin \beta n) + A_2 \varepsilon^{+an} (\cos \beta n + j \sin \beta n)$$

and

$$e_0 = A_1 + A_2;$$

hence,

$$\left.\begin{aligned} A_1 &= \frac{e_0}{1 - \varepsilon^{-2an} (\cos 2\beta n - j \sin 2\beta n)}, \\ A_2 &= - A_1 \varepsilon^{-2an} (\cos 2\beta n - j \sin 2\beta n), \end{aligned}\right\} \quad (14)$$

and the potential difference against ground is

$$E = e_0 \frac{\varepsilon^{-al}(\cos \beta l + j \sin \beta l) - \varepsilon^{-a(2n-l)} [\cos \beta (2n - l) - j \sin \beta (2n - l)]}{1 - \varepsilon^{-2an} (\cos 2\beta n - j \sin 2\beta n)}. \quad (15)$$

From equation (5), substituting (15) and (9), we have

$$I = -\sqrt{\frac{Y}{Z}}\, e_0 \frac{\varepsilon^{-al}(\cos \beta l - j \sin \beta l) + \varepsilon^{-a(2n-l)} [\cos \beta (2n - l) - j \sin \beta (2n - l)]}{1 - \varepsilon^{-2an}(\cos 2\beta n - j \sin 2\beta n)}. \quad (16)$$

Reduced to absolute terms this gives the potential difference against ground as

$$e = e_0 \sqrt{\frac{\varepsilon^{-2al} + \varepsilon^{-2a(2n-l)} - 2\varepsilon^{-2an} \cos 2\beta (n - l)}{1 + \varepsilon^{-4an} - 2\varepsilon^{-2an} \cos 2\beta n}}, \quad (17)$$

the current as

$$i = e_0 \sqrt{\frac{y}{z}} \sqrt{\frac{\varepsilon^{-2al} + \varepsilon^{-2a(2n-l)} + 2\,\varepsilon^{-2an} \cos 2\,\beta\,(n-l)}{1 + \varepsilon^{-4an} - 2\,\varepsilon^{-2an} \cos 2\,\beta n}}, \tag{18}$$

and the potential gradient, or potential difference between adjacent cylinders, is

$$e' = x_c i = e_0 x_c \sqrt{\frac{y}{z}} \sqrt{\frac{\varepsilon^{-2al} + \varepsilon^{-2a(2n-l)} + 2\,\varepsilon^{-2an} \cos 2\,\beta\,(n-l)}{1 + \varepsilon^{-4an} - 2\,\varepsilon^{-2an} \cos 2\,\beta n}}. \tag{19}$$

For an infinite length of line, $n = \infty$, that is, for a very large number of ligh+ning arrester cylinders, where ε^{-2an} is negligible, as in the case where the discharge passes from the line into the arrester without reaching the ground, equations (17), (18), (19) simplify to

$$e = e_0 \varepsilon^{-al}, \tag{20}$$

$$i = e_0 \sqrt{\frac{y}{z}}\, \varepsilon^{-al}, \tag{21}$$

and

$$e' = e_0 x_c \sqrt{\frac{y}{z}}\, \varepsilon^{-al}; \tag{22}$$

that is, are simple exponential curves.

Substituting (4) and (3) in (21) and (22) gives

$$e' = e_0 \varepsilon^{-al} \sqrt[4]{\frac{C_0^2 + \left(\frac{g}{2\,\pi f}\right)^2}{C^2\{[1 - (2\,\pi f)^2\, CL]^2 + (2\,\pi f C r)^2\}}} \tag{23}$$

and

$$i = 2\,\pi f C e'; \tag{24}$$

or, approximately, if r and g are negligible, we have

$$e' = e_0 \varepsilon^{-al} \sqrt{\frac{C_0}{C\,\{1 - (2\,\pi f)^2\, CL\}}} \tag{25}$$

and

$$i = 2\,\pi f e_0 \varepsilon^{-al} \sqrt{\frac{C C_0}{1 - (2\,\pi f)^2\, CL}}\,. \tag{26}$$

47. Assume, as example, a lightning arrester having the following constants: $L = 2 \times 10^{-8}$ henry; $C_0 = 10^{-13}$ farads;

$C = 4 \times 10^{-11}$ farads; $r = 1$ ohm; $g = 4 \times 10^{-6}$ mho; $f = 10^8 =$ 100 million cycles per second; $n = 300$ cylinders, and $e_0 = 30{,}000$ volts; then from equation (3), $x = 12.6$ ohms, $x_c = 39.7$ ohms, and $b = 62.8 \times 10^{-6}$ mhos;

from equation (1),

$$Z = 1 - 27.1\,j \text{ ohms};$$

from equation (2),

$$Y = (4 + 62.8\,j)\,10^{-6} \text{ mho};$$

from equation (4),

$$z = 27.1 \text{ ohms and } y = 62.9 \times 10^{-6} \text{ mho};$$

from equation (10),

$$\alpha = 0.0021 \text{ and } \beta = 0.0412;$$

from equation (17),

$$e = 35{,}500\sqrt{\varepsilon^{-0.0042\,l} + 0.08\,\varepsilon^{+0.0042\,l} - 0.568\cos(24.72 - 0.0824\,l)};$$

from equation (18),

$$i = 54\sqrt{\varepsilon^{-0.0042\,l} + 0.08\,\varepsilon^{+0.0042\,l} + 0.568\cos(24.72 - 0.0824\,l)},$$

and from equation (19),

$$e' = 2140\sqrt{\varepsilon^{-0.0042\,l} + 0.08\,\varepsilon^{+0.0042\,l} + 0.568\cos(24.72 - 0.0824\,l)}.$$

Hence, at $l = 0$, $e = 30{,}000$ volts, $i = 64.6$ amperes, and $e' = 2560$ volts; and at $l = 300$, $e = 0$, $i = 57.5$ amperes, and $e' = 2280$ volts.

With voltages per gap varying from 2280 to 2560, 300 gaps would, by addition, give a total voltage of about 730,000, while the actual voltage is only about one-twenty-fourth thereof; that is, the sum of the voltages of many spark-gaps in series may be many times the resultant voltage, and a lightning flash may pass possibly for miles through clouds with a total potential of only a few hundred million volts. In the above example the 300 cylinders include 7.86 complete wave-lengths of the discharge.

CHAPTER VI

ALTERNATING MAGNETIC FLUX DISTRIBUTION

48. As carrier of magnetic flux iron is used, as far as possible, since it has the highest permeability or magnetic conductivity. If the magnetic flux is alternating or otherwise changing rapidly, an e.m.f. is generated by the change of magnetic flux in the iron, and to avoid energy losses and demagnetization by the currents produced by these e.m.fs. the iron has to be subdivided in the direction in which the currents would exist, that is, at right angles to the lines of magnetic force. Hence, alternating magnetic fields and magnetic structures desired to respond very quickly to changes of m.m.f. are built of thin wires or thin iron sheets, that is, are laminated.

Since the generated e.m.fs. are proportional to the frequency of the alternating magnetism, the laminations must be finer the higher the frequency.

To fully utilize the magnetic permeability of the iron, it therefore has to be laminated so as to give, at the impressed frequency, practically uniform magnetic induction throughout its section, that is, negligible secondary currents. This, however, is no longer the case, even with the thinnest possible laminations, at extremely high frequencies, as oscillating currents, lightning discharges, etc., and under these conditions the magnetic flux distribution in the iron is not uniform, but the magnetic flux density, $\mathfrak{B}$, decreases rapidly, and lags in phase, with increasing depth below the surface of the lamination, so that ultimately hardly any magnetic flux exists in the inside of the laminations, but practically only a surface layer carries magnetic flux. The apparent permeability of the iron thus decreases at very high frequency, and this has led to the opinion that at very high frequencies iron cannot follow a magnetic cycle. There is, however, no evidence of such a "viscous hysteresis." Magnetic investigations at 100,000 to 200,000 cycles per second have given the same magnetic cycles as at low frequencies. It therefore is probable that iron follows magnetically even at the highest frequencies, traversing practically the same hysteresis cycle irrespective of the frequency, if the true m.m.f., that is, the resultant of the impressed m.m.f. and the m.m.f. of the secondary currents in the iron, is considered. Since with increasing frequency, at constant impressed m.m.f., the resultant m.m.f. decreases, due to the increase of the demagnetizing secondary currents, this simulates the effect of a viscous hysteresis.

Frequently also, for mechanical reasons, iron sheets of greater thickness than would give uniform flux density have to be used in an alternating field.

Since rapidly varying magnetic fields usually are alternating, and the subdivision of the iron is usually by lamination, it will be sufficient to consider as illustration of the method the distribution of alternating magnetic flux in iron laminations.

49. Let Fig. 13 represent the section of a lamination. The alternating magnetic flux is assumed to pass in a direction perpendicular to the plane of the paper.

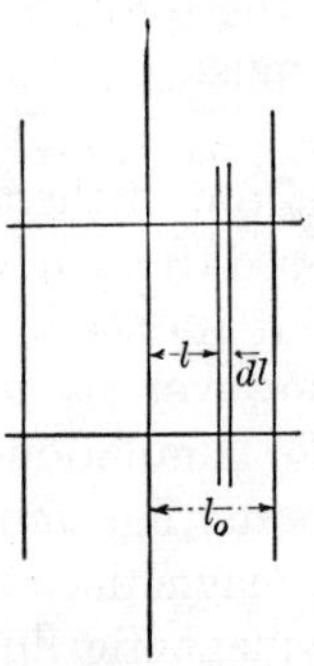

FIG. 13. Alternating magnetic flux distribution in solid iron.

Let μ = the magnetic permeability, λ = the electric conductivity, l = the distance of a layer dl from the center line of the lamination, and $2\,l_0$ = the total thickness of the lamination. If then $\underset{\cdot}{I}$ = the current density in the layer dl, and $\underset{\cdot}{E}$ = the e.m.f. per unit length generated in the zone dl by the alternating magnetic flux, we have

$$\underset{\cdot}{I} = \lambda \underset{\cdot}{E}. \tag{1}$$

The magnetic flux density $\mathcal{B}_1$ at the surface $l = l_0$ of the lamination corresponds to the impressed or external m.m.f. The density $\underset{\cdot}{\mathcal{B}}$ in the zone dl corresponds to the impressed m.m.f. plus the sum of all the m.m.fs. in the zones outside of dl, or from l to l_0.

The current in the zone dl is

$$\underset{\cdot}{I}\,dl = \lambda \underset{\cdot}{E}\,dl \tag{2}$$

and produces the m.m.f.

$$\underset{\cdot}{\mathcal{H}} = 0.4\,\pi\lambda \underset{\cdot}{E}\,dl, \tag{3}$$

which in turn would produce the magnetic flux density

$$d\underset{\cdot}{\mathcal{B}} = 0.4\,\pi\lambda\mu \underset{\cdot}{E}\,dl; \tag{4}$$

that is, the magnetic flux density $\dot{\mathcal{B}}$ at the two sides of the zone dl differs by the magnetic flux density $d\dot{\mathcal{B}}$ (equation (4)) produced by the m.m.f. in zone dl, and this gives the differential equation between $\dot{\mathcal{B}}$, E, and l,

$$\frac{d\dot{\mathcal{B}}}{dl} = 0.4\ \pi\lambda\mu\dot{E}. \tag{5}$$

The e.m.f. generated at distance l from the center of the lamination is due to the magnetic flux in the space from l to l_0. Thus the e.m.fs. at the two sides of the zone dl differ from each other by the e.m.f. generated by the magnetic flux $\dot{\mathcal{B}}dl$ in this zone.

Considering now $\dot{\mathcal{B}}$, $\dot{E}$, and $\dot{I}$ as complex quantities, the e.m.f. $d\dot{E}$, that is, the difference between the e.m.fs. at the two sides of the zone dl, is in quadrature ahead of $\dot{\mathcal{B}}dl$, and thus denoted by

$$d\dot{E} = j\ 2\ \pi f\dot{\mathcal{B}}\ 10^{-8}\ dl, \tag{6}$$

where f = the frequency of alternating magnetism.

This gives the second differential equation

$$\frac{d\dot{E}}{dl}\ j\ 2\ \pi f\dot{\mathcal{B}}\ 10^{-8}. \tag{7}$$

50. Differentiating (5) in respect to l, and substituting (7) therein, gives

$$\frac{d^2\dot{\mathcal{B}}}{dl^2} = 0.8\ j\pi^2 f\lambda\mu\ 10^{-8}\ \dot{\mathcal{B}}, \tag{8}$$

or, writing

$$c^2 = fa^2 = 0.4\ \pi^2 f\lambda\mu\ 10^{-8}, \tag{9}$$

$$a^2 = 0.4\ \pi^2\lambda\mu\ 10^{-8}, \tag{10}$$

we have

$$\frac{d^2\dot{\mathcal{B}}}{dl^2} = 2\ jc^2\dot{\mathcal{B}}. \tag{11}$$

This differential equation is integrated by

$$\dot{\mathcal{B}} = A\varepsilon^{-vl}; \tag{12}$$

this equation substituted in (11) gives

$$v^2 = 2\ jc^2; \tag{13}$$

hence,

$$v = \pm (1 + j)\,c \tag{14}$$

and

$$\dot{\mathfrak{B}} = A_1\varepsilon^{+(1+j)\,cl} + A_2\varepsilon^{+(1-j)\,cl}.$$

Since $\dot{\mathfrak{B}}$ must have the same value for $-\,l$ as for $+\,l$, being symmetrical at both sides of the center line of the lamination,

$$A_1 = A_2 = A,$$

hence,

$$\dot{\mathfrak{B}} = A\,\{\varepsilon^{+(1+j)\,cl} + \varepsilon^{-(1+j)\,cl}\}; \tag{15}$$

or, substituting

$$\varepsilon^{\pm jcl} = \cos cl \pm j \sin cl \tag{16}$$

gives

$$\dot{\mathfrak{B}} = A\,\{(\varepsilon^{+cl} + \varepsilon^{-cl}) \cos cl + j\,(\varepsilon^{+cl} - \varepsilon^{-cl}) \sin cl\}. \tag{17}$$

51. Denoting the flux density in the center of the lamination, for $l = 0$, by $\dot{\mathfrak{B}}_0$ from (17) we have

$$\dot{\mathfrak{B}}_0 = 2\,A;$$

hence,

$$A = \tfrac{1}{2}\,\dot{\mathfrak{B}}_0 \tag{18}$$

and

$$\dot{\mathfrak{B}} = \dot{\mathfrak{B}}_0 \left\{ \frac{\varepsilon^{+cl} + \varepsilon^{-cl}}{2} \cos cl + j \frac{\varepsilon^{+cl} - \varepsilon^{-cl}}{2} \sin cl \right\}. \tag{19}$$

Denoting the flux density at the outside of the lamination, for $l = l_0$, that is, the density produced by the external m.m.f., by $\dot{\mathfrak{B}}_1$, substituted in (19), we have

$$\dot{\mathfrak{B}}_1 = \dot{\mathfrak{B}}_0 \left\{ \frac{\varepsilon^{+cl_0} + \varepsilon^{-cl_0}}{2} \cos cl_0 + j \frac{\varepsilon^{+cl_0} - \varepsilon^{-cl_0}}{2} \sin cl_0 \right\}, \tag{20}$$

and substituting (20) in (19),

$$\dot{\mathfrak{B}} = \dot{\mathfrak{B}}_1 \frac{(\varepsilon^{+cl} + \varepsilon^{-cl}) \cos cl + j\,(\varepsilon^{+cl} - \varepsilon^{-cl}) \sin cl}{(\varepsilon^{+cl_0} + \varepsilon^{-cl_0}) \cos cl_0 + j\,(\varepsilon^{+cl_0} - \varepsilon^{-cl_0}) \sin cl_0}. \tag{21}$$

The mean or apparent value of the flux density, i.e., the average throughout the lamination, is

$$\dot{\mathfrak{B}}_m = \frac{1}{l_0} \int_0^{l_0} \dot{\mathfrak{B}}\, dl. \tag{22}$$

Using equation (15) as the more convenient for integration gives

$$\dot{\mathfrak{B}}_m = \frac{A}{(1+j)\,cl_0}\left[\varepsilon^{+(1+j)\,cl} - \varepsilon^{-(1+j)\,cl}\right]_0^{l_0}$$

$$= \frac{A\left(\varepsilon^{+(1+j)\,cl_0} - \varepsilon^{-(1+j)\,cl_0}\right)}{(1+j)\,cl_0}, \tag{23}$$

and substituting herein (16), (18) and (20), gives

$$\left.\begin{aligned} \dot{\mathfrak{B}}_m &= \frac{\dot{\mathfrak{B}}_0}{(1+j)\,cl_0}\left\{\frac{\varepsilon^{+cl_0}-\varepsilon^{-cl_0}}{2}\cos cl_0 + j\frac{\varepsilon^{+cl_0}+\varepsilon^{-cl_0}}{2}\sin cl_0\right\} \\ &= \frac{\dot{\mathfrak{B}}_1}{(1+j)\,cl_0}\,\frac{(\varepsilon^{+cl_0}-\varepsilon^{-cl_0})\cos cl_0 + j\,(\varepsilon^{+cl_0}+\varepsilon^{-cl_0})\sin cl_0}{(\varepsilon^{+cl_0}+\varepsilon^{-cl_0})\cos cl_0 + j\,(\varepsilon^{+cl_0}-\varepsilon^{-cl_0})\sin cl_0}. \end{aligned}\right\} \tag{24}$$

The absolute values of the flux densities are derived as square root of the sum of the squares of real and imaginary terms in equations (19), (20), (21), and (24), as

$$\mathfrak{B} = \frac{\mathfrak{B}_0}{2}\sqrt{\varepsilon^{+2\,cl} + \varepsilon^{-2\,cl} + 2\cos 2\,cl}, \tag{25}$$

$$\mathfrak{B}_1 = \frac{\mathfrak{B}_0}{2}\sqrt{\varepsilon^{+2\,cl_0} + \varepsilon^{-2\,cl_0} + 2\cos 2\,cl_0}, \tag{26}$$

$$\mathfrak{B} = \mathfrak{B}_1\sqrt{\frac{\varepsilon^{+2\,cl} + \varepsilon^{-2\,cl} + 2\cos 2cl}{\varepsilon^{+2\,cl_0} + \varepsilon^{-2\,cl_0} + 2\cos 2\,cl_0}}, \tag{27}$$

and

$$\left.\begin{aligned} \mathfrak{B}_m &= \frac{\mathfrak{B}_0}{2\,cl_0\sqrt{2}}\sqrt{\varepsilon^{+2cl_0} + \varepsilon^{-2cl_0} - 2\cos 2\,cl_0} \\ &= \frac{\mathfrak{B}_1}{cl_0\sqrt{2}}\sqrt{\frac{\varepsilon^{+2cl_0} + \varepsilon^{-2cl_0} - 2\cos 2\,cl_0}{\varepsilon^{+2cl_0} + \varepsilon^{-2cl_0} + 2\cos 2\,cl_0}}. \end{aligned}\right\} \tag{28}$$

52. Where the thickness of lamination, $2\,l_0$, or the frequency f, is so great as to give cl_0 a value sufficiently high to make ε^{-cl_0}, or the reflected wave, negligible compared with the main wave ε^{+cl_0}, the equations can be simplified by dropping ε^{-cl}. In this case the flux density, $\mathfrak{B}$, is very small or practically nothing in the interior, and reaches appreciable values only near the surface. It then is preferable to count the distance from the surface of the lamination into the interior, that is, substitute the independent variable

$$s = l_0 - l. \tag{29}$$

Dropping ε^{-cl} and ε^{-cl_0} in equation (21) gives

$$\dot{\mathfrak{B}} = \dot{\mathfrak{B}}_1 \frac{\varepsilon^{cl}(\cos cl + j \sin cl)}{\varepsilon^{cl_0}(\cos cl_0 + j \sin cl_0)}$$

$$= \dot{\mathfrak{B}}_1 \varepsilon^{-c(l_0 - l)} \{\cos c(l_0 - l) - j \sin c(l_0 - l)\};$$

hence,

$$\dot{\mathfrak{B}} = \dot{\mathfrak{B}}_1 \varepsilon^{-cs}(\cos cs - j \sin cs); \tag{30}$$

and the absolute value is

$$\mathfrak{B} = \mathfrak{B}_1 \varepsilon^{-cs}, \tag{31}$$

and at the center of the lamination,

$$\left.\begin{aligned} \dot{\mathfrak{B}}_0 &= \dot{\mathfrak{B}}_1 \varepsilon^{-cl_0}(\cos cl_0 + j \sin cl_0), \\ \mathfrak{B}_0 &= \mathfrak{B}_1 \varepsilon^{-cl_0}. \end{aligned}\right\} \tag{32}$$

From equation (24) the mean value of flux density follows when dropping ε^{-cl_0} as negligible, thus:

$$\dot{\mathfrak{B}}_m = \frac{\dot{\mathfrak{B}}_1}{(1 + j) cl_0} = \frac{(1 - j)\dot{\mathfrak{B}}_1}{2\, cl_0} \tag{33}$$

or the absolute value is

$$\mathfrak{B}_m = \frac{\mathfrak{B}}{cl_0\sqrt{2}}. \tag{34}$$

53. As seen, the preceding equations of the distribution of alternating magnetic flux in a laminated conductor are of the same form as the equations of distribution of current and voltage in a transmission line, but more special in form, that is, the attenuation constant α and the wave length constant β have the same value, c. As result, the distribution of the alternating magnetic flux in the lamina depends upon one constant only, cl_0.

The wave length is given by

$$cl_w = 2\pi;$$

hence
$$l_w = \frac{2\pi}{c}$$

and by (9)

$$l_w = \frac{2\pi}{a\sqrt{f}}$$
$$= \frac{10{,}000}{\sqrt{0.1\,\lambda\mu f}}, \tag{35}$$

and the attenuation during one wave length, or decrease of intensity of magnetism, per wave length, is

$$\epsilon^{-2\pi} = 0.0019,$$

per half-wave length it is

$$\epsilon^{-\pi} = 0.043.$$

and per quarter-wave length

$$\epsilon^{-\frac{\pi}{2}} = 0.207.$$

At the depth $\frac{l_w}{4}$ below the surface, the magnetic flux lags 90 degrees and has decreased to about 20 per cent; at the depth $\frac{l_w}{2}$ it lags 180 degrees, that is, is opposite in direction to the flux at the surface of the lamination, but is very small, the intensity being less than 5 per cent of that at the surface, and at the depth l_w the flux is again in phase with the surface flux, but its intensity is practically nil, less than 0.2 per cent of the surface intensity; that is, the penetration of alternating flux into the laminated iron is inappreciable at the depth of one wave length.

By equations (33) and (34), the total magnetic flux per unit width of lamination is

$$2\,l_0\dot{\mathfrak{B}}_m = \frac{2\dot{\mathfrak{B}}_1}{(1+j)\,c} = \frac{(1-j)\,\dot{\mathfrak{B}}_1}{c}$$

the absolute value is $2\,l_0\mathfrak{B}_m = \frac{2\,\mathfrak{B}_1}{c\sqrt{2}}$;

that is, the same as would be produced at uniform density in a thickness of lamination

$$2\,\dot{l}_p = \frac{2}{(1+j)\,c} = \frac{(1-j)}{c}$$

or absolute value,

$$2\,l_p = \frac{\sqrt{2}}{c};$$

which means that the resultant alternating magnetism in the lamination lags 45 degrees, or one-eighth wave behind the impressed m.m.f., and is equal to a uniform magnetic density penetrating to a depth

$$l_p = \frac{1}{c\sqrt{2}}. \tag{36}$$

l_p, therefore, can be called the depth of penetration of the alternating magnetism into the solid iron.

Since the only constant entering into the equation is cl_0, the distribution of alternating magnetism for all cases can be represented as function of cl_0.

If cl_0 is small, and therefore the density in the center of the lamination $\mathfrak{B}_0$ comparable with the density $\mathfrak{B}_1$ at the outside, the equations (19), (20), and (24) respectively (25), (26), and (28) have to be used; if cl_0 is large, and the flux density $\mathfrak{B}_0$ in the center of the lamination is negligible, the simpler equations (30) to (34) can be used.

54. As an example, let $\mu = 1000$ and $\lambda = 10^5$; then $a = 1.98$, and for $f = 60$ cycles per second, $c = a\sqrt{f} = 15.3$; hence, the thickness of effective layer of penetration is

$$l_p = \frac{1}{c\sqrt{2}} = 0.046\ cm = 0.018 \text{ inches.}$$

In Fig. 14 is shown, with cl as abscissas, the effective value of the magnetic flux, which from equation (25) is

$$\mathfrak{B} = \frac{\mathfrak{B}_0}{2}\sqrt{\varepsilon^{+2cl} + \varepsilon^{-2cl} + 2\cos 2cl},$$

and also the space-phase angle between $\mathfrak{B}$ and $\mathfrak{B}_0$, which from equation (19) is

$$\tan \tau_0 = \frac{\varepsilon^{cl} - \varepsilon^{-cl}}{\varepsilon^{cl} + \varepsilon^{-cl}} \tan cl. \tag{37}$$

In Fig. 15 is shown, with cs as abscissas, the effective value of the magnetic flux, which from equation (31) is

$$\mathfrak{B} = \mathfrak{B}_1 \varepsilon^{-cs},$$

and also the space-phase angle between $\mathfrak{B}$ and $\mathfrak{B}_1$, which from equation (30) is

$$\tan \tau_1 = \tan cs. \tag{38}$$

The thickness of the equivalent layer is marked in Fig. 15.

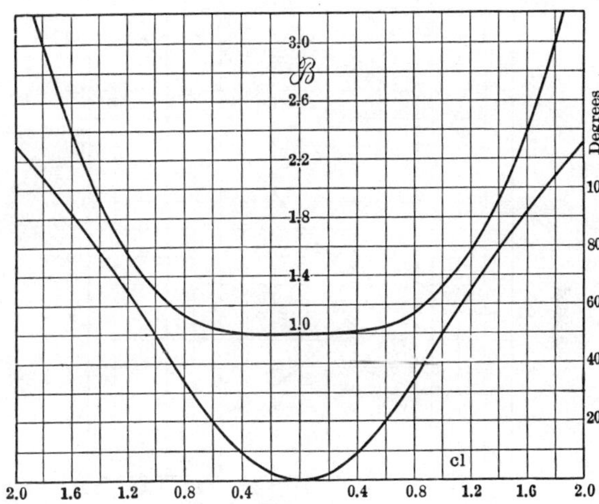

FIG. 14. Alternating magnetic flux distribution in solid iron.

As further illustrations are shown in Fig. 16 the absolute values of magnetic flux density $\mathfrak{B}$ throughout a layer of 14 mils thickness, that is, of l_0 = 0.007 inches = 0.018 cm. thickness.

For 60 cycles, by Curve I, $c = 15.3$ $cl_0 = 0.275$
For 1000 cycles, by Curve II, $c = 62.5$ $cl_0 = 1.125$
For 10,000 cycles, by Curve III, $c = 198$ $cl_0 = 3.55$

It is seen that the density in Curve I is perfectly uniform, while in Curve III practically no flux penetrates to the center.

55. The effective penetration of the alternating magnetism into the iron, or the thickness l_p of surface layer which at constant induction $\mathfrak{B}_1$ would give the same total magnetic flux as exists in the lamination, is

$$l_p = \frac{1}{(1+j)c} = \frac{1-j}{2c} \tag{39}$$

or the absolute value is

$$l_p = \frac{1}{c\sqrt{2}};$$

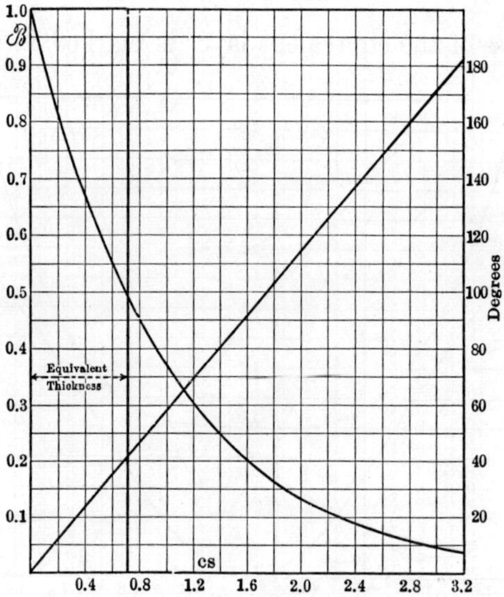

FIG. 15. Alternating magnetic flux distribution in solid iron.

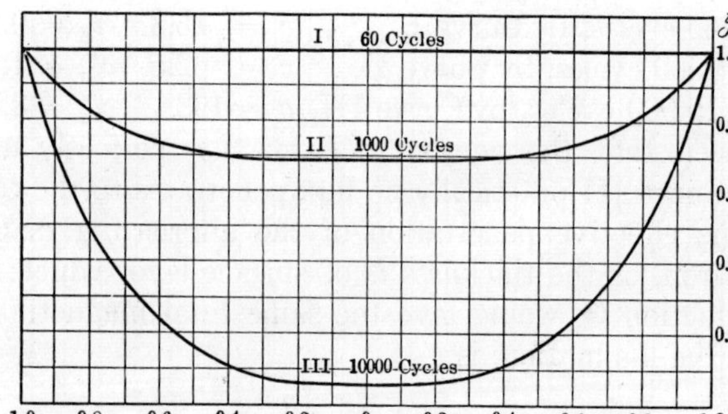

FIG. 16. Alternating magnetic flux distribution in solid iron.

hence, substituting for c from equation (9),

$$l_p = \frac{10^4}{\pi \sqrt{0.8\,\lambda \mu f}} = \frac{3570}{\sqrt{\lambda \mu f}} \tag{40}$$

that is, the penetration of an alternating magnetic flux into a solid conductor is inversely proportional to the square root of the electric conductivity, the magnetic permeability, and the frequency.

The values of penetration, l_p, in centimeters for various materials and frequencies are given below.

Frequency.	25	60	1000	10,000	10^5	10^6
Soft iron, $\mu = 1000$, $\lambda = 10^5$........	0.0714	0.0460	0.0113	0.0036	0.00113	0.00036
Cast iron, $\mu = 200$, $\lambda = 10^4$........	0.504	0.325	0.080	0.0252	0.0080	0.0025
Copper, $\mu = 1$, $\lambda = 6 \times 10^5$........	0.922	0.595	0.144	0.0461	0.0144	0.0046
Resistance alloys, $\mu = 1$, $\lambda = 10^4$....	7.14	4.60	1.13	0.357	0.113	0.036

As seen, even at frequencies as low as 25 cycles alternating magnetism does not penetrate far into solid wrought iron, but penetrates to considerable depth into cast iron. It also is interesting to note that little difference exists in the penetration into copper and into cast iron, the high conductivity of the former compensating for the higher permeability of the latter.

56. The wave length, $l_w = \frac{2\pi}{c}$, substituting for c, from equation (9), is

$$l_w = \frac{31{,}600}{\sqrt{\lambda \mu f}}; \tag{41}$$

that is, the wave length of the oscillatory transmission of alternating magnetism in solid iron is inversely proportional to the square root of the electric conductivity, the magnetic permeability, and the frequency.

Comparing this equation (41) of the wave length l_w with equation (40) of the depth of penetration l_p, it follows that the depth of penetration is about one-ninth of the wave length, or 40 degrees, or, more accurately, since

$$l_p = \frac{1}{c\sqrt{2}} \text{ and } l_w = \frac{2\pi}{c},$$

we have
$$\frac{l_p}{l_w} = \frac{1}{2\sqrt{2}\pi} = \frac{1}{8.9}, \tag{42}$$
or 40.5 degrees.

The speed of propagation is
$$S = fl_w$$
$$= \frac{31{,}600\sqrt{f}}{\sqrt{\lambda\mu}}: \tag{43}$$
that is, the speed of propagation is inversely proportional to the square root of the electric conductivity and of the magnetic permeability, but directly proportional to the square root of the frequency. This gives a curious instance of a speed which increases with the frequency. Numerical values are given below.

Frequency.	25 Cycles.	10,000 Cycles.
Soft iron, $\mu = 1000$, $\lambda = 10^5$	$S = 15.8$ cm.	316 cm.
Cast iron, $\mu = 200$, $\lambda = 10^4$	111 cm.	2230 cm.
Copper, $\mu = 1$, $\lambda = 6 \times 10^5$	204 cm.	4080 cm.

It is seen that these speeds are extremely low compared with the usual speeds of electromagnetic waves.

57. Since instead of $\dot{\mathfrak{B}}_1$, corresponding to the impressed m.m.f. and permeability μ, the mean flux density in the lamina is $\dot{\mathfrak{B}}_m$, the effect is the same as if the permeability of the material were changed from μ to
$$\mu' = \mu\frac{\dot{\mathfrak{B}}_m}{\dot{\mathfrak{B}}_1}, \tag{44}$$
and μ' can be called the *effective permeability*, which is a function of the thickness of the lamination and of the frequency, that is, a function of cl_0; μ' appears in complex form thus,
$$\mu' = \mu_1' - j\mu_2';$$
that is, the permeability is reduced and also made lagging.

For high values of cl_0, that is, thin laminations or high frequencies, from (33), we have
$$\mu' = \frac{\mu}{(1+j)\,cl_0}$$
$$= \frac{\mu}{2\,cl_0} - j\frac{\mu}{2\,cl_0}. \tag{45}$$
or, absolute
$$\mu' = \frac{\mu}{cl_0\sqrt{2}}$$

58. As illustration, for iron of 14 mils thickness, or $l_0 = 0.018$ centimeters, and the constants $\mu = 1000$ and $\lambda = 10^5$, that is $a = 1.98$, the absolute value of the effective permeability is

$$\mu' = \frac{\mu}{cl_0\sqrt{2}}$$

and

$$c = a\sqrt{f};$$

hence,

$$\mu' = \frac{\mu}{al_0\sqrt{2f}} = \frac{19{,}800}{\sqrt{f}}; \tag{46}$$

that is, the effective or apparent permeability at very high frequencies decreases inversely proportional to the square root of the frequency. In the above instance the apparent permeability is:

At low frequency, $\mu = 1000$;
at 10,000 cycles, $\mu' = 198$;
at 1,000,000 cycles, $\mu' = 19.8$;
at 100 million cycles, $\mu' = 1.98$, and
at 392 million cycles, $\mu' = 1$,

or the same as air, and at still higher frequencies the presence of iron reduces the magnetic flux.

It is interesting to note that with such a coarse lamination as a 14-mil sheet, even at the highest frequencies of millions of cycles, an appreciable apparent permeability is still left; that is, the magnetic flux is increased by the presence of iron; and the effect of iron in increasing the magnetic flux disappears only at 400 million cycles, and beyond this frequency iron lowers the magnetic flux. However, even at these frequencies, the presence of iron still exerts a great effect in the rapid damping of the oscillations by the lag of the mean magnetic flux by 45 degrees.

Obviously, in large solid pieces of iron, the permeability μ' falls below that of air even at far lower frequencies.

Where the penetration of the magnetic flux l_p is small compared with the dimensions of the iron, its shape becomes immaterial, since only the surface requires consideration, and so in this case any solid structure, no matter what shape, can be considered magnetically as its outer shell of thickness l_p when dealing with rapidly alternating magnetic fluxes.

At very high frequencies, when dealing with alternating magnetic circuits, the outer surface and not the section is, therefore, the dominating feature.

The lag of the apparent permeability represents an energy component of the e.m.f. of self-induction due to the magnetic flux, which increases with increasing frequency, and ultimately becomes equal to the reactive component.

CHAPTER VII

DISTRIBUTION OF ALTERNATING-CURRENT DENSITY IN CONDUCTOR

59. If the frequency of an alternating or oscillating current is high, or the section of the conductor which carries the current is very large, or its electric conductivity or its magnetic permeability high, the current density is not uniform throughout the conductor section, but decreases towards the interior of the conductor, due to the higher e.m.f. of self-inductance in the interior of the conductor, caused by the magnetic flux inside of the conductor. The phase of the current inside of the conductor also differs from that on the surface and lags behind it.

In consequence of this unequal current distribution in a large conductor traversed by alternating currents, the effective resistance of the conductor may be far higher than the ohmic resistance, and the conductor also contains internal inductance.

In the extreme case, where the current density in the interior of the conductor is very much lower than on the surface, or even negligible, due to this "screening effect," as it has been called, the current can be assumed to exist only in a thin surface layer of the conductor, of thickness l_p; that is, in this case the effective resistance of the conductor for alternating currents equals the ohmic resistance of a conductor section equal to the periphery of the conductor times the "depth of penetration."

Where this unequal current distribution throughout the conductor section is considerable, the conductor section is not fully utilized, but the material in the interior of the conductor is more or less wasted. It is of importance, therefore, in alternating-current circuits, especially in dealing with very large currents, or with high frequency, or materials of very high permeability, as iron, to investigate this phenomenon.

An approximate determination of this effect for the purpose of deciding whether the unequal current distribution is so small as to be negligible in its effect on the resistance of the conductor, or whether it is sufficiently large to require calculation and methods of avoiding it, is given in Dover edition Volume I, Part II, paragraph 113, p. 264.

An appreciable increase of the effective resistance over the ohmic resistance may be expected in the following cases:

(1) In the low-tension distribution of heavy alternating currents by large conductors.

(2) When using iron as conductor, as for instance iron wires in high potential transmissions for branch lines of smaller power, or steel cables for long spans in transmission lines.

(3) In the rail return of single-phase railways.

(4) When carrying very high frequencies, such as lightning discharges, high frequency oscillations, wireless telegraph currents, etc.

In the last two cases, which probably are of the greatest importance, the unequal current distribution usually is such that practically no current exists at the conductor center, and the effective resistance of the track rail even for 25-cycle alternating current thus is several times greater than the ohmic resistance, and conductors of low ohmic resistance may offer a very high effective resistance to a lightning stroke.

By subdividing the conductor into a number of smaller conductors, separated by some distance from each other, or by the use of a hollow conductor, or a flat conductor, as a bar or ribbon, the effect is reduced, and for high-frequency discharges, as lightning arrester connections, flat copper ribbon offers a very much smaller effective resistance than a round wire. Stranding the conductor, however, has no direct effect on this phenomenon, since it is due to the magnetic action of the current, and the magnetic field in the stranded conductor is the same as in a solid conductor, other things being equal. That is, while eddy currents in the conductor, due to external magnetic fields, are eliminated by stranding the conductor, this is not the case with the increase of the effective resistance by unequal current distribution. Stranding the conductor, however, may reduce unequal current distribution indirectly, especially with iron as conductor material, by reducing the effective or mean permeability of the conductor, due to the break in the magnetic circuit between the iron strands, and also by the reduction of the mean conductivity of the conductor section. For instance, if in a stranded conductor 60 per cent of the conductor section is copper, 40 per cent space between the strands, the mean conductivity is 60 per cent of that of copper. If by the subdivision of an iron conductor into strands the reluctance of the magnetic circuit is increased tenfold, this represents a reduction of the mean permeability to one-tenth. Hence, if for the conductor material proper $\mu = 1000$, $\lambda = 10^5$, and the conductor section is reduced by stranding to 60 per cent, the permeability to one-tenth, the mean values would be

$$\mu_0 = 100 \quad \text{and} \quad \lambda_0 = 0.6 \times 10^5,$$

and the factor $\sqrt{\lambda\mu}$, in the equation of current distribution, is reduced from $\sqrt{\lambda\mu} = 10{,}000$ to $\sqrt{\lambda_0\mu_0} = 2450$, or to 24.5 per cent of its previous value. In this case, however, with iron as conductor material, an investigation must be made on the current distribution in each individual conductor strand.

Since the simplest way of reducing the effect of unequal current distribution is the use of flat conductors, the most important case is the investigation of the alternating-current distribution throughout the section of the flat conductor. This also gives the solution for conductors of any shape when the conductor section is so large that the current penetrates only the surface layer, as is the case with a steel rail of a single-phase railway. Where the alternating current penetrates a short distance only into the conductor, compared with the depth of penetration the curvature of the conductor surface can be neglected, that is, the conductor surface considered as a flat surface penetrated to the same depth all over. Actually on sharp convex surfaces the current penetrates somewhat deeper, somewhat less on sharp concave surfaces, so that the error is more or less compensated.

60. In a section of a flat conductor, as shown diagrammatically in Fig. 13 page 138, let λ = the electric conductivity of conductor material; μ = the magnetic permeability of conductor material; l = the distance counted from the center line of the conductor, and $2\,l_0$ = the thickness of conductor.

Furthermore, let $\dot{E}_0$ = the impressed e.m.f. per unit length of conductor, that is, the voltage consumed per unit length in the conductor after subtracting the e.m.f. consumed by the self-inductance of the external magnetic field of the conductor; thus, if $\dot{E}_1$ = the total supply voltage per unit length of conductor and $\dot{E}_2$ = the external reactance voltage, or voltage consumed by the magnetic field outside of the conductor, between the conductors, we have

$$\dot{E}_0 = \dot{E}_1 - \dot{E}_2.$$

Let

$\dot{I} = i_1 - ji_2$ = current density in conductor element dl,

$\dot{\mathcal{B}} = b_1 - jb_2$ = magnetic density in conductor element dl,

$\dot{E}$ = e.m.f. consumed in the conductor element dl by the self-inductance due to the magnetic field inside of the conductor; then the current $\dot{I}\,dl$ in the conductor element represents the m.m.f. or field intensity,

$$d\dot{\mathcal{H}} = 0.4\,\pi \dot{I}\,dl, \tag{1}$$

which causes an increase of the magnetic density $\dot{\mathcal{B}}$ between the two sides of the conductor element dl by

$$\begin{aligned} d\dot{\mathcal{B}} &= \mu d\dot{\mathcal{H}} \\ &= 0.4\,\pi \mu \dot{I}\,dl. \end{aligned} \tag{2}$$

The e.m.f. consumed by self-inductance is proportional to the magnetic flux and to the frequency, and is 90 time-degrees ahead of the magnetic flux.

The increase of magnetic flux $\dot{\mathcal{B}}\,dl$, in the conductor element dl, therefore, causes an increase in the e.m.f. consumed by self-inductance between the two sides of the conductor element by

$$d\dot{E} = -\,2\,j\pi f \dot{\mathcal{B}}\,10^{-8}\,dl, \tag{3}$$

where f = the frequency of the impressed e.m.f.

Since the impressed e.m.f. $\dot{E}_0$ equals the sum of the e.m.f. consumed by self-inductance $\dot{E}$ and the e.m.f. consumed by the resistance of the conductor element,

$$\dot{E}_0 = \dot{E} + \frac{\dot{I}}{\lambda}\,. \tag{4}$$

Differentiating (4) gives

$$d\dot{E} = -\,\frac{1}{\lambda} d\dot{I}, \tag{5}$$

and substituting (5) in (3) gives

$$d\dot{I} = 2\, j\pi f \lambda \mathcal{B}\, 10^{-8}\, dl. \tag{6}$$

The two differential equations (6) and (2) contain $\mathcal{B}$, $\dot{I}$, and l, and by eliminating $\mathcal{B}$, give the differential equation between $\dot{I}$ and l: differentiating (6) and substituting (2) therein gives

$$\frac{d^2\dot{I}}{dl^2} = 0.8\, j\pi^2 f\, 10^{-8}\, \lambda\mu \dot{I}; \tag{7}$$

or writing

$$c^2 = a^2 f = 0.4\, \pi^2\, 10^{-8}\, \lambda\mu f, \tag{8}$$

where

$$a^2 = 0.4\, \pi^2\, 10^{-8}\, \lambda\mu, \tag{9}$$

gives

$$\frac{d^2\dot{I}}{dl^2} = 2\, jc^2 \dot{I}.\text{*} \tag{10}$$

This differential equation (10) is integrated by

$$\dot{I} = \dot{A}\varepsilon^{-vl}, \tag{11}$$

and substituting (11) in (10) gives

$$v^2 = 2\, jc^2,$$

$$v = \pm\, c\, (1 + j); \tag{12}$$

hence,

$$\dot{I} = \dot{A}_1 \varepsilon^{+c(1+j)l} + \dot{A}_2 \varepsilon^{-c(1+j)l}. \tag{13}$$

Since $\dot{I}$ gives the same value for $+l$ and for $-l$,

$$\dot{A}_1 = \dot{A}_2 = \dot{A}; \tag{14}$$

hence,

$$\dot{I} = \dot{A}\,\{\varepsilon^{+c(1+j)l} + \varepsilon^{-c(1+j)l}\}. \tag{15}$$

Substituting

$$\varepsilon^{\pm jcl} = \cos cl \pm j \sin cl \tag{16}$$

gives

$$\dot{I} = \dot{A}\{(\varepsilon^{+cl} + \varepsilon^{-cl}) \cos cl + j\, (\varepsilon^{+cl} - \varepsilon^{-cl}) \sin cl\}, \tag{17}$$

and for $l = l_0$, or at the conductor surface,

$$\dot{I}_1 = \dot{A}\{(\varepsilon^{+cl_0} + \varepsilon^{-cl_0}) \cos cl_0 + j\, (\varepsilon^{+cl_0} - \varepsilon^{-cl_0}) \sin cl_0\}. \tag{18}$$

* (10) is the same differential equation, and c has the same value as in the equation of alternating magnetic flux distribution (11) on page 139, and the alternating current distribution in a solid conductor thus is the same as the alternating magnetic flux distribution in the same conductor.

At the conductor surface, however, no e.m.f. of self-inductance due to the internal field exists, and

$$\dot{I}_1 = \lambda \dot{E}_0. \tag{19}$$

Substituting (19) in (18) gives the integration constant $\dot{A}$, and this substituted in (17) gives the distribution of current density throughout the conductor section as

$$\dot{I} = \lambda \dot{E}_0 \frac{(\varepsilon^{+cl} + \varepsilon^{-cl}) \cos cl + j (\varepsilon^{+cl} - \varepsilon^{-cl}) \sin cl}{(\varepsilon^{+cl_0} + \varepsilon^{-cl_0}) \cos cl_0 + j (\varepsilon^{+cl_0} - \varepsilon^{-cl_0}) \sin cl_0}. \tag{20}$$

The absolute value is given as the square root of the sum of squares of real and imaginary terms,

$$I = \lambda E_0 \sqrt{\frac{\varepsilon^{+2cl} + \varepsilon^{-2cl} + 2 \cos 2 cl}{\varepsilon^{+2cl_0} + \varepsilon^{-2cl_0} + 2 \cos 2 cl_0}}. \tag{21}$$

The current density in the conductor center, $l = 0$, is

$$\dot{I}_0 = \frac{2 \lambda \dot{E}_0}{(\varepsilon^{+cl_0} + \varepsilon^{-cl_0}) \cos cl_0 + j (\varepsilon^{+cl_0} - \varepsilon^{-cl_0}) \sin cl_0}, \tag{22}$$

or the absolute value is

$$I_0 = \frac{2 \lambda E_0}{\sqrt{\varepsilon^{+2cl_0} + \varepsilon^{-2cl_0} + 2 \cos 2 cl_0}}. \tag{23}$$

61. It is seen that the distribution of alternating-current density throughout a solid flat conductor gives the same equation as the distribution of alternating magnetic density through an iron rail, equations of the same character as the equation of the long distance transmission line, but more special in form.

The mean value of current density throughout the conductor section,

$$\dot{I}_m = \frac{1}{l_0} \int_0^{l_0} \dot{I}\, dl, \tag{24}$$

which is derived in the same manner as in Chapter V, § 51, is

$$\dot{I}_m = \frac{\lambda \dot{E}_0 \{ (\varepsilon^{+cl_0} - \varepsilon^{-cl_0}) \cos cl_0 + j (\varepsilon^{+cl_0} + \varepsilon^{-cl_0}) \sin cl_0 \}}{(1 + j) cl_0 \{ (\varepsilon^{+cl_0} + \varepsilon^{-cl_0}) \cos cl_0 + j (\varepsilon^{+cl_0} - \varepsilon^{-cl_0}) \sin cl_0 \}}, \tag{25}$$

and the absolute value is

$$I_m = \frac{\lambda E_0}{cl_0\sqrt{2}}\sqrt{\frac{\varepsilon^{+2\,cl_0} + \varepsilon^{-2\,cl_0} - 2\cos 2\,cl_0}{\varepsilon^{+2\,cl_0} + \varepsilon^{-2\,cl_0} + 2\cos 2\,cl_0}}. \tag{26}$$

Therefore, the increase of the effective resistance R of the conductor over the ohmic resistance R_0 is

$$\frac{R}{R_0} = \frac{\dot{I}_1}{\underset{\cdot}{I}_m}, \tag{27}$$

and by (19):

$$\frac{R}{R_0} = (1 + j)\,cl_0\,\frac{(\epsilon^{+cl_0} + \epsilon^{-cl_0})\cos cl_0 + j\,(\epsilon^{+cl_0} - \epsilon^{-cl_0})\sin cl_0}{(\epsilon^{+cl_0} - \epsilon^{-cl_0})\cos cl_0 + j\,(\epsilon^{+cl_0} + \epsilon^{-cl_0})\sin cl_0}, \tag{28}$$

or the absolute value is

$$\frac{R}{R_0} = cl_0\sqrt{2}\sqrt{\frac{\epsilon^{+2cl_0} + \epsilon^{-2cl_0} + 2\cos 2\,cl_0}{\epsilon^{+2cl_0} + \epsilon^{-2cl_0} - 2\cos 2\,cl_0}}. \tag{29}$$

62. If cl_0 is so large that ϵ^{-cl_0} can be neglected compared with ϵ^{-cl_0}, then in the center of the conductor $\underset{\cdot}{I}$ is negligible, and for values of l near to l_0, or near the surface of the conductor, from equation (20) we have

$$\underset{\cdot}{I} = \lambda\underset{\cdot}{E}_0\,\frac{\epsilon^{+cl}(\cos cl + j\sin cl)}{\epsilon^{+cl_0}(\cos cl_0 + j\sin cl_0)}$$

$$= \lambda\underset{\cdot}{E}_0\epsilon^{c(l+l_0)}\{\cos c\,(l - l_0) + j\sin c\,(l - l_0)\}.$$

Substituting

$$s = l_0 - l, \tag{30}$$

where s is the depth below the conductor surface, we have

$$\underset{\cdot}{I} = \lambda\underset{\cdot}{E}_0\epsilon^{-cs}(\cos cs - j\sin cs), \tag{31}$$

and the absolute value is

$$\underset{\cdot}{I} = \lambda\underset{\cdot}{E}_0\epsilon^{-cs}; \tag{32}$$

the mean value of current density is, from (25)

$$\underset{\cdot}{I}_m = \frac{\lambda\dot{E}_0}{(1 + j)\,cl_0} = \frac{(1 - j)\lambda\dot{E}_0}{2\,cl_0} \tag{33}$$

and the absolute value is

$$I_m = \frac{\lambda E_0}{cl_0 \sqrt{2}}; \tag{34}$$

hence, the resistance ratio—or rather impedance ratio, as $\dot{I}_m$ lags behind $\dot{E}_0$—is, since the current density at the surface, or density in the absence of a screening effect, is $\dot{I}_1 = \lambda \dot{E}_0$:

$$\frac{z}{r_0} = \frac{\dot{I}_1}{\dot{I}_m} = (1 + j)\, cl_0$$

$$= cl_0 + jcl_0, \tag{35}$$

where r_0 = ohmic resistance, Z = effective impedance of the conductor, and the absolute value is

$$\frac{z}{r_0} = cl_0 \sqrt{2}; \tag{36}$$

that is, the effective impedance Z of the conductor, as given by equation (28), and, for very thick conductors, from equation (35), appears in the form

$$Z = r_0(m_1 + jm_2), \tag{37}$$

which for very thick conductors gives for m_1 and m_2 the values

$$Z = r_0\,(cl_0 + jcl_0). \tag{38}$$

63. As the result of the unequal current distribution in the conductor, the effective resistance is increased from the ohmic resistance r_0 to the value

$$r = r_0 m_1,$$
$$r = cl_0 r_0,$$

and in addition thereto an effective reactance

$$x = r_0 m_2,$$

or

$$x = cl_0 r_0,$$

is produced in the conductor.

In the extreme case, where the current does not penetrate much below the surface of the conductor, the effective resistance and the effective reactance of the conductor are equal and are

$$r = x = cl_0 r_0$$

where r_0 is the ohmic resistance of the conductor.

It follows herefrom that only $\frac{1}{cl_0}$ of the conductor section is effective; that is, the depth of the effective layer is

$$l_p = \frac{l_0}{cl_0} = \frac{1}{c};$$

or, in other words, the effective resistance of a large conductor carrying an alternating current is the resistance of a surface layer of the depth

$$l_p = \frac{1}{c}, \tag{39}$$

and in addition thereto an effective reactance equal to the effective resistance results from the internal magnetic field of the conductor.

Substituting (8) in (39) gives

$$\left.\begin{aligned} l_p &= \frac{10^4}{\pi\sqrt{0.4\,\lambda\mu f}}, \\ \text{or} \qquad & \\ l_p &= \frac{5030}{\sqrt{\lambda\mu f}}. \end{aligned}\right\} \tag{40}$$

It follows from the above equations that in such a conductor carrying an alternating current the thickness of the conducting layer, or the depth of penetration of the current into the conductor, is directly proportional, and the effective resistance and effective internal inductance inversely proportional, to the square root of the electric conductivity, of the magnetic permeability, and of the frequency.

From equation (40) it follows that with a change of conductivity λ of the material the apparent conductance, and therewith the apparent resistance of the conductor, varies proportionally to the square root of the true conductivity or resistivity.

Curves of distribution of current density throughout the section of the conductor are identical with the curves of distribution of magnetic flux, as shown by Figs. 14, 15, 16 of Chapter VI.

64. From the "depth of penetration," the actual or effective resistance of the conductor then is given by circumference of the conductor times depth of penetration. This method, of calculating the depth of penetration, has the advantage that it applies to all sizes and shapes of conductors, solid round conductors or hollow tubes, flat ribbon and even such complex shapes as the railway rail, provided only that the depth of penetration is materially less than the depth of the conductor material.

The effective resistance of the conductor then is, per unit length:

$$r = \frac{1}{\lambda l_p l_1} \tag{41}$$

Thus, substituting (40):

$$\left.\begin{aligned} r &= \frac{\pi}{l_1}\sqrt{\frac{0.4\,\mu f}{\lambda}}\,10^{-4} \\ &= \frac{\sqrt{\frac{\mu f}{\lambda}}}{5030\,l_1} \text{ ohms per cm.} \end{aligned}\right\} \tag{42}$$

where:

l_1 = circumference of conductor (actual, that is, following all indentations, etc.).

As the true ohmic resistance of the conductor is:

$$r_0 = \frac{1}{\lambda S} \text{ ohms per cm.} \tag{43}$$

where:

$$S = \text{conductor section,}$$

the "resistance ratio," or ratio of the effective resistance of unequal current distribution, to the true ohmic resistance, is:

$$\left.\begin{aligned} \frac{r}{r_0} &= \frac{S\pi}{l_1}\sqrt{0.4\,\lambda\mu f} \\ &= \frac{S}{5030\,l_1}\sqrt{\lambda\mu f} \end{aligned}\right\} \tag{44}$$

The internal reactance of unequal current distribution equals the resistance, in the range considered:

$$x_1 = r \tag{45}$$

while at low frequency, where the current distribution is still uniform, the internal reactance is, per unit length of conductor:

$$\left.\begin{aligned} x_{10} &= \pi f \mu\, 10^{-9} \text{ henry per cm.} \\ &= \pi f\, 10^{-9} \text{ for air } (\mu = 7) \end{aligned}\right\} \tag{46}$$

65. It is interesting to calculate the depth of penetration of alternating current, for different frequencies, in different materials, to indicate what thickness of conductor may be employed.

Such values may be given for 25 cycles and 60 cycles as the machine frequencies, and for 10,000 cycles and 1,000,000 cycles as the limits of frequency, between which most high frequency oscillations, lightning discharges, etc., are found, and also for 1,000,000,000 cycles as about the highest frequencies which can be produced. The depth of penetration of alternating current in centimeters is given below.

Material	μ	λ	Penetration in cm. at				
			25 Cycles.	60 Cycles.	10,000 Cycles.	10^6 Cycles.	10^9 Cycles.
Very soft iron...	2000	1.1×10^5	0.068	0.044	3.4×10^{-3}	0.34×10^{-3}	0.011×10^{-3}
Steel rail........	1000	10^5	0.101	0.065	5.0×10^{-3}	0.5×10^{-3}	0.016×10^{-3}
Cast iron........	200	10^4	0.71	0.46	0.0355	3.55×10^{-3}	0.113×10^{-3}
Copper	1	6.2×10^5	1.28	0.82	0.064	6.4×10^{-3}	0.203×10^{-3}
Aluminum	1	3.7×10^5	1.65	1.07	0.082	8.2×10^{-3}	0.263×10^{-3}
German silver ..	1	0.33×10^5	5.53	3.57	0.276	27.6×10^{-3}	0.88×10^{-3}
Graphite	1	900	33.5	21.7	1.67	0.167	5.3×10^{-3}
Silicon	1	80	112.5	72.7	5.63	0.563	17.9×10^{-3}
Salt solu., conc.	1	0.2	2.25×10^3	1.45×10^3	112	11.2	0.36
Pure river water	1	10^{-4}	100.6×10^3	65×10^3	5030	503	16

It is interesting to note from this table that even at low machine frequencies the depth of penetration in iron is so little as to give a considerable increase of effective resistance, except when using thin iron sheets, while at lightning frequencies the depth of penetration into iron is far less than the thickness of sheets which can be mechanically produced. With copper and aluminum at machine frequencies this screening effect becomes noticeable only with larger conductors, approaching one inch in thickness, but with lightning frequencies the effect is such as to require the use of copper ribbons as conductor, and the thickness of the ribbon is immaterial; that is, increasing its thickness beyond that required for mechanical strength does not decrease the resistance, but merely wastes material. In general, all metallic conductors, at lightning frequencies give such small penetration as to give more or less increase of effective resistance, and their use for lightning protection therefore is less desirable, since they offer a greater resistance for higher frequencies, while the reverse is desirable.

Only pure river water does not show an appreciable increase of resistance even at the highest obtainable frequencies, and electrolytic conductors, as salt solution, give no screening effect within the range of lightning frequencies, while cast silicon can even at one million cycles be used in a thickness up to one-half inch without increase of effective resistance.

The maximum diameter of conductor which can be used with alternating currents without giving a serious increase of the effective resistance by unequal current distribution is given below.

At 25 cycles:

Steel wire	0.30 cm.	or 0.12 inch
Copper	2.6 cm.	or 1 inch
Aluminum	3.3 cm.	or 1.3 inches

At 60 cycles:

Steel wire	0.20 cm.	or 0.08 inch
Copper	1.6 cm.	or 0.63 inch
Aluminum	2.1 cm.	or 0.83 inch

At lightning frequencies, up to one million cycles:

Copper	0.013 cm.	or 0.005 inch
Aluminum	0.016 cm.	or 0.0065 inch
German silver	0.055 cm.	or 0.022 inch
Cast silicon	1.1 cm.	or 0.44 inch
Salt solution	22 cm.	or 8.7 inches
River water	All sizes.	

APPENDIX

Transient Unequal Current Distribution.

66. The distribution of a continuous current in a large conductor is uniform, as the magnetic field of the current inside of the conductor has no effect on the current distribution, being constant. In the moment of starting, stopping, or in any way changing a direct current in a solid conductor, the corresponding change of its internal magnetic field produces an unequal current distribution, which, however, is transient.

As in this case the distribution of current is transient in time as well as in space, the problem properly belongs in Part III, but may be discussed here, due to its close relation to the permanent alternating-current distribution in a solid conductor.

Choosing the same denotation as in the preceding paragraphs, but denoting current and e.m.f. by small letters as instantaneous values, equations (1), (2), and (4) of paragraph 61 remain the same:

$$d\,\mathcal{H} = 0.4\,\pi i\,dl, \tag{1}$$

$$d\,\mathcal{B} = 0.4\,\pi\mu i\,dl, \tag{2}$$

$$e_0 = e + \frac{i}{\lambda}, \tag{4}$$

where e_0 = voltage impressed upon the conductor (exclusive of its external magnetic field) per unit length, e = voltage consumed by the change of internal magnetic field, i = current density in conductor element dl at distance l from center line of flat conductor, μ = the magnetic permeability of the conductor, and λ = the electric conductivity of the conductor.

Equation (3), however,

$$dE = -\,2\,j\pi f \dot{B}\,10^{-8}\,dl,$$

changes to

$$de = -\,\frac{d\mathcal{B}}{dt}\,10^{-8}\,dl \tag{3}$$

when introducing the instantaneous values; that is, the integral or effective value of the e.m.f. E consumed by the magnetic flux density $\mathcal{B}$ is proportional and lags 90 time-degrees behind $\mathcal{B}$, while the instantaneous value i is proportional to the rate of change of $\mathcal{B}$, that is, to its differential quotient.

Differentiating (3) with respect to dl gives

$$\frac{d^2e}{dl^2} = -\,\frac{d}{dt}\left(\frac{d\mathcal{B}}{dl}\right)10^{-8}, \tag{5}$$

and substituting herein equation (2) gives

$$\frac{d^2e}{dl^2} = -\,0.4\,\pi\mu\,\frac{di}{dt}\,10^{-8}. \tag{6}$$

Differentiating (4) twice with respect to dl gives

$$0 = \frac{d^2e}{dl^2} + \frac{1}{\lambda}\frac{d^2i}{dl^2}, \tag{7}$$

and substituting (7) into (6) gives

$$\frac{d^2i}{dl^2} = + 0.4\, \pi\mu\lambda\, 10^{-8} \frac{di}{dt} \tag{8}$$

as the differential equation of the current density i in the conductor.

Substituting

$$c^2 = 0.4\, \pi\mu\lambda\, 10^{-8} \tag{9}$$

gives

$$\frac{d^2i}{dl^2} = c^2 \frac{di}{dt}\cdot \tag{10}$$

This equation (10) is integrated by

$$i = A + B\varepsilon^{-a^2t - bl}, \tag{11}$$

and substituting (11) in (10) gives the relation

$$b^2 = - c^2a^2;$$

hence,

$$b = \pm jca, \tag{12}$$

and substituting (12) in (11), and introducing the trigonometric expressions for the exponential functions with complex imaginary exponents,

$$i = A + \varepsilon^{-a^2t} (C_1 \cos cal + C_2 \sin cal), \tag{13}$$

where

$$C_1 = B_1 + B_2 \text{ and } C_2 = j\,(B_1 - B_2).$$

Assuming the current distribution as symmetrical with the axis of the conductor, that is, i the same for $+ l$ and for $- l$, gives

$$C_2 = 0;$$

hence,

$$i = A + C\varepsilon^{-a^2t} \cos cal \tag{13}$$

as the equation of the current distribution in the conductor.

It is, however, for $t = \infty$, or for uniform current distribution,

$$i = e_0\lambda;$$

hence, substituting in (13),

$$A = e_0 \lambda$$

and

$$i = e_0 \lambda + C\varepsilon^{-a^2 t} \cos cal. \tag{14}$$

At the surface of the conductor, or for $l = l_0$, no induction by the internal magnetic field exists, but the current has from the beginning the final value corresponding to the impressed e.m.f. e_0, that is, for $l = l_0$,

$$i = e_0 \lambda,$$

and substituting this value in (14) gives

$$e_0 \lambda = e_0 \lambda + C\varepsilon^{-a^2 t} \cos cal_0;$$

hence,

$$\cos cal_0 = 0$$

and

$$cal_0 = \frac{(2\kappa - 1)\pi}{2}, \tag{15}$$

or

$$a = \frac{(2\kappa - 1)\pi}{2cl_0}, \tag{16}$$

where κ is any integer.

There exists thus an infinite series of transient terms, exponential in the time, t, and trigonometric in the distance, l, one of fundamental frequency, and with it all the odd harmonics, and the equation of current density, from (14), thus is

$$\left.\begin{aligned} i &= e_0 \lambda + \sum_1^\infty {}_\kappa C_\kappa \varepsilon^{-(2\kappa-1)^2 a_1^2 t} \cos (2\kappa - 1)\, ca_1 l \\ &= e_0 \lambda + \sum_1^\infty {}_\kappa C_\kappa \varepsilon^{-(2\kappa-1)^2 a_1^2 t} \cos \frac{(2\kappa - 1)\pi l}{2l_0}, \end{aligned}\right\} \tag{17}$$

where

$$a_1 = \frac{\pi}{2cl_0}. \tag{18}$$

The values of the integration constants C_κ are determined by the terminal conditions, that is, by the distribution of current density at the moment of start of the transient phenomenon, or $t = 0$.

For $t = 0$,

$$i_0 = e_0 \lambda + \sum_1^{\infty}{}^{\kappa} C_\kappa \cos \frac{(2\kappa - 1)\pi l}{2 l_0}. \tag{19}$$

Assuming that the current density i_0 was uniform throughout the conductor section before the change of the circuit conditions which led to the transient phenomena — as would be expected in a direct-current circuit, — from (19) we have

$$\sum_1^{\infty}{}^{\kappa} C_\kappa \cos \frac{(2\kappa - 1)\pi l}{2 l_0} = - (e_0 \lambda - i_0) = \text{constant}, \tag{20}$$

and the coefficients C_κ of this Fourier series are derived in the usual manner of such series, thus:

$$C_\kappa = 2 \operatorname{avg}\left[- (e_0 \lambda - i_0) \cos \frac{(2\kappa - 1)\pi l}{2 l_0} \right]_{l=0}^{l=l_0}$$

$$= (-1)^\kappa \frac{4}{\pi} (e_0 \lambda - i_0), \tag{21}$$

where $\operatorname{avg}[F(x)]_{x=x_1}^{x=x_2}$ denotes the average value of the function $F(x)$ between the limits $x = x_1$ and $x = x_2$

and equation (17) then assumes the form

$$i = e_0 \lambda + \frac{4}{\pi}(e_0 \lambda - i_0) \sum_1^{\infty}{}^{\kappa} \frac{(-1)^\kappa}{2\kappa - 1} \varepsilon^{-(2\kappa - 1)^2 a_1^2 t} \cos \frac{(2\kappa - 1)\pi l}{2 l_0}. \tag{22}$$

This then is the final equation of the distribution of the current density in the conductor.

If now l_1 = width of the conductor, then the total current in the conductor, of thickness $2\, l_0$, is

$$I = l_1 \int_{-l_0}^{+l_0} i\, dl$$

$$= 2\, e_0 \lambda\, l_0 l_1 - \frac{16\, l_0 l_1}{\pi^2} (e_0 \lambda - i_0) \sum_1^{\infty}{}^{\kappa} \frac{\varepsilon^{-(2\kappa - 1)^2 a_1^2 t}}{(2\kappa - 1)^2},$$

or

$$I = 2\, l_0 l_1 e_0 \lambda \left\{ 1 - \frac{8}{\pi^2}\left(1 - \frac{i_0}{e_0 \lambda}\right) \sum_1^{\infty}{}^{\kappa} \frac{1}{(2\kappa - 1)^2} \varepsilon^{-(2\kappa - 1)^2 a_1^2 t} \right\}. \tag{23}$$

For the starting of current, that is, if the current is zero, $i_0 = 0$, in the conductor before the transient phenomenon, this gives

$$I = 2\, l_0 l_1 e_0 \lambda \left\{ 1 - \frac{8}{\pi^2} \sum_1^{\infty}{}_{\kappa} \frac{1}{(2\,\kappa - 1)^2} \varepsilon^{-(2\kappa - 1)^2 a_1{}^2 t} \right\}. \tag{24}$$

While the true ohmic resistance, r_0, per unit length of the conductor is

$$r_0 = \frac{1}{2\, l_0 l_1 \lambda}, \tag{25}$$

the apparent or effective resistance per unit length of the conductor during the transient phenomenon is

$$r = \frac{e_0}{I} = \frac{r_0}{1 - \frac{8}{\pi^2} \sum_1^{\infty}{}_{\kappa} \frac{1}{(2\kappa - 1)^2} \varepsilon^{-(2\kappa - 1)^2 a_1{}^2 t}}, \tag{26}$$

and in the first moment, for $t = 0$, is

$$r = \infty,$$

since the sum is

$$\sum_1^{\infty}{}_{\kappa} \frac{1}{(2\kappa - 1)^2} = \frac{\pi^2}{8}.$$

The effective resistance of the conductor thus decreases from ∞ at the first moment, with very great rapidity — due to the rapid convergence of the series — to its normal value.

67. As an example may be considered the apparent resistance of the rail return of a direct-current railway during the passage of a car over the track.

Assume the car moving in the direction away from the station, and the current returning through the rail, then the part of the rail behind the car carries the full current, that ahead of the car carries no current, and at the moment where the car wheel touches the rail the transient phenomenon starts in this part of the rail. The successive rail sections from the wheel contact backwards thus represent all the successive stages of the transient phenomenon from its start at the wheel contact to permanent conditions some distance back from the car.

Assume the rail section as equivalent to a conductor of 8 cm. width and 8 cm. height, or $l_1 = 8$, $l_0 = 4$, and the car speed as 40 miles per hour, or 1800 cm. per second.

Assume a steel rail and let the permeability $\mu = 1000$ and the electric conductivity $\lambda = 10^5$.

Then
$$c = \sqrt{0.4\,\pi\mu\lambda\,10^{-8}} = \sqrt{1.2566} = 1.121,$$

$$a_1 = \frac{\pi}{2\,cl_0} = 0.35,$$

$$a_1^2 = 0.122.$$

Since $i_0 = 0$, the current distribution in the conductor, by (22), is

$$i = e_0\lambda\left\{1 + \frac{4}{\pi}\sum_1^\infty{}_\kappa \frac{(-1)^\kappa}{2\,\kappa - 1}\,\varepsilon^{-0.122\,(2\kappa-1)^2 t}\cos 0.393\,(2\,\kappa - 1)\,l\right\}.$$

$$= e_0\lambda\,\{1 - 1.27\,[\varepsilon^{-0.122\,t}\cos 0.393\,l - \tfrac{1}{3}\,\varepsilon^{-1.10\,t}\cos 1.18\,l + \tfrac{1}{5}\,\varepsilon^{-3.05\,t}\cos 1.96\,l - + \ldots]\},$$

the ohmic resistance per unit length of rail is

$$r_0 = \frac{1}{2\,l_0 l_1\lambda} = 0.156 \times 10^{-6}\text{ ohms per cm.}$$

and the effective resistance per unit length of rail, by (26), is

$$r = \frac{0.156 \times 10^{-6}}{1 - 0.81\,[\varepsilon^{-0.122\,t} + \frac{1}{9}\varepsilon^{-1.10\,t} + \frac{1}{25}\,\varepsilon^{-3.05\,t} + \frac{1}{49}\,\varepsilon^{-6.0\,t} + \ldots]}.$$

At a velocity of 1800 cm. per second, the distance from the wheel contact to any point p of the rail, l', is given as function of the time t elapsed since the starting of the transient phenomenon at point p by the passage of the car wheel over it, by the expression $l' = 1800\,t$, and substituting this in the equation of the effective resistance r gives this resistance as function of the distance from the car, after passage,

$$r = \frac{0.156 \times 10^{-6}}{1 - 0.81\,[\varepsilon^{-68\times 10^{-6}\,l'} + \frac{1}{9}\,\varepsilon^{-612\times 10^{-6}\,l'} + \frac{1}{25}\,\varepsilon^{-1700\times 10^{-6}\,l'} + \ldots]}$$

ohms per cm.

As illustration is plotted in Fig. 17 the ratio of the effective resistance of the rail to the true ohmic resistance, $\frac{r}{r_0}$, and with the distance from the car wheel, in meters, as abscissas, from the equation

$$\frac{r}{r_0} = \frac{1}{1-0.81\left[\varepsilon^{-0.0068\,l'} + \frac{1}{9}\varepsilon^{-0.0612\,l'} + \frac{1}{25}\varepsilon^{-0.17\,l'} + \frac{1}{49}\varepsilon^{-0.34\,l'} + \ldots\right]}.$$

As seen from the curve, Fig. 17, the effective resistance of the rail appreciably exceeds the true resistance even at a considerable distance behind the car wheel. Integrating the excess of the effective resistance over the ohmic resistance shows that

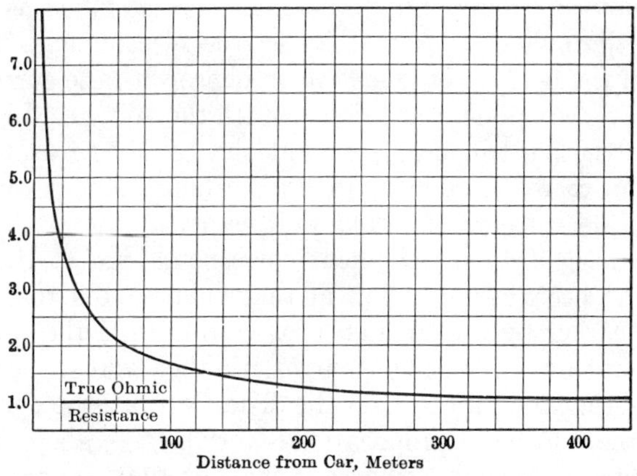

FIG. 17. Transient resistance of a direct-current railway rail return. Car speed 18 meters per second.

the excess of the effective or transient resistance over the ohmic resistance is equal to the resistance of a length of rail of about 300 meters, under the assumption made in this instance, and at a car speed of 40 miles per hour. This excess of the transient rail resistance is proportional to the car speed, thus less at lower speeds.

CHAPTER VIII

VELOCITY OF PROPAGATION OF ELECTRIC FIELD

68. In the theoretical investigation of electric circuits the velocity of propagation of the electric field through space is usually not considered, but the electric field assumed as instantaneous throughout space; that is, the electromagnetic component of the field is considered as in phase with the current, the electrostatic component as in phase with the voltage. In reality, however, the electric field starts at the conductor and propagates from there through space with a finite though very high velocity, the velocity of light; that is, at any point in space the electric field at any moment corresponds not to the condition of the electric energy flow at that moment but to that at a moment earlier by the time of propagation from the conductor to the point under consideration, or, in other words, the electric field lags the more, the greater the distance from the conductor.

Since the velocity of propagation is very high — about 3×10^{10} centimeters per second — the wave of an alternating or oscillating current even of high frequency is of considerable length; at 60 cycles the wave length is 0.5×10^{9} centimeters, and even at a hundred thousand cycles the wave length is 3 kilometers, that is, very great compared with the distance to which electric fields usually extend.

The important part of the electric field of a conductor extends to the return conductor, which usually is only a few feet distant; beyond this, the field is the differential field of conductor and return conductor. Hence, the intensity of the electric field has usually already become inappreciable at a distance very small compared with the wave length, so that within the range in which an appreciable field exists this field is practically in phase with the flow of energy in the conductor, that is, the velocity of propagation has no appreciable effect, unless the return conductor is very far distant or entirely absent, or the frequency is so high, that the distance of the return conductor is an appreciable part of the wave length.

69. Consider, for instance, a circuit representing average transmission line conditions with 6 ft. = 182 cm., between conductors, traversed by a current of $f = 10^6$, or one million cycles, such as may be produced by a nearby lightning discharge. The wave length of this current and thus of its magnetic field then would be $\frac{S}{f} = \frac{3 \times 10^{10}}{10^6} = 30{,}000$ cm. and the distance of 182 cm. between the line conductors would be $\frac{182}{30{,}000}$ or $\frac{1}{165}$ of a wave length or $\frac{360}{165} = 2.2°$. That is, the magnetic field of the current, when it reaches the return conductor, would not be in phase with the current, but $\frac{1}{165}$ of a wave length or 2.2° behind the current. The voltage induced by the magnetic field would not be in quadrature with the current, or wattless, but lag 90 + 2.2 = 92.2° behind the current, thus have an energy component equal to cos 92.2° = 3.8 per cent, giving rise to an effective resistance r_3, equal to 3.8 per cent of the reactance x. Even if at normal frequencies of 60 cycles the reactance is only equal to the ohmic resistance r_0—usually it is larger—the reactance x to 10^6 cycles would be $\frac{10^6}{60}$, = 16,700 times the ohmic resistance, and the effective resistance of magnetic radiation, r, being 3.8 per cent of this reactance, thus would be 630 times the ohmic resistance; $r_3 = 630\, r_0$. Thus, the ohmic resistance would be entirely negligible compared with the effective resistance resulting from the finite velocity of the magnetic field.

Considering, however, a high frequency oscillation of 10^6 cycles, not between the line conductors, but between line conductor and ground, and assume, under average transmission line conditions, 30 ft. as the average height of the conductor above ground. The magnetic field of the conductor then can be represented as that between the conductor and its image conductor, 30 ft. below ground, and the distance between conductor and return conductor would be 2 × 30 ft. = 1820 cm. The lag of the magnetic field, due to the finite velocity of propagation, then becomes 22°, thus quite appreciable, and the energy component of the voltage induced by the magnetic field is cos (90 + 22°) = 37 per cent.

This would give rise to an effective or radiation resistance $r_3 = 0.37\ x$. As in this case, the 60-cycle reactance usually is much larger than the ohmic resistance, assuming it as twice would make the radiation resistance $r_3 = 12{,}600\ r_0$, or more than ten thousand times the true ohmic resistance. It is true, that at these high frequencies, the ohmic resistance would be very greatly increased by unequal current distribution in the conductor. But the effective resistance of unequal current distribution increases only proportional to the square root of the frequency, and, assuming as instance conductor No. 00 B. & S. G., the effective resistance of unequal current distribution, r_1, would at 10^6 cycles be about 36 times the low frequency ohmic resistance r_0. Thus the effective resistance of magnetic radiation would still be $\frac{12{,}600}{36} = 350$ times the effective resistance of unequal current distribution; $r_3 = 350\ r_1$, and the latter, therefore, be negligible compared with the former.

It is interesting to note, that the effective resistance of radiation, r_3, does not represent energy dissipation in the conductor by conversion into heat, but energy dissipation by radiation into space, and in distinction from the "radiation resistance," which dissipates energy into space, the effective resistance of unequal current distribution may be called a "thermal resistance," as it converts electric energy into heat.

Thus in this instance of a 10^6 cycle high frequency discharge between transmission line conductor and ground, the heating of the conductor would be increased 36 fold over that produced by a low frequency current of the same amperage, by the increase of resistance by unequal current distribution in the conductor; but the total energy dissipation by the conductor would be increased by magnetic radiation still 350 times more, so that the total energy dissipation by the 10^6 cycle current would be 12,600 times greater than it would be with a low frequency current of the same value, and the attenuation or rate of decay of the current thus would be increased many thousand times, over that calculated on the assumption of constant resistance at all frequencies.

70. The finite velocity of the electric field thus requires consideration, and may even become the dominating factor in the electrical phenomena:

(*a*) In the conduction of very high frequency currents, of hundred thousands of cycles.

(*b*) In the action, propagation and dissipation of high frequency disturbances in electric circuits.

(*c*) In flattening steep wave fronts and rounding the wave shape of complex waves and sudden impulses.

(*d*) In circuits having no return circuit or no well-defined return circuit, such as the lightning stroke, the discharge path and ground circuit of the lightning arrester, the wireless antenna, etc.

(*e*) Where the electric field at considerable distance from the conductor is of importance, as in radio-telegraphy and telephony.

As illustrations of the effect of the finite velocity of the electric field may be considered in the following:

(*A*) The inductance of a finite length of an infinitely long conductor without return conductor, self-inductance as well as mutual inductance.

Such circuit is more or less approximately represented by the lightning rod, by the ground circuit of the lightning arrester, a section of wireless antenna, etc.

(*B*) The inductance of a finite length of an infinitely long conductor with return conductor, self-inductance, as well as mutual inductance.

Such for instance is the circuit of a transmission line, or the circuit between transmission line and ground.

(*C*) The capacity of a finite section of an infinitely long conductor, without return conductor as well as with return conductor.

(*D*) The mutual inductance between two finite conductors at considerable distance from each other.

Sending and receiving antennæ of radio-communication represent such pair of conductors.

(*E*) The capacity of a sphere in space.

A. INDUCTANCE OF A LENGTH l_0 OF AN INFINITELY LONG CONDUCTOR WITHOUT RETURN CONDUCTOR.

71. Such for instance is represented by a section of a lightning stroke.

The inductance of the length l_0 of a straight conductor is usually given by the equation

$$L = 2l_0 \log \frac{l'}{l_r} \times 10^{-9}, \tag{1}$$

where l' = the distance of return conductor ,l_r = the radius of the conductor, and the total length of the conductor is assumed as infinitely great compared with l_0 and l'. This is approximately the case with the conductors of a long distance transmission line.

For infinite distance l' of the return conductor, that is, a conductor without return conductor, equation (1) gives $L = \infty$; that is, a finite length of an infinitely long conductor without return conductor would have an infinite inductance L and inversely, zero capacity C.

In equation (1) the magnetic field is assumed as instantaneous, that is, the velocity of propagation of the magnetic field is neglected. Considering, however, the finite velocity of the magnetic field, the magnetic field at a distance l from the conductor and at time t corresponds to the current in the conductor at the time $t - t'$, where t' is the time required for the electric field to travel the distance l, that is, $t' = \frac{l}{S}$, where $S = 3 \times 10^{10}$ = the speed of light; or, the magnetic field at distance l and time t corresponds to the current in the conductor at the time $t - \frac{l}{S}$.

71. Representing the time t by angle $\theta = 2\pi ft$, where f= the frequency of the alternating current in the conductor, and denoting

$$a = \frac{2\pi f}{S} = \frac{2\pi}{l_w}, \tag{2}$$

where

$$l_w = \frac{S}{f} = \text{the wave length of electric field,}$$

the field at distance l and time angle θ corresponds to time angle $\theta - al$, that is, lags in time behind the current in the conductor by the phase angle al.

Let

$$i = I \cos \theta = \text{current, absolute units.} \tag{3}$$

The magnetic induction at distance l then is

$$\mathfrak{B} = \frac{2I}{l} \cos (\theta - al); \tag{4}$$

hence, the total magnetic flux surrounding the conductor of length l_0, from distance l to infinity is

$$\begin{aligned} \Phi &= l_0 \int_l^\infty \frac{2I}{l} \cos (\theta - al)\, dl \\ &= 2Il_0 \left\{ \cos \theta \int_l^\infty \frac{\cos al}{l}\, dl + \sin \theta \int_l^\infty \frac{\sin al}{l}\, dl \right\}. \end{aligned} \tag{5}$$

$\int \frac{\cos al}{l} dl$ cannot be integrated in finite form, but represents a new function which in its properties is intermediate between the sine function

$$\int \cos al\, dl = \frac{1}{a} \sin al$$

and the logarithmic function

$$\int \frac{1}{l} dl = \log l$$

and thus may be represented by a new symbol,

$$\text{sine} - \text{logarithm} = \text{sil}.$$

In the same manner, $\int \frac{\sin al}{l} dl$ is related to $\frac{1}{a} \cos al$ and to $\log l$.

Introducing therefore for these two new functions the symbols

$$\operatorname{sil} al = \int_l^\infty \frac{\cos al}{l}\, dl, \tag{6}$$

$$\operatorname{col} al = \int_l^\infty \frac{\sin al}{l}\, dl, \tag{7}$$

gives

$$\Phi = 2\, Il_0 \left\{\cos\theta \text{ sil } al + \sin\theta \text{ col } al\right\}. \tag{8}$$

The e.m.f. consumed by this magnetic flux, or e.m.f. of inductance, then is

$$e = \frac{d\Phi}{dt} = 2\,\pi f \frac{d\Phi}{d\theta};$$

hence,

$$e = 4\,\pi f Il_0 \left\{\cos\theta \text{ col } al - \sin\theta \text{ sil } al\right\}; \tag{9}$$

and since the current is

$$i = I\cos\theta,$$

the e.m.f. consumed by the magnetic field beyond distance l, or e.m.f. of inductance, contains a component in phase with the current, or *power component*,

$$e_1 = 4\,\pi f Il_0 \text{ col } al \cos\theta, \tag{10}$$

and a component in quadrature with the current, or *reactive component*,

$$e_2 = -4\,\pi f Il_0 \text{ sil } al \sin\theta, \tag{11}$$

which latter *leads* the current by a quarter period.

The reactive component e_2 is a true *self-induction,* that is, represents a surging of energy between the conductor and its electric field, but no power consumption. The effective component e_1, however, represents a power consumption

$$\begin{aligned} p &= e_1 i \\ &= 4\,\pi f I^2 l_0 \text{ col } al \cos^2\theta \end{aligned} \tag{12}$$

by the magnetic field of the conductor, due to its finite velocity; that is, it represents the *power radiated into space by the conductor.*

The energy component e_1 gives rise to an *effective resistance,*

$$r = \frac{e_1}{i} = 4\,\pi f l_0 \text{ col } al, \tag{13}$$

and the reactive component gives rise to a *reactance,*

$$x = \left[\frac{e_2}{i}\right] = 4\,\pi f l_0 \text{ sil } al, \tag{14}$$

When considering the finite velocity of propagation of the electric field, self-inductance thus is not wattless, but contains an energy component, and so can be represented by an *impedance,*

$$Z = r + jx$$
$$= 4\,\pi f l_0 \,(\text{col } al + j \text{ sil } al)\, 10^{-9} \text{ ohms.} \tag{15}$$

The *inductance* would be given by

$$L = -\frac{jZ}{2\,\pi f}$$
$$= 2\, l_0 \,\{\text{sil } al - j \text{ col } al\}\, 10^{-9} \text{ henrys,} \tag{16}$$

and the power radiated by the conductor is

$$p = i^2 r.$$

72. The functions

$$\text{sil } al = \int_l^\infty \frac{\cos al}{l}\, dl \tag{6}$$

and

$$\text{col } al = \int_l^\infty \frac{\sin al}{l}\, dl \tag{7}$$

can in general not be expressed in finite form, and so have to be recorded in tables.* Close approximations can, however, be derived for the two cases where l is very small and where l is very large compared with the wave length l_w of the electric field, and these two cases are of special interest, since the former represents the total magnetic field of the conductor, that is, its self-inductance for: $l = l_r$, and the latter the magnetic field interlinked with a distant conductor, such as the mutual inductance between sending and receiving conductor of radio telegraphy, or the induction of a lightning stroke on a transmission line.

It is

$$\left.\begin{aligned} \text{sil } 0 &= \infty, \\ \text{col } 0 &= \frac{\pi}{2}, \\ \text{sil } \infty &= 0, \\ \text{col } \infty &= 0. \end{aligned}\right\} \tag{17}$$

* Tables of these and related functions are given in the appendix.

And it can be shown that for small values of al, that is, when l is only a small fraction of a wave length, the approximations hold:

$$\left.\begin{aligned} \text{sil}\, al &= \log\frac{1}{al} - 0.5772 \\ &= \log\frac{0.5615}{al} \\ \text{col}\, al &= \frac{\pi}{2} - al \end{aligned}\right\} \qquad (18)$$

while for large values of al, the approximations hold:

$$\left.\begin{aligned} \text{sil}\, al &= -\frac{\sin al}{al} \\ \text{col}\, al &= \frac{\cos al}{al} \end{aligned}\right\} \qquad (19)$$

As seen, for larger values of al, col al has the same sign as cos al; sil al has the opposite sign of sin al.

73. As Z in (15) is the impedance and L in (16) the inductance resulting from the magnetic field of the conductor l_0, beyond the distance l, Z in (15) represents the *mutually inductive impedance* and L in (16) the *mutual inductance* of a conductor l_0 without return conductor, upon a second conductor at distance l.

In this case, if l is large compared with the wave length, we get, by substituting (19) into (15) and (16):
the *mutually inductive impedance*:

$$\begin{aligned} Z &= \frac{4\,\pi f l_0}{al}\left\{\cos al + j\sin al\right\} 10^{-9} \\ &= \frac{2\,Sl_0}{l}\left\{\cos al - j\sin al\right\} 10^{-9} \text{ ohms} \end{aligned}$$

or, absolute

$$\left.\begin{aligned} z &= \frac{2\,Sl_0}{l}\, 10^{-9} \text{ ohms} \\ &= \frac{60\, l_0}{l} \text{ ohms} \end{aligned}\right\} \qquad (20)$$

and, the *mutual inductance*:

$$\begin{aligned} M &= \frac{2\, l_0}{al}\left\{\sin al - j\cos al\right\} 10^{-9} \\ &= \frac{Sl_0}{\pi f l}\left\{\sin al - j\cos al\right\} 10^{-9} \text{ henrys} \end{aligned}$$

or, absolute:

$$\left.\begin{aligned} m &= \frac{Sl_0}{\pi f l} 10^{-9} \text{ henrys} \\ &= \frac{30\, l_0}{\pi f l} \text{ henrys} \end{aligned}\right\} \qquad (21)$$

Making, in (15) and (16):

$$l = l_r = \text{radius of the conductor,}$$

L becomes the self-inductance, and Z the self-inductive impedance of the conductor.

Since al_r always is a small quantity, equations (18) apply, and it is, substituting (2)

$$\left.\begin{aligned} \text{col}\, al_r &= \frac{\pi}{2} \\ \text{sil}\, al_r &= \log \frac{0.56}{al_r} \\ &= \log \frac{0.56\, S}{2\, \pi f l_r} \\ &= \log \frac{0.56\, S}{f l_c} \end{aligned}\right\} \qquad (21)$$

where l_c is the circumference of the conductor.

Substituting (21) into (15), gives the *self-inductive radiation impedance* of the conductor without return conductor:

$$\begin{aligned} Z &= r + jx \\ &= 4\, \pi f l_0 \left\{ \frac{\pi}{2} + j \log \frac{0.56\, S}{f l_c} \right\} 10^{-9} \text{ ohms} \end{aligned} \qquad (22)$$

comprising the *effective radiation resistance*:

$$r = 2\, \pi^2 f l_0\, 10^{-9} \text{ ohms} \qquad (23)$$

and the *effective radiation reactance*:

$$x = 4\, \pi f l_0 \log \frac{0.56\, S}{f l_c}\, 10^{-9} \text{ ohms} \qquad (24)$$

corresponding to the true *self-inductance* of the conductor without return conductor:

$$\begin{aligned} L &= \frac{x}{2\, \pi f} \\ &= 2\, l_0 \log \frac{0.56\, S}{f l_c}\, 10^{-9} \text{ henrys} \end{aligned} \qquad (25)$$

As seen, the effective radiation resistance of a conductor without return conductor is proportional to the frequency f; the inductance L decreases with increasing frequency, but logarithmically, that is, much slower than the frequency, and the reactance x thus increases somewhat slower than the frequency.

Thus per meter of conductor, or $l_0 = 100$ cm., with a conductor of No. 00 B. & S. G., of $l_c = 2.93$ cm., it is, at

$f = 100{,}000$ cycles	$f = 1{,}000{,}000$ cycles
$r = 0.195$ ohms	$r = 1.95$ ohms
$L = 2.19 \times 10^{-6}$ henry	$L = 1.73 \times 10^{-6}$ henry
$x = 1.38$ ohms	$x = 10.9$ ohms

For comparison, the true ohmic resistance of the conductor is $r_0 = 0.00026$ ohms, thus thousands of times less than the radiation resistance at these frequencies.

The power radiated by the conductor then is:

$$\begin{aligned} p &= i^2 r \\ &= 2\,\pi^2 f l_0 i^2\, 10^{-9} \text{ watts} \end{aligned} \tag{26}$$

hence, at 100 amperes, per meter length of conductor, at

$$\begin{aligned} f &= 100{,}000 \text{ cycles} : p = 1.95 \text{ kilowatts} \\ f &= 1{,}000{,}000 \text{ cycles} : p = 19.5 \text{ kilowatts} \end{aligned}$$

B. INDUCTANCE OF A LENGTH l_0 OF AN INFINITELY LONG CONDUCTOR, WITH RETURN CONDUCTOR AT DISTANCE l'.

74. Such for instance is represented by a section of a transmission line, etc. Let again

$$i = I \cos \theta = \text{current, absolute units.} \tag{27}$$

The magnetic induction, at distance l then is

$$\mathfrak{B} = \frac{2I}{l} \cos (\theta - al) \tag{28}$$

hence, the total magnetic flux surrounding the conductor of length l_0, from its surface at distance l_r (the radius of the conductor) up to the return conductor at distance l', is

$$\begin{aligned} \Phi &= l_0 \int_{l_r}^{l'} \frac{2\,I}{l} \cos (\theta - al)\, dl \\ &= 2\,I l_0 \left\{ \cos \theta \int_{l_r}^{l'} \frac{\cos al}{l}\, dl + \sin \theta \int_{l_r}^{l'} \frac{\sin al}{l}\, dl \right\} \end{aligned} \tag{29}$$

It is, however,

$$\left.\begin{aligned}&\int_{l_r}^{l'}\frac{\cos al}{l}\,dl = \int_{l_r}^{\infty}\frac{\cos al}{l}\,dl - \int_{l'}^{\infty}\frac{\cos al}{l}\,dl = \text{sil } al_r - \text{sil } al' \\ &\int_{l_r}^{l'}\frac{\sin al}{l}\,dl = \text{col } al_r - \text{col } al'\end{aligned}\right\} \quad (30)$$

Hence:

$$\Phi = 2\,Il_0\{\cos\theta(\text{sil } al_r - \text{sil } al') + \sin\theta(\text{col } al_r - \text{col } al')\} \quad (31)$$

The voltage induced by this magnetic flux in the conductor is

$$e = \frac{d\Phi}{dt}\,10^{-9} = 2\,\pi f\,\frac{d\Phi}{d\theta}\,10^{-9} \text{ volts}$$

$$= 4\,\pi f I l_0\,\{\cos\theta\,(\text{col } al_r - \text{col } al') - \sin\theta\,(\sin al_r - \sin al')\}\,10^{-9} \quad (32)$$

and the effective *self-inductive radiation impedance* of the conductor is

$$\frac{e}{I} = 4\,\pi f l_0\{\cos\theta\,(\text{col } al_r - \text{col } al') - \sin\theta\,(\text{sil } al_r - \text{sil } al')\}10^{-9} \text{ ohms}$$

or, in symbolic expression,

$$Z = r + jx$$

$$= 4\,\pi f l_0\{(\text{col } al_r - \text{col } al') + j(\text{sil } al_r - \text{sil } al')\}10^{-9} \text{ ohms} \quad (33)$$

and, the *effective radiation resistance:*

$$r = 4\,\pi f l_0\,(\text{col } al_r - \text{col } al')10^{-9} \text{ ohms} \quad (34)$$

effective radiation self-inductance:

$$L = 2\,l_0\,(\text{sil } al_r - \text{sil } al') \text{ henrys} \quad (35)$$

Since even at a billion cycles, the wave length $l_w = 30$ cm. is large compared with the conductor radius l_r of a transmission line, al_r is very small, and we can therefore substitute (18), and get:

$$r = 4\,\pi f l_0 \left\{\frac{\pi}{2} - \text{col } al'\right\} 10^{-9} \text{ ohms} \quad (36)$$

$$L = 2\,l_0 \left\{\log\frac{0.56}{al_r} - \text{sil } al'\right\} 10^{-9} \text{ henrys} \quad (37)$$

75. Under transmission line conditions, for all except the highest frequencies al' is so small as to permit the approximations (18). This gives

$$r = 4\,\pi f l_0 \alpha l'\ 10^{-9} \text{ ohms} \tag{38}$$

$$L = 2\,l_0 \left\{ \log \frac{1}{al_r} - \log \frac{1}{al'} \right\} 10^{-9} \text{ henrys}$$

$$= 2\,l_0 \log \frac{l'}{l}\ 10^{-9} \text{ henrys} \tag{39}$$

Equation (39) is the usual inductance formula, derived without considering the finite velocity of the elcetric field.

Substituting (2) into (38) gives

$$r = \frac{8\,\pi^2 f^2 l' l_0}{S} 10^{-9} \text{ ohms}$$

$$= 2.63\, f^2 l' l_0\ 10^{-18} \text{ ohms} \tag{40}$$

The effective radiation resistance r thus is proportional to the square of the frequency, and proportional to the distance from the return conductor, for all frequencies of a wave length large compared with the distance from the return conductor. For such frequencies, the finite velocity of the field has no appreciable effect yet on the inductance of the conductor.

At $al = 0.1$ the error of approximations (18) is still less than 0.01 per cent.

and

$$al' < 0.1$$

gives, by (2):

$$f < \frac{S}{2\pi l'}$$

hence:

$$f < \frac{4770 \times 10^6}{l'}$$

At $l' = 6$ ft. $= 182$ cm., or 6 ft. distance between transmission line conductors, equations (39) and (40) thus are applicable up to frequencies of

$$f = \frac{4770}{l'} = 26.2 \text{ million cycles,}$$

thus for all frequencies which come into engineering consideration.

In a high-frequency oscillation between line and ground, as

due for instance to lightning, assuming the average height of the line as 30 ft., it is: $l' = 2 \times 30$ ft. $= 1820$ cm., hence,

$$f = 2.62 \text{ million cycles.}$$

However, even far beyond these frequencies, equations (39) and (40) are still approximately applicable.

76. As an instance may be calculated the effective radiation resistance per kilometer of long distance transmission conductor of No. 00 B. & S. G.,

(*a*) Against the return conductor, at 6 ft. = 182 cm. distance;

(*b*) Against the ground, at elevation of 30 ft. = 910 cm.

The true ohmic resistance is

$$r_0 = 0.257$$

The radiation resistance is, by (40),

$$\begin{aligned} r &= 2.63\ f^2 l'\ 10^{-18} \text{ ohms per cm.} \\ &= 0.263\ f^2 l'\ 10^{-12} \text{ ohms per km.} \end{aligned}$$

hence for

$f =$	25	60	10^3
(*a*) $l' - 182;\ r =$	3.0×10^{-6}	17.3×10^{-6}	48×10^{-6}
(*b*) $l' = 1820;\ r =$	30×10^{-6}	173×10^{-6}	480×10^{-6}

10^4	10^5	10^6	10^7	10^8 cycles
0.0048	0.48	48	4800	225,000 ohms*
0.048	4.8	480	22,500*	194,000* ohms.

By (40), the radiation resistance r is proportional to the square of the frequency, that is, $\frac{r_0}{f^2}$ is constant, up to those frequencies, where $al' = \frac{2\pi f l'}{S}$ becomes appreciable compared with the quarter wave length $\frac{\pi}{2}$, that is, as long as f is well below the value

$$f_0 = \frac{S}{4l'}$$

or, as long as the distance of the return conductor, l', is well below

$$l' = \frac{S}{4f}$$

that is, a quarter wave length.

* In these values, the more complete equation (36) had to be used.

If the return conductor is at a distance equal to a quarter wave length,

$$l' = \frac{S}{4f}$$

then in equation (36), col $al' = 0$, and this equation becomes

$$r = 2\,\pi^2 f l_0\, 10^{-9}$$
$$= 19.7\, f l_0\, 10^{-9} \text{ ohms}$$

which is the equation of the radiation resistance of a conductor without return conductor (23). That is,

The radiation resistance of a conductor with return conductor at quarter wave length distance, is the same as that of a conductor without return conductor.

The radiation resistance of a conductor is a maximum with the return conductor at such distance l', which makes col al' in (36) a negative maximum. This is for $al' = \pi$, hence,

$$l' = \frac{S}{2\pi}$$
$$r = 23.2\, f l_0\, 10^{-9} \text{ ohms}$$

That is, the radiation resistance of a conductor is a maximum with the return conductor at half a wave length distance.

C. CAPACITY OF A LENGTH l_0 OF AN INFINITELY LONG CONDUCTOR.

77. Such for instance is represented by a section of a transmission line, etc.

Let

$$e = E \cos \theta$$
$$= \text{voltage impressed upon conductor of radius } l_r. \qquad (41)$$

Then

$$G = \frac{de}{dl}$$
$$= \text{dielectric gradient at distance } l \text{ from conductor} \qquad (42)$$

$$K = \frac{G}{4\pi S^2}$$
$$= \frac{1}{4\pi S^2}\frac{de}{dl} \qquad (43)$$

$$= \text{dielectric field intensity, and}$$

$$\begin{aligned} D &= kK \\ &= \frac{k}{4\pi S^2}\frac{de}{dl} \qquad (44) \\ &= \text{dielectric density} \\ \Psi &= 2\,\pi l l_0 D \\ &= \frac{kll_0}{2S^2}\frac{de}{dl} \qquad (45) \\ &= \text{dielectric flux, where} \\ 2\,\pi l l_0 &= \text{section traversed by flux.} \end{aligned}$$

Thus,

$$\begin{aligned} \frac{de}{dl} &= \frac{2S^2}{kll_0}\Psi \qquad (46) \\ &= \text{dielectric gradient, and} \\ e &= \int_{l_r}^{l'} \frac{de}{dl}\,dl \\ &= \frac{2S^2}{kl_0}\int_{l_r}^{l'} \frac{\Psi}{l}\,dl \qquad (47) \\ &= \text{voltage impressed upon conductor, where:} \\ l' &= \text{distance of return conductor.} \end{aligned}$$

It is, however, by definition (see "Electric Discharges, Waves and Impulses," Chapter XI, Dover edition Volume II, Part II),

$$\Psi = Ce \qquad (48)$$

where

$$C = \text{capacity}$$

The dielectric flux Ψ, at distance l, lags behind the voltage impressed upon the conductor (41), by the phase angle al, thus is:

$$\Psi = CE \cos (\theta - al) \qquad (49)$$

Substituting (41) and (49) into (47), and cancelling by E, give

$$\begin{aligned} \cos\theta &= \frac{2S^2C}{kl_0}\int_{l_r}^{l'} \frac{\cos(\theta - al)}{l}\,dl \\ &= \frac{2S^2C}{kl_0}\left\{\cos\theta \operatorname{sil} al + \sin\theta \operatorname{col} al\right\}_{l_r}^{l'} \qquad (50) \end{aligned}$$

thus,

$$\frac{1}{C} = \frac{2S^2}{kl_0}\left\{(\operatorname{sil} al_r - \operatorname{sil} al') + \tan\theta\,(\operatorname{col} al_r - \operatorname{col} al')\right\}10^{-9} \qquad (51)$$

where the $\tan\theta$ represents the quadrature component of capacity, or effective dielectric radiation resistance, and the 10^{-9} reduces from absolute units to farads.

It then is, in complex expression:

$$\frac{1}{C} = \frac{2S^2}{kl_0}\left\{(\text{sil } al_r - \text{sil } al') + j(\text{col } al_r - \text{col } al')\right\} 10^{-9} \qquad (52)$$

78. The *dielectric radiation impedance* then is

$$Z = \frac{-j}{2\pi fC} = \frac{S^2}{\pi fkl_0}\left\{(\text{col } al_r - \text{col } al') - j(\text{sil } al_r - \text{sil } al')\right\} 10^{-9} \text{ ohms} \qquad (53)$$

Since by (33) the *magnetic radiation impedance* is

$$Z' = 2\,\pi fLj = 4\,\pi fl_0\left\{(\text{col } al_r - \text{col } al') + j(\text{sil } al_r - \text{sil } al')\right\} 10^{-9} \text{ ohms} \qquad (33)$$

it is, for the absolute values,

$$\frac{z'}{z} = 4\pi^2 f^2 LC = \frac{4\pi^2 f^2 kl_0^2}{S^2} \qquad (54)$$

or, in air, for $k = 1$,

$$LC = \frac{l_0^2}{S^2} \qquad (55)$$

and, per unit length, or $l_0 = 1$,

$$LC = \frac{1}{S^2} \qquad (56)$$

where L and C are the radiation inductance and radiation capacity, including the wattless or reactive components, the true inductance and capacity, as well as the energy components, the effective magnetic and dielectric radiation resistances.

Equations (55) and (56) are the same equations which apply to the values of L and C calculated without considering the finite velocity of the field. Thus the capacity can be calculated from the inductance, and inversely, even in the general case.

It is, by (18),

$$\left.\begin{aligned} \text{sil } al_r &= \log \frac{0.56}{al_r} \\ \text{col } al_r &= \frac{\pi}{2} \end{aligned}\right\} \qquad (57)$$

since al_r is very small.

Substituting (57) into (53) gives, as the *dielectric radiation impedance,*

$$Z = \frac{S^2}{\pi f k l_0}\left\{\left(\frac{\pi}{2} - \operatorname{col} al'\right) - j\left(\log\frac{0.56}{al_r} - \operatorname{sil} al'\right)\right\}10^{-9}\text{ ohms} \quad (58)$$

79. If l' is small compared with a quarter wave length, it is, by (18),

$$\left.\begin{aligned}\operatorname{sil} al' &= \log\frac{0.56}{al'}\\ \operatorname{col} al' &= \frac{\pi}{2} - al'\end{aligned}\right\} \quad (59)$$

hence

$$\frac{1}{C} = \frac{2S^2}{kl_0}\left\{\log\frac{l'}{l_r} + jal'\right\}10^{-9}$$

and, substituting (2),

$$\frac{1}{C} = \frac{2S^2}{kl_0}\left\{\log\frac{l'}{l_r} + j\frac{2\pi f l'}{S}\right\}10^{-9} \quad (60)$$

$$= \left\{\frac{2S^2}{kl_0}\log\frac{l'}{l_r} + \frac{4\pi f l' S}{kl_0}\right\}10^{-9} \quad (61)$$

If

$$l' = \infty.,$$

that is, *in a conductor without return conductor,* such as the vertical discharge path of a lightning arrester, it is, substituting (57) and (2) into (52):

$$\frac{1}{C} = \frac{2S^2}{kl_0}\left\{\log\frac{0.56}{2\pi f l_r} + j\frac{\pi}{2}\right\}10^{-9} \quad (62)$$

The capacity reactance, or *dielectric radiation impedance of the conductor without return conductor,* then is,

$$Z = \frac{-j}{2\pi f C}$$

$$Z = \frac{S^2}{\pi f k l_0}\left\{\frac{\pi}{2} - j\log\frac{0.56}{2\pi f l_r}\right\}10^{-9} \quad (63)$$

while the *dielectric radiation impedance of the conductor with return conductor at distance* l', is, by (60):

$$Z = \frac{S^2}{\pi f k l_0}\left\{\frac{2\pi f l'}{S} - j\log\frac{l'}{l_r}\right\}10^{-9}$$

$$= \left\{\frac{2Sl'}{kl_r} - j\frac{S^2}{\pi f k l_0}\log\frac{l'}{l_r}\right\}10^{-9} \quad (64)$$

Thus comprising an *effective dielectric radiation resistance:*

Conductor with return conductor,

$$r = \frac{2Sl'}{kl_0} 10^{-9} \text{ ohms} \tag{65}$$

Conductor without return conductor,

$$r = \frac{S^2}{2fkl_0} 10^{-9} \text{ ohms} \tag{66}$$

and an *effective dielectric radiation* (*capacity-*) *reactance*:
Conductor with return conductor,

$$x = \frac{S^2}{\pi fkl_0} \log \frac{l'}{l_r} 10^{-9} \text{ ohms} \tag{67}$$

Conductor without return conductor,

$$x = \frac{S^2}{\pi fkl_0} \log \frac{0.56\, S}{2\,\pi fl_r} 10^{-9} \text{ ohms} \tag{68}$$

An effective capacity:

$$C = \frac{1}{2\pi fx} \tag{69}$$

Conductor with return conductor:

$$C = \frac{kl_0\, 10^9}{2S^2 \log \frac{l'}{l_r}} \text{ farads} \tag{70}$$

Conductor without return conductor:

$$C = \frac{kl_0\, 10^9}{2S^2 \log \frac{0.56S}{2\pi fl_r}} \text{ farads} \tag{71}$$

It is interesting to note that the values (70) and (67) of the capacity and the capacity reactance of a conductor with return conductor at distance l', is the same as derived without considering the velocity of propagation of the electric field. That is, the finite velocity of the electric field does not change the equation of the capacity (nor that of the inductance), as long as the return conductor is well within a quarter-wave distance.

As value of the *attenuation constant of dielectric radiation* then follows:

$$u_2 = \frac{g}{C} = \frac{2\pi fg}{b} = \frac{2\pi fr}{x}$$

and by (65) to (68) conductor with return conductor:

$$u_2 = \frac{4\pi^2 f^2 l' k}{S \log \frac{l'}{l_r}} \tag{72}$$

Conductor without return conductor:

$$u_2 = \frac{\pi^2 f}{\log \frac{0.56\, S}{2\,\pi f l_r}} \tag{73}$$

While the *attenuation constant of magnetic radiation* is, from (38) and (39), conductor with return conductor:

$$u_1 = \frac{r}{L} = \frac{4\,\pi^2 f^2 l'}{S \log \frac{l'}{l_r}} \tag{74}$$

and, from (23) and (25), conductor without return conductor:

$$u_1 = \frac{r}{L} = \frac{\pi^2 f}{\log \frac{0.56\ S}{2\,\pi f l_r}} \tag{75}$$

that is, the same values as the attenuation constant of dielectric radiation, as was to be expected.

It is interesting to note, that the effective dielectric capacity resistance, of the conductor with return conductor (65) is again proportional to the distance of the return conductor, l', like the effective resistance of magnetic radiation. As the dielectric capacity reactance (68) is inverse proportional to the frequency, the capacity current is (approximately) proportional to the frequency, and the power consumption by dielectric radiation, $i^2 r$, thus (approximately) proportional to the square of the frequency, just like the power consumption by magnetic radiation, in *B*.

Numerical values are given in the next chapter, and in Part III of this volume.

It is interesting to note, that the expressions of inductance L and capacity C, (39) and (70), are the same as the values of L and C, calculated without considering the finite velocity of the electric field, that is, are the "low frequency values" of external inductance and capacity. Thus,

As long as the distance of the return conductor is small compared with a quarter wave length, electric radiation due to the finite velocity of the electric field, does not affect or change the values of external inductance L and of capacity C, but causes energy dissipation by electromagnetic and by dielectric or electrostatic radiation, which is represented by effective resistances. The electromagnetic radiation resistance is represented by a series resistance, proportional to the square of the frequency; the electrostatic resistance by a shunted resistance in series with the capacity. It is independent of the frequency and the power consumed by either resistance thus is proportional to the square of the frequency.

D. MUTUAL INDUCTANCE OF TWO CONDUCTORS OF FINITE LENGTH AT CONSIDERABLE DISTANCE FROM EACH OTHER.

80. Such for instance are sending and receiving antennæ of radio-communication. Or lightning stroke and transmission line.

The electric field of an infinitely long conductor without return conductor decreases inversely proportional to the distance l, and therefore is represented by

$$\psi = \frac{\Psi}{l} \tag{76}$$

where Φ is the intensity of the field at unit distance from the conductor.

The electric field of a conductor of finite length l_0 decreases inversely proportional to the distance l and also proportionally to the angle φ subtended by the conductor l_0 from the distance l,

$$\psi = \frac{\varphi}{\pi} \frac{\Psi}{l}. \tag{77}$$

Since this angle φ, for great distances l, is given by

$$\varphi = \frac{l_0}{l} \tag{78}$$

the electric field of a conductor without return conductor, of finite length l_0, at great distances l, is represented by

$$\psi = \frac{l_0 \Psi}{\pi l^2}. \tag{79}$$

Since the electric field of the return conductor is opposite to that of the conductor, it follows that the electric field of an infinitely long conductor, with the return conductor at distance l_1, is, by (76),

$$\psi = \frac{\Psi}{l - l'} - \frac{\Psi}{l + l'} \tag{80}$$

where $l' = l_1 \cos \tau$ is the projection of the distance l_1 between the conductors upon the direction l, that is, l' is the difference in the distance of the two conductors from the point l.

For large distances l, equation (80), becomes

$$\psi = \frac{l'\Psi}{l^2}. \tag{81}$$

In the same manner, from equation (79) it follows that the decrease of the electric field of the conductor of finite length l_0 with its return conductor at the distance l_1, that is, of a rectangular circuit of the dimensions of l_0 and l_1, is:

$$\psi = \frac{l_0\Psi}{\pi\left(l - \frac{l'}{2}\right)^2} - \frac{l_0\Psi}{\pi\left(l + \frac{l'}{2}\right)^2};$$

hence,

$$\psi = \frac{2\, l_0 l'\Psi}{\pi l^3} \tag{82}$$

81. Let l_1 and l_2 be the lengths of two conductors without return conductors, at distance l_0 from each other.

By equation (79), the electric field of a conductor of length l_1, without return conductor, at distance l, is given by:

$$\psi = \frac{l_1\Psi}{\pi l^2} \tag{79}$$

With the current

$$i = I \cos \theta \tag{83}$$

in the conductor l_1, the magnetic field at unit distance is

$$\Psi = 2\, i \tag{84}$$

and the electromagnetic component of the field at distance l thus is

$$\begin{aligned} \mathcal{B} &= \frac{2\, l_1 i}{\pi l^2} \\ &= \frac{2\, l_1 I \cos(\theta - al)}{\pi l^2} \end{aligned} \tag{85}$$

where the al represents the finite velocity of propagation from the conductor l_1 to the distance l.

The magnetic flux intercepted by the receiving conductor of length l_2, at distance l_0, then is:

$$\Phi = \int_{l_0}^{\infty} \frac{2\, l_1 l_2 I \cos(\theta - al)}{\pi l^2} dl$$

$$= \frac{2\, l_1 l_2 I}{\pi} \left\{ \cos\theta \int_{l_0}^{\infty} \frac{\cos al}{l^2} dl + \sin\theta \int_{l_0}^{\infty} \frac{\sin al}{l^2} dl \right\}. \tag{86}$$

These integrals can not be integrated in finite form, but represent new functions, and as such may be denoted by

$$a \operatorname{coll} al = \int_{l}^{\infty} \frac{\cos al}{l^2} dl \tag{87}$$

$$a \operatorname{sill} al = \int_{l}^{\infty} \frac{\sin al}{l^2} dl \tag{88}$$

These functions are further discussed in the appendix.

82. Substituting (87) and (88) into (86), gives:

$$\Phi = \frac{2\, a l_1 l_2 I}{\pi} \{\cos\theta \operatorname{coll} al_0 + \sin\theta \operatorname{sill} al_0\} \tag{89}$$

and the *mutual inductance*,

$$M = \frac{\Phi}{I} \tag{90}$$

thus is given, in symbolic representation:

$$M = \frac{2a}{\pi} l_1 l_2 \{\operatorname{coll} al_0 - j \operatorname{sill} al_0\}\ 10^{-9} \text{ henry.} \tag{91}$$

By partial integration, coll and sill can be reduced to sil and col, thusly:

$$a \operatorname{coll} al = \int_{l}^{\infty} \frac{\cos al}{l^2} dl = -\int_{l}^{\infty} \cos al\, d\left(\frac{1}{l}\right) = \frac{\cos al}{l} - a \operatorname{col} al \tag{92}$$

$$a \operatorname{sill} al = \int_{l}^{\infty} \frac{\sin al}{l^2} dl = -\int_{l}^{\infty} \sin al\, d\left(\frac{1}{l}\right) = \frac{\sin al}{l} + a \operatorname{sil} al. \tag{93}$$

For large values of l, by equations (20), col al approximates $\frac{\cos al}{al}$, and sil al approximates $-\frac{\sin al}{al}$, that is, coll al and sill al approximate zero, with increasing l, at a higher rate than do sil al and col al, as was to be expected.

Substituting (82) and (83) into (81) gives

$$M = \frac{2}{\pi} l_1 l_2 \left\{ \left(\frac{\cos a l_0}{l_0} - a \operatorname{col} a l_0 \right) - j \left(\frac{\sin a l_0}{l_0} + a \operatorname{sil} a l_0 \right) \right\} 10^{-9} \text{ henry} \quad (94)$$

From the mutual inductance M follows the *mutual impedance*

$$\begin{aligned} Z &= 4 j\pi f M \\ &= 4 a f l_1 l_2 \{ \operatorname{sill} a l_0 + j \operatorname{coll} a l_0 \} 10^{-9} \\ &= 4 f l_1 l_2 \left\{ \left(\frac{\sin a l_0}{l_0} + a \operatorname{sil} a l_0 \right) + j \left(\frac{\cos a l_0}{l_0} - a \operatorname{col} a l_0 \right) \right\} 10^{-9} \text{ ohms} \quad (95) \end{aligned}$$

or, absolute,

$$\begin{aligned} z &= 4 a f l_1 l_2 \sqrt{\operatorname{sill}^2 a l_0 + \operatorname{coll}^2 a l_0}\, 10^{-9} \\ &= 4 f l_1 l_2 \sqrt{\left(\frac{\sin a l_0}{l_0} + a \operatorname{sil} a l_0 \right)^2 + \left(\frac{\cos a l_0}{l_0} - a \operatorname{col} a l_0 \right)^2}\, 10^{-9} \text{ ohms} \quad (96) \end{aligned}$$

For large values of $a l_0$, it is:

$$\operatorname{sill} a l_0 = -\frac{\sin a l_0}{(a l_0)^2} \quad (97)$$

$$\operatorname{coll} a l_0 = \frac{\cos a l_0}{(a l_0)^2}. \quad (98)$$

thus, substituted into (86):

$$\begin{aligned} z &= 4 f \frac{l_1 l_2}{a l_0^2} 10^{-9} \\ &= \frac{2 S l_1 l_2}{\pi l_0^2} 10^{-9} \text{ ohms.} \quad (99) \end{aligned}$$

Thus, for instance:

$$\begin{aligned} l_1 = l_2 &= 50 \text{ ft.} = 1515 \text{ cm.} \\ l_0 &= 10 \text{ miles} = 1.61 \times 10^6 \text{ cm.} \\ f &= 500{,}000 \text{ cycles} \\ z &= 53 \times 10^{-6} \end{aligned}$$

and, if

$$\begin{aligned} i_1 &= 10 \text{ amps.} \\ e_2 &= z\, i_1 = 0.53 \text{ millivolts.} \end{aligned}$$

E. CAPACITY OF A SPHERE IN SPACE.

83. Let

l_0 = radius of the sphere.
$e_0 = E \cos \theta$ = voltage of the sphere.
e = voltage at distance l from centre of sphere.

The voltage gradient at distance l then is:

$$G = \frac{de}{dl} \tag{100}$$

and the electrostatic field, in electromagnetic units:

$$K = \frac{G}{4\pi S^2}$$
$$= \frac{1}{4\pi S^2}\frac{de}{dl} \tag{101}$$

thus the dielectric flux:

$$\Psi = 4\pi l^2 K$$
$$= \frac{l^2}{S^2}\frac{de}{dl} \tag{102}$$

Let C = capacity of the sphere. The dielectric flux then is

$$\Psi = Ce \tag{103}$$

and lags by angle al behind the voltage, due to the finite velocity of the field (more correctly, by $a(l - l_0)$, but l_0 may be neglected against l).

Thus the flux at distance l is

$$\Psi = CE \cos(\theta - al) \tag{104}$$

Substituting (94) into (92), and resolving, gives:

$$\frac{de}{dl} = \frac{CES^2 \cos(\theta - al)}{l^2} \tag{105}$$

and, integrated from l_0 to ∞:

$$e = E \cos\theta = CES^2 \int_{l_0}^{\infty} \frac{\cos(\theta - al)}{l^2} dl \tag{106}$$

hence,

$$\frac{1}{C} = S^2 \left\{ \int_{l_0}^{\infty} \frac{\cos al}{l^2}\, dl + \tan\theta \int_{l_0}^{\infty} \frac{\sin al}{l^2}\, dl \right\} 10^{-9} \tag{107}$$

where the 10^{-9} reduces the farads.

The integrals are the same found in the case of mutual inductance, D, equations (87), (88), and reduce to sil and col by equations (92) and (93).

Writing them symbolically, (107) becomes:

$$\frac{1}{C} = S^2 \left\{ \left(\frac{\cos al_0}{l_0} - a \operatorname{col} al_0 \right) - j \left(\frac{\sin al_0}{l_0} + a \operatorname{sil} al_0 \right) \right\} 10^{-9} \tag{108}$$

If, as usual, l_0 is small compared with the wave length, it is:

$$\left.\begin{aligned} \cos al_0 &= 1 \\ \operatorname{col} al_0 &= \frac{\pi}{2} \\ \sin al_0 &= 0 \\ \operatorname{sil} al_0 &= \log \frac{0.56}{al_0} \end{aligned}\right\} \tag{109}$$

thus,

$$\frac{1}{C} = S^2 \left\{ \left(\frac{1}{l_0} - \frac{a\pi}{2} \right) - ja \log \frac{0.56}{al_0} \right\} 10^{-9} \tag{110}$$

and, substituting (2):

$$\frac{1}{C} = S^2 \left\{ \left(\frac{1}{l_0} - \frac{\pi^2 f}{S} \right) - j \frac{2\pi f}{S} \log \frac{0.56\, S}{2\pi f l_0} \right\} 10^{-9} \tag{111}$$

For $a = O$, or infinite velocity of the electric field, (110) becomes

$$C = \frac{l_0}{S^2} \tag{112}$$

which is the usual expression of the capacity of a sphere in space.

CHAPTER IX

HIGH-FREQUENCY CONDUCTORS

84. As the result of the phenomena discussed in the preceding chapters, conductors intended to convey currents of very high frequency, as lightning discharges, high frequency oscillations of transmission lines, the currents used in wireless telegraphy, etc., cannot be calculated by the use of the constants derived at low frequency, but effective resistance and inductance, and therewith the power consumed by the conductor, and the voltage drop, may be of an entirely different magnitude from the values which would be found by using the usual values of resistance and inductance. In conductors such as are used in the connections and the discharge path of lightning arresters and surge protectors, the unequal current distribution in the conductor (Chapter VII) and the power and voltage consumed by electric radiation, due to the finite velocity of the electric field (Chapter VIII), require consideration.

The true ohmic resistance in high frequency conductors is usually entirely negligible compared with the effective resistance resulting from the unequal current distribution, and still greater may be, at very high frequency, the effective resistance representing the power radiated into space by the conductor. The total effective resistance, or resistance representing the power consumed by the current in the conductor, thus comprises the true ohmic resistance, the effective resistance of unequal current distribution, and the effective resistance of radiation.

The power consumed by the effective resistance of unequal current distribution in the conductor is converted into heat in the conductor, and this resistance thus may be called the "thermal resistance" of the conductor, to distinguish it from the radiation resistance. The power consumed by the radiation resistance is not converted into heat in the conductor, but is dissipated in the space surrounding the conductor, or in any other conductor on which the electric wave impinges. That is, at very high frequency, the total power consumed by the effective resistance of the conductor does not appear as heating of the conductor, but a large part of it may be sent out into space as electric radiation, which accounts for the power exerted upon bodies near the path of a lightning stroke, as "side discharge."

It demonstrates that safety from lighting is not given by merely affording a discharge path, but while discharging through such path, most of the energy of the lightning may be communicated by radiation to other bodies.

The inductance is reduced by the unequal current distribution in the conductor, which, by deflecting most of the current into the outer layer of the conductor, reduces or practically eliminates the magnetic field inside of the conductor. The lag of the magnetic field in space, behind the current in the conductor, due to the finite velocity of radiation, also reduces the inductance to less than that from the conductor surface to a distance of one-half wave. An exact determination of the inductance is, however, not possible; the inductance is represented by the electromagnetic field of the conductor, and this depends upon the presence and location of other conductors, etc., in space, on the length of the conductor, and the distance from the return conductor. Since very high frequency currents, as lightning discharges, frequently have no return conductor, but the capacity at the end of the discharge path returns the current as "displacement current," the extent and distribution of the magnetic field is indeterminate. If, however, the conductor under consideration is a small part of the total discharge — as the ground connection of a lightning arrester, a small part of the discharge path from cloud to ground — and the frequency very high, so that the wave length is relatively short, and the space covered by the first half wave thus is known to be free of effective return conductors, the magnitude of the inductance can be calculated with fair approximation by assuming the conductor as a finite section of a conductor without return conductor.

85. In long distance transmission lines and other electric power circuits, disturbances leading to the appearance of high frequency currents may be either between the lines—such as caused by switching, sudden changes of load, spark discharges or short circuits between conductors, etc. Or they may be between line and ground, such as caused by lightning, by arcing grounds, short circuits to ground, with grounded neutral, etc.

In the former case, high frequency currents between the line conductors, the electric field is essentially contained between the line conductors, in a space which is usually practically free of other conductors. The effect of the finite velocity of the field on the inductance or rather the impedance of the conductor, the radiation resistance, etc., can be well approximated by the equations of a finite section of an infinitely long conductor, having its return conductor at a distance equal to the distance between the transmission conductors (Chapter VIII).

In the case of a high frequency disturbance between line and ground, all the line conductors may share in the conduction, that is carry current simultaneously in the same direction, as frequently the case with lightning discharges, etc. The impedance then is the joint impedance of all the line conductors (about one-third that of one conductor, with a three-wire line); the field is between the line and the ground, and usually fairly free of conducting bodies. Thus the radiation resistance, etc., can be calculated under the assumption of the image conductor as return conductor, that is a return conductor at a distance equal to twice the height of the line.

If the high frequency disturbance originates between one line conductor and the ground, as usual with arcing grounds, and occasionally with lightning discharges, etc., the high frequency field is between this conductor and the ground as return conductor, but the other line conductors (and other parallel circuits, as telephone lines, etc.) are within the high frequency field, and currents are induced in them by mutual induction. These currents, being essentially in reverse direction to the inducing current, act as partial return current, and the constants, as radiation resistance, etc., are intermediate between those of the two cases previously discussed.

With regards to unequal current distribution in the conductor, obviously the existence and location of the return conductor is of no moment.

In many cases therefore, for the two extremes—low frequency, where unequal current distribution and radiation are negligible, and very high frequency, where the current traverses only the outer layer and the total effect, contained within one wave length, is within a moderate distance of the conductor—the constants can be calculated; but for the intermediary case, of moderately high frequency, the conductor constants may be anywhere between the two limits, i.e., the low frequency values and the values corresponding to an infinitely long conductor without return conductor.

Since, however, the magnitude of the conductor constants, as derived from the approximate equations of unequal current distribution and of radiation, are usually very different from the low frequency values, their determination is of interest even in the case of intermediate frequency, as indicating an upper limit of the conductor constants.

86. Using the following symbols, namely,

l = the length of conductor,

A = the sectional area,

l_1 = the circumference at conductor surface, that is, following all the indentations of the conductor,

l_2 = the shortest circumference of the conductor, that is, circumference without following its indentations,

l_r = the radius of the conductor,

l' = the distance from the return conductor,

λ = the conductivity of conductor material,

μ = the permeability of conductor material,

f = the frequency,

S = the speed of light $= 3 \times 10^{10}$ cm., and

$$a = \frac{2\pi f}{S} = 2.09 f\, 10^{-10} = \text{ the wave length constant,} \qquad (1)$$

At low frequency, the current density throughout the conductor section is uniform, and its resistance is the true ohmic resistance:

$$r_0 = \frac{l}{\lambda A} \text{ ohms.} \qquad (1)$$

The external reactance, that is, reactance due to the magnetic field outside of the conductor, is at low frequency, where the finite velocity of the magnetic field can be neglected, given by:

$$x_0 = 4\, \pi f l \log \frac{l'}{l_r}\, 10^{-9} \text{ ohms} \qquad (2)$$

or, reducing to common logarithms, by multiplying with log 10 = 2.3026:

$$x_0 = 9.21\,\pi f l \lg \frac{l'}{l_r}\, 10^{-9} \text{ ohms} \qquad (3)$$

where lg may denote the common, log the natural logarithm.

In addition to the external reactance, there exists an internal reactance, due to the magnetic flux inside of the conductor. At low frequency, where the current density in the conductor is uniform, this is:

$$x_0' = \pi f \mu l\, 10^{-9} \text{ ohms} \qquad (4)$$

and the total low frequency impedance thus is:

$$\left.\begin{aligned} Z_0 &= r_0 + j(x_0 + x_0') \\ &= l\left\{\frac{1}{\lambda A} + j\pi f\left(4 \lg \frac{l'}{l_r} + \mu\right) 10^{-9}\right\} \\ &= l\left\{\frac{1}{\lambda A} + j\pi f\left(9.21 \lg \frac{l'}{l_r} + \mu\right) 10^{-9}\right\} \text{ ohms} \end{aligned}\right\} \qquad (5)$$

and the low frequency inductance:

$$\left.\begin{aligned} L_0 &= \frac{x_0 + x_0'}{2\,\pi f} \\ &= l\left(2 \log \frac{l'}{l_r} + \frac{\mu}{2}\right) 10^{-9} \\ &= l\left(4.6 \lg \frac{l'}{l_r} + \frac{\mu}{2}\right) 10^{-9} \text{ henry} \end{aligned}\right\} \qquad (6)$$

The magnetic field of the current surrounds this current and fills all the space outside thereof, up to the return current. Some of the magnetic field due to the current in the interior and the center of a conductor carrying current, thus is inside of the conductor, while all the magnetic field of the current in the outer layer of the conductor is outside of it. Therefore, more magnetic field surrounds the current in the interior of the conductor than the current in its outer layer, and the inductance therefore increases from the outer layer of the conductor towards its interior, by the "internal magnetic field." In the interior of the conductor, the reactance voltage thus is higher than on the outside. At low frequency, with moderate size of conductor, this difference is inappreciable in its effect. At higher frequencies, however, the higher reactance in the interior of the conductor, due to this internal magnetic field, causes the current density to decrease towards the interior of the conductor, and the current to lag, until finally the current flows practically only through a thin layer of the conductor surface.

As the result hereof, the effective resistance of the conductor is increased, due to the uneconomical use of the conductor material caused by the lower current density in the interior, and due to the phase displacement between the currents in the successive layers of the conductor, which results in the sum of the currents in the successive layers of the conductor being larger than the resultant current. Due to this unequal current distribution, the internal reactance of the conductor is decreased, as less current penetrates to the interior of the conductor, and thus produces less magnetic field inside of the conductor.

The equivalent *depth of penetration* of the current into the conductor, from Chapter VII, (40), is

$$l_p = \frac{10^4}{\pi\sqrt{0.4\,\lambda\mu f}} = \frac{5030}{\sqrt{\lambda\mu f}}; \tag{7}$$

hence, the effective *resistance of unequal current distribution*, or *thermal resistance* of the conductor, is, approximately,

$$r_1 = \frac{l}{\lambda l_p l_1} = \frac{l\pi}{l_1}\sqrt{\frac{0.4\,\mu f}{\lambda}}\,10^{-4} = \frac{1.98 l}{l_1}\sqrt{\frac{\mu f}{\lambda}}\,10^{-4} \text{ ohms}, \tag{8}$$

The *effective reactance of the internal flux*, at high frequency, approaches the value:

$$x_1 = r_1 = \frac{1.98\,l}{l_1}\sqrt{\frac{\mu f}{\lambda}}\,10^{-4} \text{ ohms}. \tag{9}$$

87. The effective resistance resulting from the finite velocity of the electric field, or *radiation resistance*, by assuming the conductor as a section of an infinitely long *conductor without return conductor*, from Chapter VIII, (23), is

$$r_2 = 2\,l\pi^2 f\,10^{-9} = 1.97\,lf\,10^{-8} \text{ ohms}, \tag{10}$$

and the *effective reactance of the external field* of a finite section of an infinitely long round conductor without return conductor, from Chapter VIII, (25), is

$$x_2 = 4\,\pi f l\left(\log\frac{1}{a l_r} - 0.5772\right)10^{-9} \text{ ohms}. \tag{11}$$

and, substituting (7),

$$\left.\begin{aligned} x_2 &= 4\pi fl\left(\log\frac{S}{2\pi fl_r} - 0.5772\right)10^{-9}\text{ ohms} \\ &= 4\pi fl\,(21.72 - \log l_r f)\,10^{-9}\text{ ohms}\end{aligned}\right\} \tag{12}$$

or, substituting the common logarithm: log = 2.303 lg, gives:

$$x_2 = 2.89\,fl(9.45 - \text{lg}\ l_r f)\,10^{-8}\text{ ohms} \tag{13}$$

Assuming now that the external magnetic field of a conductor of any shape is equal to that of a round conductor having the same minimum circumference, as is approximately the case, that is, substituting:

$$l_2 = 2\,\pi l_r$$

in equations (12) and (13), gives

$$\left.\begin{aligned} x_2 &= 4\pi fl\left(\log\frac{S}{l_2 f} - 0.5772\right)10^{-9}\text{ ohms} \\ &= 2.89fl\,(10.25 - \text{lg}\ l_2 f)\,10^{-8}\text{ ohms}\end{aligned}\right\} \tag{14}$$

While the case of a conductor without return conductor may be approximated under some conditions, such as lightning discharges, under other conditions, such as high frequency disturbances in transmission lines, the case of a conductor with return conductor at finite distance l' is more representative.

The effective radiation resistance and reactance of a section of an infinitely long *conductor with return conductor at distance* l', are, by Chapter VIII (42), and by (45) (44):

Radiation resistance,

$$r_3 = 4\pi fl\left(\frac{\pi}{2} - \text{col}\ al'\right)10^{-9}\text{ ohms}$$

or, substituting (1),

$$r_3 = 4\pi fl\left(\frac{\pi}{2} - \text{col}\ \frac{2\pi fl'}{S}\right)10^{-9}\text{ ohms} \tag{15}$$

and, if the distance of the return conductor, l', is small compared to the wave length, this becomes

$$\left.\begin{aligned} r_3 &= \frac{8\pi^2 f^2 l' l}{S}\,10^{-9}\text{ ohms} \\ &= 2.63\,f^2 l' l\,10^{-18}\text{ ohms}\end{aligned}\right\} \tag{16}$$

Radiation reactance,

$$x_3 = 4\pi fl\left(\log\frac{1}{al_r} - 0.5772 - \text{sil}\ al'\right)10^{-9}\text{ ohms} \tag{17}$$

or, substituting (7),

$$\left.\begin{aligned} x_3 &= 4\pi f l\left(\log\frac{S}{2\pi f l_r} - 0.5772 - \operatorname{sil}\frac{2\pi f l'}{S}\right)10^{-9}\text{ ohms} \\ &= 4\pi f l\left(21.72 - \log l_r f - \operatorname{sil}\frac{2\pi f l'}{S}\right)10^{-9}\text{ ohms}\end{aligned}\right\} \quad (18)$$

and, if the distance of the return conductor, l′, is small compared to the wave length, (17) becomes the ordinary low frequency reactance formula:

$$x_3 = 4\pi f l \log\frac{l'}{l_r}\,10^{-9}\text{ ohms} = 4\pi f l\log\frac{2\pi l'}{l_2}\,10^{-9}\text{ ohms} \quad (19)$$

88. The total *impedance* of a conductor for high frequencies is therefore:

Conductor without return conductor:

$$Z = (r_1 + r_2) + j\,(x_1 + x_2)$$

$$= l\left\{\left[\frac{\pi}{l_1}\sqrt{\frac{0.4\mu f}{\lambda}}\,10^{-4} + 2\pi^2 f\,10^{-9}\right] + j\left[\frac{\pi}{l_1}\sqrt{\frac{0.4\mu f}{\lambda}}\,10^{-4} + 4\pi f\left(\log\frac{S}{l_2 f} - 0.5772\right)10^{-9}\right]\right\}\text{ ohms}$$

$$= l\left\{\left[\frac{1.98}{l_1}\sqrt{\frac{\mu f}{\lambda}}\,10^{-4} + 1.97 f\,10^{-8}\right] + j\left[\frac{1.98}{l_1}\sqrt{\frac{\mu f}{\lambda}}\,10^{-4} + 2.89 f\,(10.25 - \lg l_2 f)\,10^{-8}\right]\right\}\text{ ohms} \quad (20)$$

Conductor with return conductor at distance l′:

$$Z = (r_1 + r_3) + j\,(x_1 + x_3)$$

$$= l\left\{\left[\frac{\pi}{l_1}\sqrt{\frac{0.4\mu f}{\lambda}}\,10^{-4} + \frac{8\pi^2 f^2 l'}{S}\,10^{-9}\right] + j\left[\frac{\pi}{l_1}\sqrt{\frac{0.4\mu f}{\lambda}}\,10^{-4} + 4\pi f\log\frac{l'}{l_r}\,10^{-9}\right]\right\}$$

$$= l\left\{\left[\frac{1.98}{l_1}\sqrt{\frac{\mu f}{\lambda}}\,10^{-4} + 2.63 f^2 l'\,10^{-18}\right] + j\left[\frac{1.98}{l_1}\sqrt{\frac{\mu f}{\lambda}}\,10^{-4} + 2.89 f\lg\frac{l'}{l_r}\,10^{-8}\right]\right\}\text{ ohms} \quad (21)$$

or, if l' is of the same or higher magnitude as the wave length,

$$Z = (r_1 + r_3) + j\,(x_1 + x_3)$$

$$= l\left\{\left[\frac{\pi}{l_1}\sqrt{\frac{0.4\mu f}{\lambda}}\,10^{-4} + 4\pi f\left(\frac{\pi}{2} - \operatorname{col}\frac{2\pi f l'}{S}\right)10^{-9}\right] + j\left[\frac{\pi}{l_1}\sqrt{\frac{0.4\mu f}{\lambda}}\,10^{-4} + 4\pi f\left(\log\frac{S}{2\pi f l_r} - 0.5772 - \operatorname{sil}\frac{2\pi f l'}{S}\right)10^{-9}\right]\right\}$$

$$= l\left\{\left[\frac{1.98}{l_1}\sqrt{\frac{\mu f}{\lambda}}\,10^{-4} + 4\pi f\left(\frac{\pi}{2} - \operatorname{col}\frac{2\pi f l'}{S}\right)10^{-9}\right] + j\left[\frac{1.98}{l_1}\sqrt{\frac{\mu f}{\lambda}}\,10^{-4} + 4\pi f\left(21.72 - \log l_r f - \operatorname{sil}\frac{2\pi f l'}{S}\right)10^{-9}\right]\right\}\text{ ohms} \quad (22)$$

The *inductance* $L = \dfrac{x}{2\pi f}$ is:

Conductor without return conductor,

$$L = l\left\{\frac{1}{l_1}\sqrt{\frac{0.1\mu}{\lambda f}}\,10^{-4} + 2\left(\log\frac{S}{l_2 f} - 0.5772\right)10^{-9}\right\}\text{ henry} \quad (23)$$

Conductor with return conductor at distance l',

$$L = l\left\{\frac{1}{l_1}\sqrt{\frac{0.1\mu}{\lambda f}}\,10^{-4} + 2\log\frac{l'}{l_r}\,10^{-9}\right\}\text{ henry} \quad (24)$$

or

$$\left.\begin{aligned} L &= l\left\{\frac{1}{l_1}\sqrt{\frac{0.1\mu}{\lambda f}}\,10^{-4} + 2\left(\log\frac{S}{2\pi f l_r} - 0.5772 - \operatorname{sil}\frac{2\pi f l'}{S}\right)10^{-9}\right\}\text{ henry} \\ &= l\left\{\frac{0.316}{l_1}\sqrt{\frac{\mu}{\lambda f}}\,10^{-4} + 2\left(21.72 - \log l_r f - \operatorname{sil}\frac{2\pi f l'}{S}\right)10^{-9}\right\}\text{ henry}\end{aligned}\right\} \quad (25)$$

89. As an instance may be considered the high frequency impedance of a copper wire No. 00 B. & S. G., that is of the radius,

$$l_r = 0.1825\text{ in.} = 0.463\text{ cm.}$$

under the three conditions:

(*a*) Return conductor at $l' = 6$ ft. $= 182$ cm. distance, corresponding to transmission line conductors oscillating against each other.

(*b*) With the ground as return conductor, at 30 ft. distance, that is, $l' = 2 \times 30$ ft. $= 1820$ cm., corresponding to a transmission line conductor oscillating against ground.

(*c*) No return conductor, corresponding to the vertical discharge path of a lightning stroke.

With copper as conductor material, it is:

$$\lambda = 6.2 \times 10^5$$
$$\mu = 1$$

It is then, by the preceding equations, per meter length of conductor, or

$$l_0 = 100 \text{ cm.}$$

Low frequency values:
true ohmic resistance, (1),

$$r_0 = \frac{100}{\lambda \pi l_r^2} = 0.24 \times 10^{-3} \text{ ohms}$$

external reactance, (2):

$$x_0 = 0.4\, \pi f \log \frac{l'}{l_r}\, 10^{-6} \text{ ohms}$$

hence,

(*a*)	$l' = 182$:	$x_0 = 7.5\, f\, 10^{-6}$ ohms
(*b*)	$= 1820$:	$x_0 = 10.4\, f\, 10^{-6}$ ohms
(*c*)	$= \infty$:	$x_0 = \infty$

internal reactance, (4):

$$x_0' = 0.1\, \pi f\, 10^{-6} = 0.314\, f\, 10^{-6} \text{ ohms}$$

High frequency values:
thermal resistance, (8):

$$r_1 = \frac{100\,\pi}{l_1}\sqrt{\frac{0.4\,\mu f}{\lambda}}\, 10^{-4} = \frac{50}{l_r}\sqrt{\frac{0.4\,\mu f}{\lambda}}\, 10^{-4} = 8.65\sqrt{f}\; 10^{-6} \text{ ohms}$$

internal reactance, (9):

$$x_1 = r_1 = 8.65\sqrt{f}\, 10^{-6} \text{ ohms}$$

radiation resistance,
(*a*) and (*b*) (16):

$$r_3 = \frac{0.8\,\pi^2 f^2 l'}{S}\,10^{-6} = 263\,f^2 l'\,10^{-18}\text{ ohms}$$

(*a*) $l' = 182$: $r_3 = 0.048\,f^2\,10^{-12}$ ohms
(*b*) $l' = 1820$: $r_3 = 0.48\,f^2\,10^{-12}$ ohms
(*c*) $l' = \infty$ (10):

$$r_2 = 0.2\,\pi^2 f\,10^{-6} = 1.97\,f\,10^{-6}\text{ ohms}$$

radiation reactance,
(*a*) and (*b*) (19):

$$x_3 = 0.4\,\pi f \log\frac{l'}{l_r}\,10^{-6} = \mathcal{H}_0$$

(*a*) $l' = 182$: $x_3 = 7.5\,f\,10^{-6}$ ohms
(*b*) $l' = 1820$: $x_3 = 10.4\,f\,10^{-6}$ ohms
(*c*) $l' = \infty$ (14):

$$x_2 = 0.4\,\pi f\left(\log\frac{S}{l_2 f} - 0.5772\right)10^{-6} = f(28.5 - 2.89\,\lg f)10^{-6}\text{ ohms}$$

For r_3 and x_3, for $f = 10^8$, $l' = 182$, and for $f = 10^7$ and 10^8, $l' = 1820$, the more complete equations (15) and (18) must be used, as l' exceeds a quarter wave length.

Table I gives numerical values, from 1 cycle to 10^8 cycles, of r, x, Z, cos a and the resistance ratios. These values are plotted in Fig. 18, in logarithmic scale.

90. The low frequency values of resistance r_0 and external and internal reactance $x_0 + x_0'$, have no existence at the higher frequencies. But as they are the values calculated by the usual formulas, they are given in Table I for comparison with the true effective high frequency values. The values r_2 and x_2, though given for all frequencies, have a meaning only for the very high frequencies, 10^4 and higher, since at lower frequencies the condition of a conductor without return conductor can hardly be realized, as any conductor within a quarter wave length would act more or less as effective return conductor.

As seen from the equations, and illustrated in Table I, the thermal resistance of the conductor, that is, the resistance which converts electric energy into heat in the conductor, is the true ohmic resistance at low frequencies, but with increasing frequency

TABLE I.—CIRCUIT CONSTANTS OF COPPER WIRE NO. 00 B. & S. G. PER METER LENGTH OF CONDUCTOR.

(a) Return conductor at 6 ft. = 182 cm. distance.
(b) Return conductor at 2 × 30 ft. = 1820 cm. distance.
(c) No return conductor, conductivity: $\lambda = 6.2 \times 10^5$.

		Cycles, f =	1	10	10^2	10^3	10^4	10^5	10^6	10^7	10^8
Low Frequency		r_0 =	0.000240	0.000240	0.000240	0.000240	0.000240	0.000240	0.000240	0.000240	0.000240
		x_{10} =	0.0000003	0.000003	0.00003	0.0003	0.003	0.03	0.3	3	30
	(a)	x_0 =	0.0000075	0.000075	0.00075	0.0075	0.075	0.75	7.5	75	750
	(b)		0.0000104	0.000104	0.00104	0.0104	0.104	1.04	10.4	104	1040
	(a)	z_0 =	0.000240	0.000252	0.000816	0.0078	0.078	0.78	7.8	78	780
	(b)		0.000240	0.000263	0.001097	0.0107	0.107	1.07	10.7	107	1070
	(a)	$\cos \omega_0$ =	1.000	0.952	0.293	0.0308	0.0031	0.00031	0.000031	0.000003	0.0000003
	(b)		1.000	0.912	2.218	0.225	0.0022	0.00022	0.000022	0.000002	0.0000002
High Frequency		r_1 =	0.000240	0.000240	0.000240	0.000273	0.000865	0.00273	0.00865	0.0273	0.0865
		x_1 =	0.0000003	0.000003	0.000030	0.000273	0.000865	0.00273	0.00865	0.0273	0.0865
	(a)	r_3 =	....	....	....	....	0.0000048	0.00048	0.048	4.8	225
	(b)		....	....	....	....	0.000048	0.0048	0.48	22.5	194
	(c)	r_2 =	0.000002	0.000020	0.000197	0.00197	0.0197	0.197	1.97	19.7	197
	(a)	x_3 =	0.0000075	0.000075	0.00075	0.0075	0.075	0.75	7.5	75	490
	(b)		0.0000104	0.000104	0.00104	0.0104	0.104	1.04	10.4	78	505
	(c)	x_2 =	0.0000285	0.000256	0.00227	0.0198	0.169	1.40	11.2	83	515
	(a)	z =	0.000240	0.000252	0.000816	0.00778	0.759	0.753	7.5	75.2	539
	(b)		0.000240	0.000263	0.001097	0.01068	0.1049	1.043	10.4	81.2	541
	(c)		0.000242	0.000367	0.00234	0.0201	0.170	1.42	11.4	85.3	551
	(a)	$\cos \omega$ =	1.000	0.952	0.293	0.0351	0.0115	0.0043	0.0076	0.0642	0.418
	(b)		1.000	0.912	0.218	0.0257	0.0087	0.0072	0.047	0.277	0.358
	(c)		1.000	0.710	0.186	0.112	0.121	0.141	0.173	0.232	0.358
	(a)	$\frac{r_1 + r_3}{r_0}$ =	1.000	1.000	1.000	1.14	3.62	13.4	236	20,100	937,000
	(b)		1.000	1.000	1.000	1.14	3.80	31.3	2030	94,000	808,000
	(c)	$\frac{r_1 + r_2}{r_0}$ =	1.008	1.082	1.82	9.35	85.5	832	8250	82,000	820,000

begins to rise due to unequal current distribution in the conductor at about 1000 cycles in copper wire number 00 B. & S. G., and approaches proportionality with the square root of the frequency, hence reaches values many times the ohmic resistance, at very high frequencies.

The radiation resistance of the conductor without return conductor, r_2, is proportional to the frequency, but the radiation resistance of the conductor with return conductor, r_3, is proportional to the square of the frequency, hence very small until high frequencies are reached—10,000 to 100,000 cycles. The radiation resistance r_3 of the conductor with return conductor then increases very rapidly and reaches values many thousand times the ohmic resistance. At the very highest frequencies, many millions of cycles, its rate of increase becomes less again, and it approaches proportionality with the first power of the frequency, and approaches the value of radiation resistance r_2, of the conductor without return conductor, at frequencies of a wave length comparable with the distance of the return conductor. The radiation resistance r_3 of the conductor with return conductor is the larger, the greater the distance of the return conductor, and is proportional to this distance, within the range in which it is proportional to the square of the frequency. The radiation resistance of the conductor without return conductor, at the very highest frequencies, is the same as that of the conductor with return conductor, but, being proportional to the frequency, with decreasing frequency, it decreases at a lesser rate, and would even at commercial machine frequencies still be appreciable, if at such frequencies the conditions of a conductor without return conductor could be realized.

The total effective resistance of a conductor under transmission line conditions, that is, with return conductor at finite distance, is at low frequencies constant and is the true ohmic resistance. With increasing frequency, it begins to increase first slowly—at about 1000 cycles under transmission line conditions—and approaches proportionality to the square root of the frequency, as the result of the screening effect of the unequal current distribution in the conductor. Then the increase becomes more rapid, due to the appearance of the radiation resistance—at about 100,000 cycles under transmission line conditions—and reaches proportionality with the square of the frequency, at values many thousand times the ohmic resistance. Finally, at the very highest frequencies—10 million cycles—the rate of increase becomes less again, and approaches proportionality with the frequency.

91. It is interesting to note that the external reactance of the conductor with return conductor, or radiation reactance x_3, has up to very high frequencies, millions of cycles, the same value x_0 as calculated by the low frequency formula, that is, by neglecting the finite velocity of the field, hence is proportional to the frequency. The internal reactance x_1 is $= x_0'$, and proportional to the frequency at low frequencies, but drops behind x_0' due to unequal current distribution in the conductor, and approaches proportionality with the square root of the frequency. As, however, the internal reactance is a small part of the total reactance, it follows, that the total reactance of the conductor and thus also the absolute value of the impedance (for all higher frequencies, where the reactance preponderates) can be calculated by the usual low frequency reactance formula, which neglects the finite velocity of the field. Hence, the inductance L of the conductor can be assumed as constant for all frequencies up to millions of cycles; it decreases only very slowly by the decreasing internal reactance of unequal current distribution. Only at the very highest frequencies, where the wave length is comparable with the distance of the return conductor, the inductance L decreases, and the reactance $x_1 + x_3$ increases less then proportional to the frequency.

In a conductor without return conductor, the reactance at the very highest frequencies is approximately the same as in a conductor with return conductor. With decreasing frequency, however, $x_1 + x_2$ decreases less than proportional to the frequency, that is, the inductance L increases—and becomes infinity for zero frequency, if such were possible.

Without considering unequal current distribution in the conductor and the finite velocity of electric field, the power factor $\cos \omega$ would steadily decrease, from unity at very low frequencies, to zero at infinite frequency. Due to the increase of the effective resistance, the power factor $\cos \omega$ first decreases, from unity at low frequency, down to a minimum at some high frequency, and then increases again to high values at very high frequencies. The minimum value of the power factor is the lower and occurs at the higher frequencies, the shorter the distance of the return conductor is. Thus with the return conductor at 6 ft. distance, the power factor is 0.43 per cent at 100,000 cycles; with the return conductor at 60 ft. distance, it is 0.72 per cent; in the conductor without return conductor the power factor is 11.2 per cent at 1000 cycles.

92. It is of interest to determine the effect of size, shape and material on the high-frequency constants of a conductor.

These high-frequency constants are, per unit length of conductor:

Internal constants: Thermal resistance and internal reactance:

$$r_1 = x_1 = \frac{\pi}{l_1}\sqrt{\frac{0.4\,\mu f}{\lambda}}\,10^{-4} \text{ ohms per cm.} \qquad (8)\ (9)$$

External constants: Radiation resistance:

$$r_3 = \frac{8\,\pi^2 f^2 l'}{\lambda}\,10^{-9} \text{ ohms per cm.} \qquad (16)$$

External reactance:

$$x_3 = 4\,\pi f \log \frac{2\pi l'}{l_2}\,10^{-9} \text{ ohms per cm.} \qquad (19)$$

These approximations hold for all but the very highest, and very low frequencies, that is, are correct within the frequency range with lower limit of about 1000 to 10,000 cycles, and upper limit of about 10 million cycles. Thus they apply for all those high frequencies which are of importance in the disturbances occurring in industrial circuits, with the exception of the lowest harmonics of low-frequency surges.

The constants of the conductor material enter the equations only as the ratio $\left(\frac{\mu}{\lambda}\right)$, permeability to conductivity, in the internal constants r_1 and x_1. Thus higher permeability has the same effect in increasing the thermal resistance as lower conductivity, and for instance, a cast silicon rod of permeability $\mu = 1$, and conductivity $\lambda = 55$, has the same high-frequency resistance and reactance, as a rod of the same size, of wrought iron, of permeability $\mu = 2000$ and conductivity $\lambda = 1.1 \times 10^5$, that is, of the same $\frac{\mu}{\lambda} = 0.0182$, though the latter has 2000 times the conductivity of the former.

Provided, however, that size of conductor and frequency are such as to fulfill the conditions under which equations (8) and (9) are applicable, which is, that the conductor is large compared with the depth of penetration of the current into the conductor:

$$l_p = \frac{10^4}{\pi\sqrt{0.4\ \lambda\mu f}}.$$

Thus an iron rod of 2 inches (5 cm.) diameter has at one million cycles the same thermal resistance as a silicon rod of the same size: 0.17 ohms per meter, since the depth of penetration is $l_p =$ 0.00034 cm. for iron, 0.68 cm. for silicon, thus in either case small compared with the radius of the conductor $l_r = 2.5$ cm.

At 10,000 cycles, however, the iron rod has the thermal resistance and internal reactance $r_1 = x_1 = 0.017$ ohms per meter, the penetration being $l_p = 0.0034$ cm., thus small. For the silicon rod, however, at 10,000 cycles the penetration is $l_p = 6.8$ cm., thus at the radius $l_r = 2.5$ cm. formulas (8) and (9) do not apply any more, but it is approximately (that is neglecting unequal current distribution): $r_1 = 0.093$ ohms per meter, or 5.5 times the resistance of the iron rod, while the internal reactance is $x_1 =$ 0.031 ohms per meter, hence 80 per cent. higher than that of the iron rod of the same size.

93. In the equations of the external constants, the radiation resistance and reactance, the material constants of the conductor do not enter, and the radiation resistance and the external reactance thus are independent of the conductor material.

The dimensional constants of the conductor, size and shape, enter the equation only as the circumference of the conductor l_1, l_2, that is, only the circumference of the conductor counts in high-frequency conduction, and all conductors of the same material, regardless of size and shape, have the same high-frequency resistances and reactances as long as they have the same conductor circumference. Thus a solid copper rod or a thin copper cylinder of the same outer diameter as the rod, or a flat copper ribbon of a circumference equal to that of the rod, are equally good high-frequency conductors, though the hollow cylinder or the ribbon may contain only a small part of the material contained in the solid copper rod. Provided, however, that the thickness or depth of the conductor (the thickness of wall of the hollow cylinder, half the thickness of the copper ribbon) is larger than the depth of penetration of current into the conductor, which is

$$l_p = \frac{10^4}{\pi\sqrt{0.4\,\lambda\mu f}}$$

In the expression of the radiation resistance, r_3, neither the material nor the dimensions of the conductor enter, that is, the radiation resistance of a conductor is independent of size, shape or material of the conductor and depends only on frequency and distance of the return conductor.

Thus a thin steel wire or a wet string have the same radiation resistance as a large copper bar. Obviously, the thermal resistance of the former is much larger and the total effective resistance thus would be larger except at those very high frequencies, at which the radiation resistance dominates.

94. As illustration may be calculated, for frequencies from 10 thousand to 10 million cycles and for 60 cycles, the resistances and reactances and thus the total impedance, the power factor, the voltage drop per meter at 100 amperes, of various conductors, for the three conditions:

(*a*) High frequency between conductors 6 ft. apart:

$$l' = 182 \text{ cm.}$$

(*b*) High frequency between conductor and ground 30 ft. below conductor:

$$l' = 1820 \text{ cm.}$$

(*c*) High-frequency discharge through vertical conductor without return conductor:

$$l' = \infty$$

For the conductors:

(1) Copper wire No. 00 B. & S. G.

$$l_r = 0.463 \text{ cm.} \quad \lambda = 6.2 \times 10^5$$

(2) Iron wire of the same size as (1): $\mu = 2000$

$$\lambda = 1.1 \times 10^5$$

(3) Copper ribbon of thickness equal to twice the depth of penetration at 10,000 cycles, and the same amount of material as (1)

(4) Iron ribbon of the same size as (3).

(5) Two inch iron pipe, ⅛ inch thick.

This gives the depth of penetration at $f = 10^4$ cycles:

for copper, $$l_p = \frac{6.4}{\sqrt{f}} = 0.064 \text{ cm.}$$

for iron, $$l_p = \frac{0.34}{\sqrt{f}} = 0.0034 \text{ cm.}$$

It thus is:

(1) and (2) copper wire: $l_1 = l_2 = 2\,\pi l_r = 2.91$ cm.

(3) The area of the copper wire is $A = \pi l_r^2 = 0.672$ cm.^{2} Twice the depth of penetration is: $2\,l_p = 0.128$ cm., hence the thickness of the copper ribbon of equal weight with the wire is 0.128 cm. = 0.05 inches, the width is $l_3 = 5.25$ cm. or about two inches. The circumference then is:

$$l_1 = l_2 = 10.75 \text{ cm. or 3.7 times that of the wire.}$$

(4) The same dimensions as (3).

(5) $l_r = 1$ inch $= 2.54$ cm.

$l_1 = l_2 = 16$ cm.

$A = 4.6$ cm.^{2} = 6.85 that of (1) to (4).

Table II gives the values of r_0, r_1, r_3, r_2, x_0, x_1, x_3, x_2, $r_1 + r_3$, $r_1 + r_2$, $x_1 + x_3$, $x_1 + x_2$, z, $\cos \omega$, e and l_p, for $f = 10^6$ cycles.

Table III gives the values of $r_1 + r_3$ or $r_1 + r_2$, z, $\cos \omega$ and e, for $f = 60$, 10^4, 10^5, 10^6, 10^7 cycles for the five kinds of conductors.

95. It is interesting to compare in Table III, the constants of the first four conductors, as they have the same section, representing about average section of transmission conductors, but represent two shapes, round wire and thin flat ribbon, and two kinds of material, copper—high conductivity and non-magnetic—and iron—magnetic material of medium conductivity.

As seen, the effect of conductor shape and conductor material is very great at machine frequencies, 60 cycles, but becomes small and almost negligible at extremely high frequencies. This is rather against the usual assumption.

TABLE II.—CIRCUIT CONSTANTS FOR $f = 10^6$ = ONE MILLION CYCLES, PER METER LENGTH OF CONDUCTOR.

(*a*) Return conductor at 6 ft. = 182 cm. distance.
(*b*) Return conductor at 2 × 30 ft. = 1820 cm. distance.
(*c*) No return conductor.

		(1) Copper Wire.	(2) Iron Wire.	(3) Copper Ribbon.	(4) Iron Ribbon.	(5) Iron Pipe.	
$r_0 = \frac{100}{A\lambda} =$		0.00024	0.00135	0.00024	0.00135	0.000197	
$x_0 = 0.4\,\pi f\left(\log\frac{2\,\pi f l'}{l_2} + \frac{l_0}{l_2}\frac{\mu}{4}\right)10^{-6} =$	(*a*) (*b*)	7.8 10.8	635 638	3.05 5.95	173 176	81 84	Ohms Ohms
$r_1 = \frac{100\,\pi}{l_1}\sqrt{\frac{0.4\,\mu f}{\lambda}}\,10^{-4} =$		0.00865	0.92	0.00233	0.248	0.167	Ohms
$r_3 = \frac{0.8\,\pi^2 f^2 l'}{S}\,10^{-6} =$	(*a*) (*b*)	0.048 0.48	0.048 0.48	0.048 0.48	0.048 0.48	0.048 0.48	Ohms Ohms
$r_2 = 0.2\,\pi^2 f\,10^{-6} =$	(*c*)	1.97	1.97	1.97	1.97	1.97	Ohms
$x_1 = \frac{100\,\pi}{l_1}\sqrt{\frac{0.4\,\mu f}{\lambda}}\,10^{-4} =$		0.00865	0.92	0.00233	0.248	0.167	Ohms
$x_3 = 0.4\,\pi f\log\frac{2\pi f l'}{l_2}\,10^{-6} =$	(*a*) (*b*)	7.5 10.4	7.5 10.4	2.97 5.87	2.97 5.87	2.47 5.37	Ohms Ohms
$x_2 = 0.4\,\pi f\left(\log\frac{S}{l_2 f} - 0.5772\right)10^{-6} =$	(*c*)	11.2	11.2	9.5	9.5	8.9	Ohms

$r = r_1 + r_3 =$	(a)	0.0567	0.97	0.0503	0.296	0.215	Ohms
	(b)	0.489	1.40	0.482	0.728	0.647	Ohms
$r = r_1 + r_2 =$	(c)	1.98	2.89	1.97	2.22	2.14	Ohms
$x = x_1 + x_3 =$	(a)	7.5	8.42	2.97	3.22	2.64	Ohms
	(b)	10.4	11.32	5.87	6.12	5.54	Ohms
$x = x_1 + x_2 =$	(c)	11.2	12.12	9.50	9.75	9.07	Ohms
$z = \sqrt{r^2 + x^2} =$	(a)	7.5	8.46	2.97	3.23	2.65	Ohms
	(b)	10.4	11.4	5.89	6.16	5.58	Ohms
	(c)	11.4	12.45	9.7	10.0	9.3	Ohms
$\cos \omega = \frac{r}{z} =$	(a)	0.76	11.5	1.7	9.2	8.1	Per cent.
	(b)	4.7	12.3	8.2	11.8	11.6	Per cent.
	(c)	17.3	23.3	20.3	22.2	23.0	Per cent.
$e = 100\,z =$	(a)	750	846	296	323	265	Volts
	(b)	1040	1140	589	616	558	Volts
	(c)	1140	1245	970	1000	930	Volts
$l_p = \frac{10^1}{\pi \sqrt{0.4\lambda\mu f}} =$		0.0064	0.00034	0.0064	0.00034	0.00034	Cm.

TABLE III.

	Conductors 6 Ft. apart.					Conductor 30 Ft. above Ground.					Conductor without Return Conductor.				
Frequency, cycles:	60	10^4	10^5	10^6	10^7	60	10^4	10^5	10^6	10^7	60	10^4	10^5	10^6	10^7
Resistance, r, in ohms per meter;															
(1) Copper wire	0.00024	0.00087	0.0052	0.0567	4.827	0.00024	0.00091	0.0075	0.489	22.5	....	0.0206	0.200	1.98	19.7
(2) Iron wire	0.0072	0.092	0.292	0.97	7.72	0.0072	0.092	0.297	1.40	25.4	....	0.112	0.489	2.89	20.0
(3) Copper ribbon	0.00024	0.00024	0.00122	0.0503	4.81	0.00024	0.00028	0.0055	0.482	22.5	....	0.0199	0.198	1.97	19.7
(4) Iron ribbon	0.0019	0.0248	0.079	0.296	5.58	0.0019	0.0248	0.083	0.728	23.3	....	0.0435	0.265	2.22	20.5
(5) Iron pipe	0.0013	0.0167	0.053	0.215	5.33	0.0013	0.0167	0.058	0.647	23.0	....	0.0364	0.250	2.14	20.2
Impedance, z, in ohms per meter.															
(1) Copper wire	0.00053	0.076	0.753	7.51	75	0.00069	0.105	1.043	10.41	81	....	0.115	1.14	11.4	85
(2) Iron wire	0.0105	0.191	1.083	8.46	78	0.0106	0.218	1.37	11.4	85	....	0.233	1.49	12.45	88
(3) Copper ribbon	0.00030	0.030	0.298	2.97	30	0.00043	0.059	0.588	5.89	49	....	0.097	0.97	9.7	72
(4) Iron ribbon	0.0028	0.060	0.315	3.23	31	0.0030	0.087	0.67	6.16	51	....	0.127	1.06	10.0	74
(5) Iron pipe	0.00195	0.0446	0.305	2.65	26	0.0021	0.072	0.592	5.58	47	....	0.112	0.97	9.3	70
Power factor, cos ω, in per cent.															
(1) Copper wire	45.3	1.14	0.69	0.76	6.4	34.8	0.87	0.72	4.7	27.8	....	17.9	17.5	17.3	23.2
(2) Iron wire	69.5	48.2	27.01	11.5	9.9	68.0	42.2	21.6	12.3	30.0	....	48.0	32.7	23.3	22.7
(3) Copper ribbon	80.0	0.8	0.4	1.7	16.0	53.5	0.48	0.93	8.2	46.0	....	20.5	20.4	20.3	27.3
(4) Iron ribbon	68.0	41.3	25.1	9.2	18.0	63.3	28.6	12.4	11.8	45.8	....	34.3	25.0	22.2	27.7
(5) Iron pipe	66.7	37.5	17.4	8.1	20.4	62.0	23.2	9.8	11.6	49.0	....	32.5	25.8	23.0	29.0
Voltage, e, in volts per meter;															
(1) Copper wire	0.053	7.6	75	751	7500	0.069	10.5	104	1041	8100	....	11.5	114	1140	8500
(2) Iron wire	1.05	19.1	108	846	7800	1.06	21.8	137	1140	8500	....	23.3	149	1245	8800
(3) Copper ribbon	0.03	3.0	30	297	3000	0.043	5.9	59	589	4900	....	9.7	97	970	7200
(4) Iron ribbon	0.28	6.0	36	323	3100	0.30	8.7	67	616	5100	....	12.7	106	1000	7400
(5) Iron pipe	0.195	4.5	30	265	2600	0.21	7.2	59	558	4700	....	11.2	97	930	7000

The reason is, at machine frequencies, the unequal current distribution, or the screening effect of the internal magnetic field, is still practically absent in copper conductors, even in round wires of medium size, and it is practically complete in iron conductors, even in ribbon of $\frac{1}{20}$ inch thickness, while at very high frequencies the effect of radiation preponderates, which is independent of the material, and the radiation resistance even independent of the shape of the conductor.

Thus under transmission line conditions, first and second of Table III, at 60 cycles the impedance, and hence the voltage drop in the iron conductor is from 7 to 20 times that of the copper conductor; at 10,000 cycles the voltage drop in the iron conductor is only from 1.5 to 2.5 times that in the copper conductor; the difference has decreased to from 14 per cent. to 44 per cent. at 100,000 cycles, 5 per cent. to 12.5 per cent. at one million cycles, while at 10 million cycles the voltage drop in the iron conductor is only 3 to 5 per cent. higher, thus practically the same and the only difference is that due to the conductor shape.

The effective resistance, and thus the power consumption in the iron conductor at 60 cycles is from 8 to 30 times that of the copper conductor, but with increasing frequency the difference in the effective resistance increases to from 88 to 106 times at 10,000 cycles, reaches a maximum and then decreases again, and is from 15 to 65 times that of the copper conductor at 100,000 cycles, only $1\frac{1}{2}$ to 17 times at a million cycles, while at 10 million cycles and above all the differences in the effective resistance practically disappear.

As the result, the power factor of the conductor—being the same, 100 per cent., at extremely low frequency and not much different, and fairly high at machine frequency—decreases with increasing frequency, reaches a minimum and then increases again to considerable values at extremely high frequency—where the high radiation resistance comes into play. The difference between iron and copper, however, is that the minimum value of the power factor, at medium high frequencies, is very low in copper, a fraction of 1 per cent., while in the iron conductor the power factor always retains considerable values, the minimum being 10 or more times that of the copper conductor. Thus an oscillation in an iron conductor must die out at a much faster rate than in a copper conductor and the liability of the formation of a continual or cumulative oscillation may exist in copper conductors but hardly in iron conductors.

The effect of the shape of the conductor on the impedance or voltage drop is fairly uniform throughout the entire frequency range, the voltage drop being the smaller, the larger the circumference.

With regard to the effective resistance, however, the effect of the conductor shape is considerable already at very low frequencies in iron conductors, but still absent with copper conductors, due to the absence of the screening effect in copper at low frequency. With increasing frequency, the difference appears in the effective resistance of the copper conductor also, with the appearance of unequal current distribution, and the ratio of the resistance of the round conductor to that of the flat conductor approaches the same value in copper as in iron. With the approach of very high frequency, however, the difference decreases again, with the appearance of radiation effect, and finally vanishes.

96. Thus to convey currents of extremely high frequency, an iron conductor is almost as good as a copper conductor of the same shape and cross section. As iron is very much cheaper than copper, it follows that in high-frequency conduction an iron conductor under the conditions of Table III, should be better than a copper conductor of the same cost and the same general shape, due to the larger size or rather circumference of the iron conductor.

There is, however, a material advantage at extremely high frequencies as well as at moderately high frequencies, in lower voltage drop at the same current resulting from such a shape conductor as gives maximum circumference, such as ribbon or hollow conductor. This advantage of ribbon or hollow tube, over the solid round conductor, exists also in the resistance and thus power consumption at medium high frequencies, but not at extremely high frequencies, but at the latter, in power consumption all conductors, regardless of size, shape or material, are practically equal.

With the thickness of ribbon conductor considered in Table III, of about $\frac{1}{20}$ inch, which is about the smallest mechanically permissible under usual conditions, the screening effect even in copper conductors is practically complete already at 10,000 cycles,

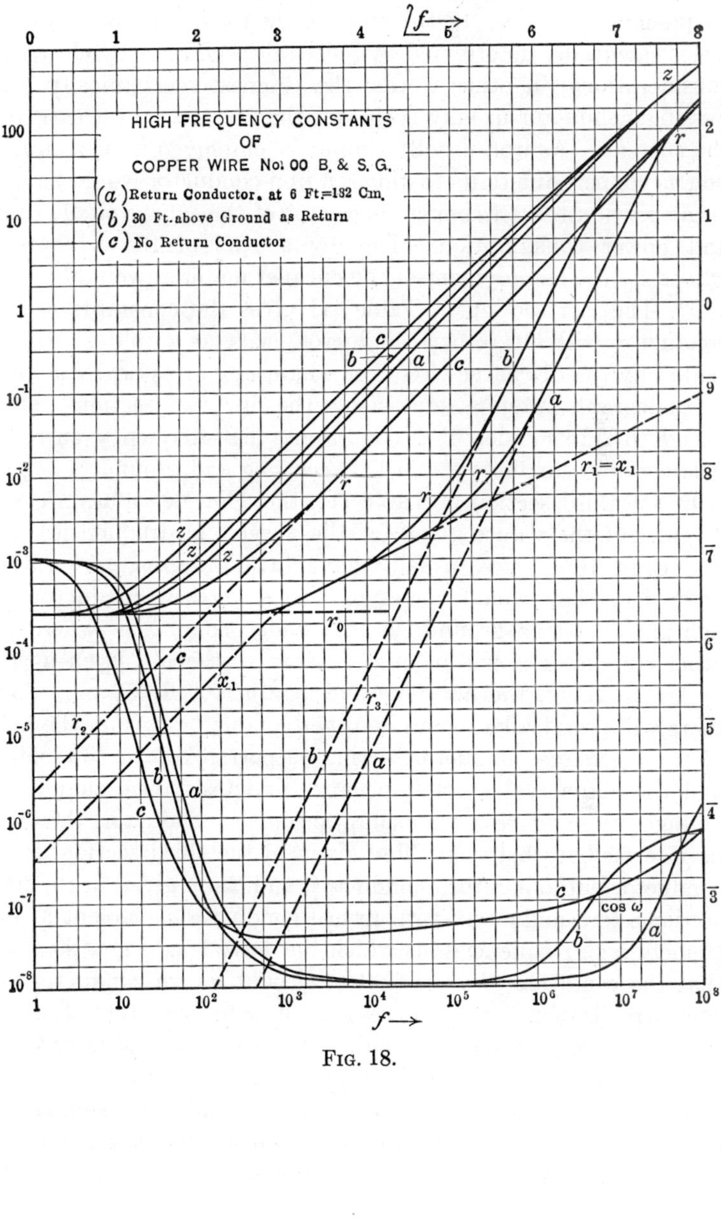
HIGH FREQUENCY CONSTANTS
OF
COPPER WIRE No. 00 B. & S. G.
(a) Return Conductor, at 6 Ft.=182 Cm.
(b) 30 Ft. above Ground as Return
(c) No Return Conductor
cos ω
f→

FIG. 18.

that is, the depth of penetration less than one-half the thickness of the conductor. Herefrom follows, that in the design of a high frequency conductor, the thickness of the ribbon or hollow cylinder is essentially determined by mechanical and not by electrical considerations; in other words, the thinnest mechanically permissible conductor usually is thicker than necessary for carrying the current. As iron usually cannot be employed in as thin ribbon as copper, due to its rusting, an iron conductor would have a larger section than a copper conductor of the same voltage drop and power consumption. Thereby a part of the advantage gained by the employment of the cheaper material would be lost.

97. The last section of Table III gives the constants of the conductor without return conductor, such as would be represented by the discharge circuit of a lightning arrester, by a wireless telegraph antenna, etc.; while the first two sections correspond to transmission line conditions, high-frequency currents between line conductors and between line and ground.

In the third section of Table III, the 60-cycle values are not given, and the values given for the lower high frequencies, 10^4 and even 10^5 cycles, usually have little meaning, are rarely realizable; they would correspond to a vertical conductor, as lightning arrester ground circuit, under conditions where no other conductor is within quarter-wave distance. Even at 10^5 cycles, however, the quarter-wave length is still 750 m. Thus there will practically always be other conductors within the field of the discharge conductor, acting as partial return conductor, and the actual values of impedance and resistance, that is, of voltage drop and power consumption in the conductor, thus will be intermediate between those given in the third section, for a conductor without return conductor, and those given in the first sections, for conductors with return conductors. Except at extremely high frequencies, at which the wave length gets so short that the condition of a conductor without return conductor becomes realizable. It is interesting to note, therefore, that at extremely high frequencies, the constants of the conductor without return conductor approach those of the conductor with return conductor. At lower high frequencies, impedance and resistance, and thus voltage drop and power consumption, of the return less conductor are much higher than those of the conductor with return, the more so, the lower the frequency. This is due to the considerable effect exerted already at low frequencies by electrical radiation.

However, while the case of the conductor without return conductor is not realizable at low and medium high frequencies in industrial circuits, it probably is more or less realized by the lightning discharge between ground and cloud, and the constants given in the third section of Table III would probably approximately represent the conditions met in the conductors of lightning rods such as used for the protection of buildings against lightning.

It is interesting to note that with such a conductor without return conductor, the power factor is always fairly high, even with copper as conductor material. The impedance and thus the voltage drop does not differ much from those of the conductor with return conductor. The resistance, however, and thus the power consumption are much higher, sometimes, in copper conductors, more than a hundred times as large, showing the large amount of energy radiated by the conductor—which reappears more or less destructively as "induced lightning stroke" in objects in the neighborhood of the lightning stroke.

PART III

TRANSIENTS IN TIME AND SPACE

CHAPTER I

GENERAL EQUATIONS

1. Considering the flow of electric power in a circuit. Electric power p can be resolved in two components, one component, proportional to the magnetic effects, called the current i, and one component, proportional to the dielectric effects, called the voltage e:

$$p = ei$$

There may be energy dissipation, and energy storage in the electric circuit, and either may depend on the voltage, or on the current, giving four constants r, g, L and C, representing respectively the energy dissipation and the energy storage depending on current and on voltage respectively.

The rate of energy storage can not be proportional to the current i or voltage e, but only to their rate of change, $\frac{di}{dt}$ and $\frac{de}{dt}$: if the rate of energy storage depended on the current i itself, then at constant current i, energy storage would constantly take place, and the amount of stored energy continuously increase, at constant condition of the circuit, which obviously is impossible.

Energy dissipation however, in its simplest form, would be proportional to the current itself, respectively the voltage. Thus we have:

Energy dissipation: ri

ge

Energy storage: $L\frac{di}{dt}$

$C\frac{de}{dt}$

The energy relation of an electric circuit thus can be characterized by four constants, namely:

r = effective resistance, representing the power or rate of energy consumption depending upon the current, i^2r; or the power component of the e.m.f. consumed in the circuit, that is, with an alternating current, the voltage, ir, in phase with the current.

L = effective inductance, representing the energy storage depending upon the current, $\frac{i^2L}{2}$, as electromagnetic component of the electric field; or the voltage generated due to the change of the current, $L\frac{di}{dt}$, that is, with an alternating current, the reactive voltage consumed in the circuit jxi, where $x = 2\,\pi fL$ and f = frequency.

g = effective (shunted) conductance, representing the power or rate of energy consumption depending upon the voltage, e^2g; or the power component of the current consumed in the circuit, that is, with an alternating voltage, the current, eg, in phase with the voltage.

C = effective capacity, representing the energy storage depending upon the voltage, $\frac{e^2C}{2}$, as electrostatic component of the electric field; or the current consumed by a change of the voltage, $C\frac{de}{dt}$, that is, with an alternating voltage, the (leading) reactive current consumed in the circuit, jbe, where $b = 2\,\pi fC$ and f = frequency.

In the investigation of electric circuits, these four constants, r, L, g, C, usually are assumed as located separately from each other, or localized. Although this assumption can never be perfectly correct, — for instance, every resistor has some inductance and every reactor has some resistance, — nevertheless in most cases it is permissible and necessary, and only in some classes of phenomena, and in some kinds of circuits, such as high-frequency phenomena, voltage and current distribution in long-distance, high-potential circuits, cables, telephone circuits, etc., this assumption is not permissible, but r, L, g, C must be treated as distributed throughout the circuit.

In the case of a circuit with distributed resistance, inductance, conductance, and capacity, as r, L, g, C, are denoted the effective resistance, inductance, conductance, and capacity, respectively, per unit length of circuit. The unit of length of the circuit may be chosen as is convenient, thus: the centimeter in the high-frequency oscillation over the multigap lightning arrester circuit, or a mile or kilometer in a long-distance transmission circuit or high-potential cable, or the distance of the velocity of light, 300,000 km., as most convenient in studying the laws of electric waves, etc.

The permanent values of current and e.m.f. in such circuits of distributed constants have, for alternating-current circuits, been investigated in Part II, where it was shown that they can be treated as transient phenomena in space, of the complex variables, current $\dot{I}$ and e.m.f. $\dot{E}$.

Transient phenomena in circuits with distributed constants, and, therefore, the general investigation of such circuits, leads to transient phenomena of two independent variables, time t and space or distance l; that is, these phenomena are transient in time and in space.

The difficulty met in studying such phenomena is that they are not alternating functions of time, and therefore can no longer be represented by the complex quantity.

It is possible, however, to derive from the constants of the circuit, r, L, g, C, and without any assumption whatever regarding current, voltage, etc., general equations of the electric circuits, and to derive some results and conclusions from such equations.

These general equations of the electric circuit are based on the single assumption that the constants r, L, g, C remain constant with the time t and distance l, that is, are the same for every unit length of the circuit or of the section of the circuit to which the equations apply. Where the circuit constants change, as where another circuit joins the circuit in question, the integration constants in the equations also change correspondingly.

Special cases of these general equations then are all the phenomena of direct currents, alternating currents, discharges of reactive coils, high-frequency oscillations, etc., and the difference between these different circuits is due merely to different values of the integration constants.

2. In a circuit or a section of a circuit containing distributed resistance, inductance, conductance, and capacity, as a transmission line, cable, high-potential coil of a transformer, telephone or telegraph circuit, etc., let r = the effective resistance per unit length of circuit; L = the effective inductance per unit length of circuit; g = the effective shunted conductance per unit length of circuit; C = the effective capacity per unit length of circuit; t = the time, l = the distance, from some starting point; e = the voltage, and i = the current at any point l and at any time t; then e and i are functions of the time t and the distance l.

In an element dl of the circuit, the voltage e changes, by de, by the voltage consumed by the resistance of the circuit element, $ri\,dl$, and by the voltage consumed by the inductance of the circuit element, $L\frac{di}{dt}dl$. Hence,

$$\frac{de}{dl} = ri + L\frac{di}{dt}. \tag{1}$$

In this circuit element dl the current i changes, by di, by the current consumed by the conductance of the circuit element, $ge\,dl$, and by the current consumed by the capacity of the circuit element, $C\frac{de}{dt}dl$. Hence,

$$\frac{di}{dl} = ge + C\frac{de}{dt}. \tag{2}$$

Differentiating (1) with respect to t and (2) with respect to l, and substituting then (1) into (2), gives

$$\frac{d^2i}{dl^2} = rgi + (rC + gL)\frac{di}{dt} + LC\frac{d^2i}{dt^2}, \tag{3}$$

and in the same manner,

$$\frac{d^2e}{dl^2} = rge + (rC + gL)\frac{de}{dt} + LC\frac{d^2e}{dt^2}. \tag{4}$$

These differential equations, of the second order, of current i and voltage e are identical; that is, in an electric circuit current and e.m.f. are represented by the same equations, which differ by the integration constants only, which are derived from the terminal conditions of the problem.

These differential equations are linear functions in the dependent variable and its derivates, and as the general exponential function is the only integral of such a differential equation, that is, is the only function linearly related to its derivates, these equations are integrated by the exponential function. That is:

Equation (3) is integrated by terms of the form

$$i = A\varepsilon^{-al-bt}. \tag{5}$$

Substituting (5) in (3) gives the identity

$$\begin{aligned} a^2 &= rg - (rC + gL)\,b + LCb^2 \\ &= (bL - r)\,(bC - g). \end{aligned} \tag{6}$$

In the terms of the form (5) the relation (6) thus must exist between the coefficients of l and t.

Substituting (5) into (1) gives

$$\frac{de}{dl} = (r - bL)\,A\varepsilon^{-al-bt}, \tag{7}$$

and, integrated,

$$e = \frac{bL - r}{a} A\varepsilon^{-al-bt}. \tag{8}$$

Or, substituting (5) in (8), and then substituting:

$$a = \sqrt{(bL - r)(bC - g)}$$

gives:

$$z = \frac{bL - r}{a} = \sqrt{\frac{bL - r}{bC - g}} \tag{9}$$

as what may be called the "surge impedance," or "natural impedance" of the circuit, and

$$e = \sqrt{\frac{bL - r}{bC - g}}\,A\epsilon^{-al-bt} \tag{10}$$

or:

$$e = zi \tag{11}$$

The integration constant of (8) would be a function of t, and since it must fulfill equation (4), must also have the form (5) for the special value $a = 0$, hence, by (6), $b = \frac{r}{L}$ or $b = \frac{g}{C}$ and therefore can be dropped.

In their most general form the *equations of the electric circuit* are

$$i = \sum\nolimits^n \left\{ A_n \varepsilon^{-a_nl - b_nt} \right\}, \tag{12}$$

$$e = \sum\nolimits^n \left\{ \frac{b_nL - r}{a_n} A_n \varepsilon^{-a_nl - b_nt} \right\}, \tag{13}$$

$$a_n^2 - (b_nL - r)\,(b_nC - g) = 0, \tag{14}$$

where A_n and a_n and b_n are integration constants, the last two being related to each other by the equation (14).

3. These pairs of integration constants, A_n and (a_n, b_n), are determinated by the terminal conditions of the problem.

Some such terminal conditions, for instance, are:

Current i and voltage e given as a function of time at one point l_0 of the circuit — at the generating station feeding into the circuit or at the receiving end of the transmission line.

Current i given at one point, voltage e at another point — as voltage at the generator end, current at the receiving end of the line.

Voltage given at one point and the impedance, that is, the complex ratio $\frac{\text{volts}}{\text{amperes}}$, at another point, as voltage at the generator end, load at the receiving end of the circuit.

Current and voltage given at one time t_0 as function of the distance l, as distribution of voltage and current in the circuit at the starting moment of an oscillation, etc.

Other frequent terminal conditions are:

Current zero at all times at one point l_0, as the open end of the circuit.

Voltage zero at all times at one point l_0, as the grounded or the short-circuited end of the circuit.

Current and voltage at all times at one point l_0 of the circuit, equal to current and voltage at one point of another circuit, as the connecting point of one circuit with another one.

As illustration, some of these cases will be discussed below.

The quantities i and e must always be real; but since a_n and b_n appear in the exponent of the exponential function, a_n and b_n may be complex quantities, in which case the integration constants A_n must be such complex quantities that by combining the different exponential terms of the same index n, that is, corresponding to the different pairs of a and b derived from the same equation (13), the imaginary terms in A_n and $\frac{b_n L - r}{a_n} A_n$ cancel.

In the exponential function

$$\varepsilon^{-al-bt},$$

writing

$$a = h + jk \quad \text{and} \quad b = p + jq, \tag{15}$$

we have

$$\varepsilon^{-al-bt} = \varepsilon^{-hl-pt}\varepsilon^{-j(kl+qt)},$$

and the latter term resolves into trigonometric functions of the angle

$$kl + qt.$$

$$kl + qt = \text{constant} \tag{16}$$

therefore gives the relation between l and t for constant phase of the oscillation or alternation of the current or voltage.

h and p thus are the coefficients of the transient, k and q the coefficients of the periodic term.

With change of time t the phase thus changes in position l in the circuit, that is, moves along the circuit.

Differentiating (16) with respect to t gives

$$k\frac{dl}{dt} + q = 0,$$

or

$$\frac{dl}{dt} = -\frac{q}{k}; \tag{17}$$

that is, the phase of the oscillation or alternation moves along the circuit with the speed $-\frac{q}{k}$, or, in other words,

$$S = -\frac{q}{k} \tag{18}$$

is the speed of *propagation of the electric phenomenon in the circuit*, and the phenomenon may be considered a wave motion.

(If no energy losses occur, $r = 0$, $g = 0$, in a straight conductor in a medium of unit magnetic and dielectric constant, that is, unit permeability and unit inductive capacity, S is the velocity of light.)

4. Since (14) is a quadratic equation, several pairs or corresponding values of a and b exist, which, in the most general case, are complex imaginary. The terms with conjugate complex imaginary values of a and b then have to be combined for the elimination of their imaginary form, and thereby trigonometric functions appear; that is, several terms in the equations (12) and

(13), which correspond to the same equation (14), and thus can be said to form a group, can be combined with each other.

Such a group of terms, of the same index n, is defined by the equation (14),

$$a_n^2 = (b_n L - r)(b_n C - g).$$

For convenience the index n may be dropped in the investigation of a group of terms of current and voltage, thus:

$$a^2 = (bL - r)(bC - g), \tag{19}$$

and the following substitutions may be made:

$$a = a_1 \sqrt{LC}, \tag{20}$$

$$\left.\begin{aligned} a &= h + jk, \\ a_1 &= h_1 + jk_1, \\ b &= p + jq, \end{aligned}\right\} \tag{21}$$

from which

$$\begin{aligned} h &= h_1 \sqrt{LC} \text{ and} \\ k &= k_1 \sqrt{LC}. \end{aligned} \tag{22}$$

Substituting (18) in (19),

$$(h_1 + jk_1)^2 = \left[(p + jq) - \frac{r}{L}\right]\left[(p + jq) - \frac{g}{C}\right]. \tag{23}$$

Carrying out and separating the real and the imaginary terms, equation (23) resolves into the two equations thus:

and

$$\left.\begin{aligned} h_1^2 - k_1^2 &= \left(p - \frac{r}{L}\right)\left(p - \frac{g}{C}\right) - q^2 \\ h_1 k_1 &= q\left(2\,p - \frac{r}{L} - \frac{g}{C}\right). \end{aligned}\right\} \tag{24}$$

Substituting

$$u = \frac{1}{2}\left(\frac{r}{L} + \frac{g}{C}\right) \tag{25}$$

$$m = \frac{1}{2}\left(\frac{r}{L} - \frac{g}{C}\right), \tag{26}$$

and $$p = s + u \tag{27}$$

into (24) gives

$$\left.\begin{aligned} h_1^2 - k_1^2 &= s^2 - q^2 - m^2, \\ \text{or} \qquad h_1 k_1 &= sq, \end{aligned}\right\} \tag{28}$$

and

$$\left.\begin{aligned} s^2 - q^2 &= h_1^2 - k_1^2 + m^2, \\ \text{or} \qquad sq &= h_1 k_1. \end{aligned}\right\} \tag{29}$$

Adding four times the square of the second equation to the square of the first equation of (28) and (29) respectively, gives

$$\left.\begin{aligned} h_1^2 + k_1^2 &= \sqrt{(s^2 - q^2 - m^2)^2 + 4\, s^2 q^2} \\ &= \sqrt{(s^2 + q^2 - m^2)^2 + 4\, q^2 m^2} \\ &= R_1^2 \end{aligned}\right\} \tag{30}$$

and

$$\left.\begin{aligned} s^2 + q^2 &= \sqrt{(h_1^2 - k_1^2 + m^2)^2 + 4\, h_1^2 k_1^2} \\ &= \sqrt{(h_1^2 + k_1^2 + m^2)^2 - 4\, k_1^2 m^2} \\ &= R_0^2, \end{aligned}\right\} \tag{31}$$

and substituting (22), gives, by (28), (29) and (30), (31)

$$\left.\begin{aligned} h &= \sqrt{LC}\sqrt{\frac{1}{2}\{R_1^2 + s^2 - q^2 - m^2\}}, \\ k &= \sqrt{LC}\sqrt{\frac{1}{2}\{R_1^2 - s^2 + q^2 + m^2\}}, \\ R_1^2 &= \sqrt{(s^2 + q^2 - m^2)^2 + 4\, q^2 m^2}, \end{aligned}\right\} \tag{32}$$

or

$$\left.\begin{aligned} s &= \frac{1}{\sqrt{LC}}\sqrt{\frac{1}{2}\{R_2^2 + h^2 - k^2 + LCm^2\}}, \\ q &= \frac{1}{\sqrt{LC}}\sqrt{\frac{1}{2}\{R_2^2 - h^2 + k^2 - LCm^2\}}, \\ R_2^2 &= \sqrt{(h^2 + k^2 + LCm^2)^2 - 4\, LCk^2 m^2}. \end{aligned}\right\} \tag{33}$$

If, however $(+h + jk)$ and $(u + s + jq)$ satisfy equation (19), then any other one of the expressions

$$(\pm h \pm jk) \text{ and } (u \pm s \pm jq)$$

also satisfies equation (19), providing also the second equation of (28) or (29) is satisfied,

$$hk = LC\ sq; \tag{34}$$

that is, if s and q have the same sign, h and k must have the same sign, and inversely, if s and q have opposite signs, h and k must have opposite signs.

This then gives the corresponding values of a and b:

$$\left.\begin{array}{lll}
(1) & a = +h + jk & b = u - s - jq \\
 & \quad\; +h - jk & \quad\; u - s + jq \\
(2) & a = -h - jk & b = u - s - jq \\
 & \quad\; -h + jk & \quad\; u - s + jq \\
(3) & a = -h + jk & b = u + s - jq \\
 & \quad\; -h - jk & \quad\; u + s + jq \\
(4) & a = +h - jk & b = u + s - jq \\
 & \quad\; +h + jk & \quad\; u + s + jq
\end{array}\right\} \tag{35}$$

or eight pairs of corresponding values of a and b.

p is called the *attenuation constant,* since it represents the decrease of the electrical effect with the time.

u is called the *dissipation constant,* since it represents the dissipation of electrical energy in resistance and conductance.

m is called the *distortion constant,* since on it depends the distortion of the electric circuit, that is, the displacement between current and voltage, as will be seen hereafter.

s is called the *energy transfer constant,* since it represents the energy transfer, as will be seen hereafter.

5. Substituting the values (1) of (35) into one group of terms of equations (12) and (13),

$$\left.\begin{array}{l}
i = A\varepsilon^{-al-bt} \\
\text{and} \\
e = \dfrac{bL - r}{a} A\varepsilon^{-al-bt},
\end{array}\right\} \tag{36}$$

gives

$$i_1 = A_1\varepsilon^{-(h+jk)l-(u-s-jq)t} + A_1'\varepsilon^{-(h-jk)l-(u-s+jq)t}$$
$$= \varepsilon^{-hl-(u-s)t}\left\{A_1\varepsilon^{-jkl+jqt} + A_1'\varepsilon^{+jkl-jqt}\right\},$$

and substituting for the exponential functions with imaginary exponents their trigonometric expressions by the equation

$$\varepsilon^{\pm jx} = \cos x \pm j \sin x$$

gives

$$i_1 = \varepsilon^{-hl-(u-s)t}\{A_1[\cos(qt - kl) + j\sin(qt - kl)] + A_1'[\cos(qt - kl) - j\sin(qt - kl)]\}$$
$$= \varepsilon^{-hl-(u-s)t}\{(A_1 + A_1')\cos(qt - kl) + j(A_1 - A_1')\sin(qt - kl)\};$$

hence, A_1 and A_1' must be conjugate complex imaginary quantities, and writing

$$\left.\begin{aligned} C_1 &= A_1 + A_1' \\ \text{and} \qquad C_1' &= j(A_1 - A_1') \end{aligned}\right\} \tag{37}$$

gives

$$i_1 = \varepsilon^{-hl-(u-s)t}\{C_1\cos(qt - kl) + C_1'\sin(qt - kl)\}. \tag{38}$$

Substituting in the same manner in the equation of e, in (36), gives

$$e_1 = \frac{(u - s - jq)L - r}{h + jk}A_1\varepsilon^{-(h+jk)l-(u-s-jq)t} + \frac{(u - s + jq)L - r}{h - jk}A_1'\varepsilon^{-(h-jk)l-(u-s+jq)t}$$
$$= \varepsilon^{-hl-(u-s)t}\left\{\frac{[(u - s - jq)L - r](h - jk)}{h^2 + k^2}A_1\varepsilon^{+j(qt-kl)} + \frac{[(u - s + jq)L - r](h + jk)}{h^2 + k^2}A_1'\varepsilon^{-j(qt-kl)}\right\};$$

hence expanding, and substituting the trigonometric expressions

$$e_1 = \varepsilon^{-hl-(u-s)t}\left\{\left(\frac{[(u-s)L-r]h-qkL}{h^2+k^2} - j\frac{[(u-s)L-r]k+qhL}{h^2+k^2}\right)A_1[\cos(qt-kl)+j\sin(qt-kl)] + \left(\frac{[(u-s)L-r]h-qkL}{h^2+k^2} + j\frac{[(u-s)L-r]k+qkL}{h^2+k^2}\right)A_1'[\cos(qt-kl)-j\sin(qt-kl)]\right\}, \tag{39}$$

and introducing the denotations

$$\left.\begin{aligned} c_1 &= \frac{qkL + h\,[r - (u - s)\,L]}{h^2 + k^2} = \frac{qk + h\,(m + s)}{h^2 + k^2}\,L \\ c_1{}' &= \frac{k\,[r - (u - s)\,L] - qhL}{h^2 + k^2} = \frac{k\,(m + s) - qh}{h^2 + k^2}\,L, \end{aligned}\right\} \qquad (40)$$

and substituting (40) in (39), gives

$$\begin{aligned} e_1 &= \varepsilon^{-hl - (u - s)\,t}\{(-c_1 + jc_1{}')\,A_1\,[\cos\,(qt - kl) + j\,\sin\,(qt - kl)] \\ &\quad + (-c_1 - jc_1{}')\,A_1{}'\,[\cos\,(qt - kl) - j\,\sin\,(qt - kl)]\} \\ &= \varepsilon^{-hl - (u - s)\,t}\{[-c_1\,(A_1 + A_1{}') + jc_1{}'\,(A_1 - A_1{}')]\cos\,(qt - kl) \\ &\quad + [-jc_1(A_1 - A_1{}') - c_1{}'(A_1 + A_1{}')]\sin\,(qt - kl)\}. \qquad (41) \end{aligned}$$

Substituting the denotations (37) into (41) gives

$$\begin{aligned} e_1 &= \varepsilon^{-hl - (u - s)\,t}\{(c_1{}'C_1{}' - c_1C_1)\cos\,(qt - kl) \\ &\quad - (c_1C_1{}' + c_1{}'C_1)\sin\,(qt - kl)\}. \qquad (42) \end{aligned}$$

The second group of values of a and b in equation (35) differs from the first one merely by the reversal of the signs of h and k, and the values i_2 and e_2 thus are derived from those of i_1 and e_1 by reversing the signs of h and k.

Leaving then the same denotations c_1 and $c_1{}'$ would reverse the sign of e_2, or, by reversing the sign of the integration constants C, that is, substituting

$$\left.\begin{aligned} C_2 &= -(A_2 + A_2{}') \\ \text{and} \qquad C_2{}' &= -j\,(A_2 - A_2{}'), \end{aligned}\right\} \qquad (43)$$

the sign of i_2 reverses; that is,

$$i_2 = -\varepsilon^{+hl - (u - s)\,t}\{C_2\cos\,(qt + kl) + C_2{}'\sin\,(qt + kl)\} \qquad (44)$$

and

$$\begin{aligned} e_2 &= \varepsilon^{+hl - (u - s)\,t}\{(c_1{}'C_2{}' - c_1C_2)\cos\,(qt + kl) \\ &\quad - (c_1C_2{}' + c_1{}'C_2)\sin\,(qt + kl)\}. \qquad (45) \end{aligned}$$

The third group in equation (35) differs from the first one by the reversal of the signs of h and s, and its values i_3 and e_3 therefore are derived from i_1 and e_1 by reversing the signs of h and s.

Introducing the denotations

$$\left.\begin{aligned} c_2 &= \frac{qk - h\,(m-s)}{h^2 + k^2} L, \\ c_2' &= \frac{k\,(m-s) + qh}{h^2 + k^2} L \end{aligned}\right\} \tag{46}$$

and

$$\left.\begin{aligned} C_3 &= A_3 + A_3', \\ C_3' &= j\,(A_3 - A_3'), \end{aligned}\right\} \tag{47}$$

gives

$$i_3 = \varepsilon^{+hl-(u+s)t}\left\{C_3 \cos(qt - kl) + C_3' \sin(qt - kl)\right\} \tag{48}$$

and

$$\begin{aligned} e_3 = \varepsilon^{+hl-(u+s)t}\{&(c_2'C_3' - c_2C_3)\cos(qt - kl) \\ &- (c_2C_3' + c_2'C_3)\sin(qt - kl)\}. \end{aligned} \tag{49}$$

The fourth group in (35) follows from the third group by the reversal of the signs h and k, and retaining the denotations c_2 and c_2', but introducing the integration constants,

$$\left.\begin{aligned} C_4 &= -\,(A_4 + A_4') \\ &\text{and} \\ C_4' &= -\,j\,(A_4 - A_4'), \end{aligned}\right\} \tag{50}$$

gives

$$i_4 = -\,\varepsilon^{-hl-(u+s)t}\left\{C_4 \cos(qt + kl) + C_4' \sin(qt + kl)\right\} \tag{51}$$

and

$$\begin{aligned} e_4 = \varepsilon^{-hl-(u+s)t}\{&(c_2'C_4' - c_2C_4)\cos(qt + kl) \\ &- (c_2C_4' + c_2'C_4)\sin(qt + kl)\}. \end{aligned} \tag{52}$$

6. This then gives as the general expression of the *equations of the electric circuit:*

$$\left.\begin{aligned} i = \textstyle\sum\big[&\varepsilon^{-hl-(u-s)t}\{C_1 \cos(qt - kl) + C_1' \sin(qt - kl)\}\ (i_1) \\ &- \varepsilon^{+hl-(u-s)t}\{C_2 \cos(qt + kl) + C_2' \sin(qt + kl)\}\ (i_2) \\ &+ \varepsilon^{+hl-(u+s)t}\{C_3 \cos(qt - kl) + C_3' \sin(qt - kl)\}\ (i_3) \\ &- \varepsilon^{-hl-(u+s)t}\{C_4 \cos(qt + kl) + C_4' \sin(qt + kl)\}\big]\ (i_4) \end{aligned}\right\} \tag{53}$$

and

$$\left.\begin{aligned} e = \sum [& \varepsilon^{-hl-(u-s)t}\{(c_1'C_1' - c_1C_1)\cos(qt - kl) \\ & \quad - (c_1'C_1 + c_1C_1')\sin(qt - kl)\}\,(e_1) \\ & + \varepsilon^{+hl-(u-s)t}\{(c_1'C_2' - c_1C_2)\cos(qt + kl) \\ & \quad - (c_1'C_2 + c_1C_2')\sin(qt + kl)\}\,(e_2) \\ & + \varepsilon^{+hl-(u+s)t}\{(c_2'C_3' - c_2C_3)\cos(qt - kl) \\ & \quad - (c_2'C_3 + c_2C_3')\sin(qt - kl)\}\,(e_3) \\ & + \varepsilon^{-hl-(u+s)t}\{(c_2'C_4' - c_2C_4)\cos(qt + kl) \\ & \quad - (c_2'C_4 + c_2C_4')\sin(qt + kl)\}]\,(e_4), \end{aligned}\right\} \tag{54}$$

where $C_1, C_1', C_2, C_2', C_3, C_3', C_4, C_4'$ and two of the four values s, q, h, k are integration constants, depending on the terminal conditions, and

$$\left.\begin{aligned} c_1 &= \frac{qk + h(m + s)}{h^2 + k^2} L, \\ c_1' &= \frac{k(m + s) - qh}{h^2 + k^2} L, \\ c_2 &= \frac{qk - h(m - s)}{h^2 + k^2} L, \\ c_2' &= \frac{k(m - s) + qh}{h^2 + k^2} L, \end{aligned}\right\} \tag{55}$$

and

$$\left.\begin{aligned} u &= \frac{1}{2}\left(\frac{r}{L} + \frac{g}{C}\right), \\ m &= \frac{1}{2}\left(\frac{r}{L} - \frac{g}{C}\right), \end{aligned}\right\} \tag{56}$$

and h, k and s, q are related by the equations

$$\left.\begin{aligned} h &= \sqrt{LC}\sqrt{\tfrac{1}{2}\{R_1^2 + s^2 - q^2 - m^2\}}, \\ k &= \sqrt{LC}\sqrt{\tfrac{1}{2}\{R_1^2 - s^2 + q^2 + m^2\}}, \end{aligned}\right\} \tag{57}$$

and

$$R_1^2 = \sqrt{(s^2 + q^2 - m^2)^2 + 4\,q^2m^2};$$

hence,

$$h^2 + k^2 = LCR_1^2, \tag{58}$$

or

$$\left.\begin{aligned} s &= \frac{1}{\sqrt{LC}}\sqrt{\tfrac{1}{2}\left\{R_2^{\,2} + h^2 - k^2 + LCm^2\right\}}, \\ q &= \frac{1}{\sqrt{LC}}\sqrt{\tfrac{1}{2}\left\{R_2^{\,2} - h^2 + k^2 - LCm^2\right\}}, \\ &\text{and} \\ R_2^{\,2} &= \sqrt{(h^2 + k^2 + LCm^2)^2 - 4\,LCk^2m^2}; \end{aligned}\right\} \tag{59}$$

hence,

$$s^2 + q^2 = \frac{R_2^{\,2}}{LC}. \tag{60}$$

Writing

$$D\,(qt \pm kl) = C\cos\,(qt \pm kl) + C'\sin\,(qt \pm kl) \tag{61}$$

and

$$\begin{aligned} H\,(qt \pm kl) = {} & (c'C' - cC)\cos\,(qt \pm kl) \\ & - (c'C + cC')\sin\,(qt \pm kl), \end{aligned} \tag{62}$$

equations (50) and (51) can be written thus:

$$\left.\begin{aligned} i = \sum\big[&\varepsilon^{-hl-(u-s)t}\,D_1\,(qt-kl) - \varepsilon^{+hl-(u-s)t}\,D_2(qt+kl) && (i') \\ &+ \varepsilon^{+hl-(u+s)t}\,D_3\,(qt-kl) - \varepsilon^{-hl-(u+s)t}\,D_4(qt+kl)\big] && (i'') \end{aligned}\right\} \tag{63}$$

and

$$\left.\begin{aligned} e = \sum\big[&\varepsilon^{-hl-(u-s)t}\,H_1\,(qt-kl) + \varepsilon^{+hl-(u-s)t}\,H_2(qt+kl) && (e') \\ &+ \varepsilon^{+hl-(u+s)t}\,H_3\,(qt-kl) + \varepsilon^{-hl-(u+s)t}\,H_4(qt+kl)\big] && (e'') \end{aligned}\right\} \tag{64}$$

CHAPTER II

DISCUSSION OF SPECIAL CASES

7. The general equations of the electric circuit, (12) and (13) of Chapter I, consist of groups of terms of the form:

$$i = A\epsilon^{-al\ -bt} \tag{1}$$

$$\left.\begin{aligned} e &= \frac{bL - r}{a} A\epsilon^{-al-bt} \\ &= \sqrt{\frac{bL - r}{bC - g}} A\epsilon^{-al-bt} \\ &= zi \end{aligned}\right\} \tag{2}$$

where

$$z = \sqrt{\frac{bL - r}{bC - g}} \tag{3}$$

is the "surge impedance," or "natural impedance" of the circuit and a and b are related by the equation (14) of Chapter I:

$$\left.\begin{aligned} a^2 &= b^2LC - b\,(gL + rC) + rg \\ &= (bL - r)(bC - g) \end{aligned}\right\} \tag{4}$$

These equations must represent every existing electric circuit and every circuit which can be imagined, from the lightning stroke to the house bell and from the underground cable or transmission line to the incandescent lamp, under the only condition, that r, L, g and C are constant, or can be assumed as such with sufficient approximation.

The difference between all existing circuits thus consists merely in the difference in value of the constants A, and the constants a and b, the latter being related to each other by the equation (4).

To illustrate this, some special cases may be considered.

In general, A, a and b are general numbers, that is, complex imaginary quantities, but as such must be of such form that in the final form of i and e the imaginary terms eliminate. Thus, whenever a term of the form $X + jY$ exists, another term must exist of the form $X - jY$.

I. Special Case $b = 0$: Permanents

8. $b = 0$ means, that the electrical phenomenon is not a function of time, that is, is not transient, periodic or varying, but is constant or *permanent.*

By (4) it is:

$$a = \pm \sqrt{rg} \tag{5}$$

That is, two values of a exist, either of which gives a term equation (1), (2), and any combination of these terms thus also satisfies the differential equations.

In this case, by (3):

$$z = \pm \sqrt{\frac{r}{g}} \tag{6}$$

hence, the general equation of a *permanent* is:

$$\left.\begin{aligned} i &= A_1 \epsilon^{-l\sqrt{rg}} + A_2 \epsilon^{+l\sqrt{rg}} \\ e &= \sqrt{\frac{r}{g}} \left\{ A_1 \epsilon^{-l\sqrt{rg}} - A_2 \epsilon^{+l\sqrt{rg}} \right\} \end{aligned}\right\} \tag{7}$$

These equations do not contain L and C, that is, inductance and capacity have no effect in a *permanent,* that is, on an electrical phenomenon, which does not vary in time.

Equations (7) are the equations of a direct current circuit having distributed leakage, such as a metallic conductor submerged in water, or the current flow in the armor of a cable laid in the ground, or the current flow in the rail return of a direct current railway, etc.

r is the series resistance per unit length, g the shunted or leakage conductance per unit length of circuit.

Where the leakage conductance is not uniformly distributed, but varies, the numerical values in (7) change wherever the circuit constants change, just as would be the case if the resistance r of the conductor changed. If the leakage conductance g is not uniformly distributed, but localized periodically in space—as at the ties of the railroad track—when dealing with a sufficient circuit length, the assumption of uniformity would be justified as approximation.

9. (*a*) If the conductor is of infinite length—that is, of such great length, that the current which reaches the end of the conductor, is negligible compared with the current entering the conductor—it is:

$$A_2 = 0$$

since otherwise the second term of equation (18) would become infinite for $l = \infty$.

This gives:

$$\left.\begin{aligned} i &= A\epsilon^{-l\sqrt{rg}} \\ e &= A\sqrt{\frac{r}{g}}\,\epsilon^{-l\sqrt{rg}} \end{aligned}\right\} \tag{8}$$

or,

$$e = \sqrt{\frac{r}{g}}\,i \tag{9}$$

that is, a conductor of infinite length (or very great length) of series resistance r and shunted conductance g, has the effective resistance $r_0 = \sqrt{\frac{r}{g}}$.

It is interesting to note, that at a change of r or of g the effective resistance r_0, and thus the current flowing into the conductor at constant impressed voltage, or the voltage consumed at constant current, changes much less than r or g.

(*b*) If the circuit is open at $l = l_0$, it is:

$$i = A_1\epsilon^{-l_0\sqrt{rg}} + A_2\epsilon^{+l_0\sqrt{rg}} = 0$$

hence, if

$$A = A_1\epsilon^{-l_0\sqrt{rg}} = -A_2\epsilon^{+l_0\sqrt{rg}}$$

it is

$$\left.\begin{aligned} i &= A\left\{\epsilon^{+(l_0 - l)\sqrt{rg}} - \epsilon^{-(l_0 - l)\sqrt{rg}}\right\} \\ e &= A\sqrt{\frac{r}{g}}\left\{\epsilon^{+(l_0 - l)\sqrt{rg}} + \epsilon^{-(l_0 - l)\sqrt{rg}}\right\} \end{aligned}\right\} \tag{10}$$

(*c*) If the circuit is closed upon itself at $l = l_0$, it is,

$$e = \sqrt{\frac{r}{g}}\left\{A_1\epsilon^{-l_0\sqrt{rg}} - A_2\epsilon^{+l_0\sqrt{rg}}\right\} = 0$$

hence, if

$$A = A_1\epsilon^{-l_0\sqrt{rg}} = A_2\epsilon^{+l_0\sqrt{rg}}$$

it is

$$\left.\begin{aligned} i &= A\left\{\epsilon^{+(l_0 - l)\sqrt{rg}} + \epsilon^{-(l_0 - l)\sqrt{rg}}\right\} \\ e &= A\sqrt{\frac{r}{g}}\left\{\epsilon^{+(l_0 - l)\sqrt{rg}} - \epsilon^{-(l_0 - l)\sqrt{rg}}\right\} \end{aligned}\right\} \tag{11}$$

If, in (11), $l_0 = 0$, that is, the circuit is closed at the starting point, it is

$$\left.\begin{aligned} i &= A\{\epsilon^{-l\sqrt{rg}} + \epsilon^{+l\sqrt{rg}}\} \\ e &= A\sqrt{\frac{r}{g}}\{\epsilon^{-l\sqrt{rg}} - \epsilon^{+l\sqrt{rg}}\} \end{aligned}\right\}$$

or, counting the distance in opposite direction, that is, changing the sign of l:

$$\left.\begin{aligned} i &= A\{\epsilon^{+l\sqrt{rg}} + \epsilon^{-l\sqrt{rg}}\} \\ e &= A\sqrt{\frac{r}{g}}\{\epsilon^{+l\sqrt{rg}} - \epsilon^{-l\sqrt{rg}}\} \end{aligned}\right\} \tag{12}$$

Assuming now l to be infinitely small,

$$l \cong 0$$

we get, by

$$\begin{aligned} \epsilon^{\pm s} &= 1 \pm s + \pm \\ i &= 2A \\ e &= 2A\sqrt{\frac{r}{g}}\, l\sqrt{rg} = 2Arl = rli \end{aligned}$$

$rl = r_0$ is, however, the total resistance of the circuit, and the equations (23), for infinitely small l, thus assume the form:

$$e = r_0 i \tag{13}$$

that is, the equation of the direct current circuit with massed constants, which so appears as special case of the general direct current circuit.

10. (*d*) If the circuit, at $l = l_0$, is closed by a resistance r_0, it is:

$$\left|\frac{e}{i}\right|_{l = l_0} = r_0$$

hence,

$$\frac{A_1\epsilon^{-l_0\sqrt{rg}} - A_2\epsilon^{+l_0\sqrt{rg}}}{A_1\epsilon^{-l_0\sqrt{rg}} + A_2\epsilon^{+l_0\sqrt{rg}}} = \frac{r_0}{\sqrt{\frac{r}{g}}}$$

$$\frac{A_2\epsilon^{+l_0\sqrt{rg}}}{A_1\epsilon^{-l_0\sqrt{rg}}} = \frac{\sqrt{\frac{r}{g}} - r_0}{\sqrt{\frac{r}{g}} + r_0}$$

or,

$$A_2 = A_1 \epsilon^{-2l_0\sqrt{rg}} \frac{\sqrt{\frac{r}{g}} - r_0}{\sqrt{\frac{r}{g}} + r_0}$$

$$\left.\begin{aligned} i &= A \left\{ \epsilon^{-l\sqrt{rg}} - \frac{r_0 - \sqrt{\frac{r}{g}}}{r_0 + \sqrt{\frac{r}{g}}} \epsilon^{-(2l_0 - l)\sqrt{rg}} \right\} \\ e &= A\sqrt{\frac{r}{g}} \left\{ \epsilon^{-l\sqrt{rg}} + \frac{r_0 - \sqrt{\frac{r}{g}}}{r_0 + \sqrt{\frac{r}{g}}} \epsilon^{-(2l_0 - l)\sqrt{rg}} \right\} \end{aligned}\right\} \quad (14)$$

These equations (12) and (14) can be written in various different forms. They are interesting in showing in a direct current circuit features which usually are considered as characteristic of alternating currents, that is, of wave motion.

The first term of (12) or (14) is the outflowing or main current respectively voltage, the second term is the reflected one.

At the end of the circuit with distributed constants, reflection occurs at the resistance r_0.

If $r_0 > \sqrt{\frac{r}{g}}$, the coefficient of the second term is positive, and partial reflection of current occurs, while the return voltages add itself to the incoming voltage.

If $r_0 < \sqrt{\frac{r}{g}}$, reflection of voltage occurs, while the return current adds itself to the incoming current.

If $r_0 = \sqrt{\frac{r}{g}}$, the second term vanishes, and the equations (14) become those of (8), of an infinitely long conductor. That is:

A resistance r_0, equal to the effective resistance (surge impedance) $\sqrt{\frac{r}{g}}$ of a direct current circuit of distributed constants, passes current and voltage without reflection. A higher resistance r_0 partially reflects the voltage—completely so for $r_0 = \infty$, or open circuit. A lower resistance r_0 partially reflects the current—completely so for $r_0 = 0$, or short circuit.

$\sqrt{\frac{r}{g}}$ thus takes in direct current circuits the same position as the surge impedance in alternating current or transient circuits.

II. Special Case: $a = 0$

11. By equation (4), this gives two solutions:

$$b = \frac{r}{L} \text{ and } b = \frac{g}{C}$$

(a)
$$b = \frac{r}{L}$$

substituted in (1) and (2) gives:

$$\left.\begin{aligned} i &= A\epsilon^{-\frac{r}{L}t} \\ e &= 0 \end{aligned}\right\} \qquad (15)$$

that is, the inductive discharge of a circuit closed upon itself.

(b) Substituting in equations (1) and (2):

$$B = \frac{bL - r}{a} A$$

hence,

$$i = B \frac{a}{bL - r} \epsilon^{-al-bt}$$

$$e = B\epsilon^{-al-bt}$$

and then substituting,

$$a = 0; b = \frac{g}{C}$$

gives

$$\left.\begin{aligned} i &= 0 \\ e &= B\epsilon^{-\frac{g}{C}t} \end{aligned}\right\} \qquad (16)$$

that is, if the condenser C is shunted by the conductance g, at voltage e on the condenser and thus also on the conductance g, the current in the external or supply circuit is zero, that is, the current in the conductance g is equal and opposite to that in the condenser:

$$i_1 = ge = gB\epsilon^{-\frac{g}{C}t} \qquad (17)$$

(16) and (17) thus are the equations of the condenser discharge through the conductance g.

III. Special Case: $l \cong 0$: Massed Constants

12. In most electrical circuits, the length of the circuit is so short, that at the rate of change of the electrical phenomena, no appreciable difference exists between the different parts of the circuit as the result of the finite velocity of propagation. Or in other words, the current is the same throughout the entire circuit or circuit section.

In general, therefore, the electrical constants, resistance, inductance, capacity, conductance, can be assumed as massed together locally, and not as distributed along the circuit.

The case of distributed constants mainly requires consideration only in case of circuits of such length, that the length of the circuit is an appreciable part of the wave length of the current, or the length of the impulse, etc. This is the case:

1. In circuits of great length, such as transmission lines.
2. When dealing with transients of very short duration, or very high frequency, such as high frequency oscillations, switching impulses, etc.

The equations of the circuit of massed constants—the usual alternating or direct current circuit—should thus be derived from the equations (1) and (2) by the condition.

$$l = 0$$

In any electric circuit, there must be a point of zero potential difference. Before substituting $l = 0$, into equations (12) and (13), that is, considering only an infinitely short part of the circuit at $l = 0$, these equations must be so modified as to bring the zero point of potential difference within this part of the circuit.

Thus, we first have to substitute in (1) and (2):

$$e = 0 \text{ at } l = 0$$

by equations (4) it is:

$$a = \pm\sqrt{(bL - r)(bC - g)} \tag{18}$$

that is, to every value of b correspond two values of a, equal numerically, but of opposite signs.

Substituting these in equations (1) and (2), gives

$$\left.\begin{aligned} i &= A_1\epsilon^{-al-bt} + A_2\epsilon^{+al-bt} \\ e &= \frac{bL - r}{a}\{A_1\epsilon^{-al-bt} - A_2\epsilon^{+al-bt}\} \end{aligned}\right\} \qquad (19)$$

and for $e = 0$ at $l = 0$:

$$A_2 = A_1 = A$$

thus,

$$\left.\begin{aligned} i &= A\epsilon^{-bt}(\epsilon^{-al} + \epsilon^{+al}) \\ e &= \frac{bL - r}{a} A\epsilon^{-bt}(\epsilon^{-al} - \epsilon^{+al}) \end{aligned}\right\} \qquad (20)$$

substituting now, for infinitely small l,

$$\epsilon^{\pm al} = 1 \pm al$$

gives, as the *general equation of the circuit with massed constants*:

$$\left.\begin{aligned} i &= B\epsilon^{-bt} \\ e &= (r_0 - bL_0)B\epsilon^{-bt} \end{aligned}\right\} \qquad (21)$$

where

$$B = 2A$$
$$r_0 = lr = \text{total resistance of circuit}$$
$$L_0 = bL = \text{total inductance of circuit.}$$

The equations of voltage, in (21), may also be written:

$$\left.\begin{aligned} e &= r_0B\epsilon^{-bt} - L_0bB\epsilon^{-bt} \\ &= r_0i - L_0\frac{di}{dt} \end{aligned}\right\} \qquad (22)$$

which is the equation of the inductive circuit with massed constants.

Or

$$e = (r_0 - bL_0)i \qquad (23)$$

(*a*) *D. C.: Direct Currents.*

13. $b = 0$, that is, the electric effect is a permanent, does not vary with the time. Substituted in equations (21) gives:

$$\left.\begin{aligned} i &= B \\ e &= r_0B \end{aligned}\right\} \qquad (24)$$

the equations of the direct current circuit.

(*b*) *I. C.: Impulse Currents.*

$$b = \text{real}$$

that is, as function of time, the electrical effect is not periodic, but is transient, or is an *impulse.*

Solving equation (4) for b, gives:

$$b = u \mp s \tag{25}$$

where

$$s = \sqrt{\frac{a^2}{LC} + m^2} \tag{26}$$

and

$$\left.\begin{aligned} u &= \frac{1}{2}\left(\frac{r}{L} + \frac{g}{C}\right) \\ m &= \frac{1}{2}\left(\frac{r}{L} - \frac{g}{C}\right) \end{aligned}\right\} \tag{27}$$

Substituting these values into the general equation of massed constants, (21), gives:

$$i = B_1\epsilon^{-(u-s)t} + B_2\epsilon^{-(u+s)t}$$

or:

$$i = \epsilon^{-ut}\{B_1\epsilon^{+st} + B_2\epsilon^{-st}\} \tag{28}$$

$$e = B_1[r - (u - s) L]\epsilon^{-(u-s)t} + B_2[r - (u + s) L]\epsilon^{-(u+s)t}$$

or:

$$e = \epsilon^{-ut}\{(r - uL)(B_1\epsilon^{+st} + B_2\epsilon^{-st}) + sL (B_1\epsilon^{+st} - B_2\epsilon^{-st})\} \tag{29}$$

Assuming $i = 0$ for $t = 0$, that is, counting the time from the zero value of current, gives:

$$B_1 = - B_2 = B$$

hence,

$$\left.\begin{aligned} i &= B\epsilon^{-ut}(\epsilon^{+st} - \epsilon^{-st}) \\ e &= B\epsilon^{-ut}\{(r - uL)(\epsilon^{+st} - \epsilon^{-st}) + sL(\epsilon^{+st} + \epsilon^{-st})\} \end{aligned}\right\} \tag{30}$$

In this case, by (25) and (26), it must be:

$$u^2 \geqq s^2 \geqq m^2 \tag{31}$$

14. (*ba*) In the Special Case, that

$$s = u$$

it is, by (28) and (29),

$$\left.\begin{aligned} i &= B_1 + B_2\epsilon^{-2ut} \\ e &= rB_1 + B_2 (r - 2\,uL)\epsilon^{-2ut} \end{aligned}\right\} \tag{32}$$

or, substituting for u, gives:

$$\left.\begin{aligned} i &= B_1 + B_2\epsilon^{-\frac{r}{L}t}\epsilon^{-\frac{g}{C}t} \\ e &= rB_1 + \frac{gL}{C}B_2\epsilon^{-\frac{r}{L}t}\epsilon^{-\frac{g}{C}t} \end{aligned}\right\} \quad (33)$$

This is the general equation of a direct current circuit having inductance, resistance, capacity and conductance, including permanent as well as transient term.

If $i = 0$ at $t = 0$, it is

$$B_1 = -B_2 = B$$

hence,

$$\left.\begin{aligned} i &= B\left\{1 - \epsilon^{-\frac{r}{L}t}\epsilon^{-\frac{g}{C}t}\right\} \\ e &= rB\left\{1 - \frac{gL}{rC}\epsilon^{-\frac{r}{L}t}\epsilon^{-\frac{g}{C}t}\right\} \end{aligned}\right\} \quad (34)$$

These are the general equations of a direct current with starting transient.

For $$g = 0$$

that is, no losses in the condenser circuit, equations (34) assume the usual form of the direct current starting transient:

$$\left.\begin{aligned} i &= B\left(1 - \epsilon^{-\frac{r}{L}t}\right) \\ e &= rB \end{aligned}\right\} \quad (35)$$

(*c*) *A. C.: Alternating Currents.*

$$b = \pm jq$$

15. That is, as function of time, the electrical effect is periodic (since the exponential function with imaginary exponent is the trigonometric function), or alternating.

Substituted in equations (21), this gives:

$$\left.\begin{aligned} i &= B_1\epsilon^{+jqt} + B_2\epsilon^{-jqt} \\ e &= (r + jqL)\,B_1\epsilon^{+jqt} + (r - jqL)\,B_2\epsilon^{-jqt} \end{aligned}\right\} \quad (36)$$

and, substituting,

$$\epsilon^{\pm jqt} = \cos qt \pm j \sin qt$$

into (36), gives

$$\left.\begin{aligned} i &= A_1 \cos qt + A_2 \sin qt \\ e &= (rA_1 + qLA_2) \cos qt + (rA_2 - qLA_1) \sin qt \end{aligned}\right\} \quad (37)$$

where

$$\left.\begin{aligned} A_1 &= B_1 + B_2 \\ A_2 &= j\,(B_1 - B_2) \end{aligned}\right\} \quad (38)$$

substituting,

$$\left.\begin{aligned} qt &= \phi \\ q &= 2\,\pi f \\ 2\,\pi f L &= x \end{aligned}\right\} \quad (39)$$

gives

$$\left.\begin{aligned} i &= A_1 \cos \phi + A_2 \sin \phi \\ e &= (rA_1 + xA_2) \cos \phi + (rA_2 - xA_1) \sin \phi \\ &= r\,(A_1 \cos \phi + A_2 \sin \phi) + x\,(A_2 \cos \phi - A_1 \sin \phi) \end{aligned}\right\} \quad (40)$$

(*d*) *O. C.: Oscillating Currents.*

16. $$b = p \pm jq$$

In the general case of the circuit of massed constants, where b is a general number or complex quantity, it is, substituting in (21)

$$\left.\begin{aligned} i &= \epsilon^{-pt}\{B_1\epsilon^{+jqt} + B_2\epsilon^{-jqt}\} \\ e &= \epsilon^{-pt}\{r - pL + jqL)B_1\epsilon^{+jqt} + (r - pL - jqL)B_2\epsilon^{-jqt}\} \end{aligned}\right\} \quad (41)$$

and, substituting again the trigonometric function for the imaginary exponential function in (41), gives, in the same manner as in (*c*)

$$\left.\begin{aligned} i &= \epsilon^{-pt}\{A_1 \cos qt + A_2 \sin qt\} \\ e &= \epsilon^{-pt}\{[(r - pL)A_1 + qLA_2] \cos qt + [(r - pL)A_2 \\ &\qquad - qLA_1] \sin qt\} \\ &= \epsilon^{-pt}\{(r - pL)(A_1 \cos qt + A_2 \sin qt) + \\ &\qquad qL\,(A_2 \cos qt - A_1 \sin qt)\} \end{aligned}\right\} \quad (42)$$

These are the equations of the oscillating currents and voltages, in a circuit of massed constants, consisting of the transient term ϵ^{-pt}, and the alternating or periodic term. The latter is the same as with the alternating currents, equations (37) or (40), except that in (42), in the equations of voltage $(r - pL)$ takes the place of the resistance r in (37). That is, the effective resistance, with oscillating currents, is lowered by the negative resistance of energy return, pL.

Substituting $p = 0$ in (42) gives (37).

IV. Special Case: Impulse Currents.

$$b = \text{real}.$$

17. By the equation (4)

$$a^2 = b^2LC - b\,(rC + gL) + rg,$$

to every value of b, correspond two values of a, equal numerically but opposite in sign:

$$\pm\, a$$

To every value of a correspond two values of b:

$$b = u \pm s \tag{43}$$

where

$$\left.\begin{aligned} s &= \sqrt{m^2 + \frac{a^2}{LC}} \\ u &= \frac{1}{2}\left(\frac{r}{L} + \frac{g}{C}\right) \\ m &= \frac{1}{2}\left(\frac{r}{L} - \frac{g}{C}\right) \end{aligned}\right\} \tag{44}$$

As b is assumed to be real, it must be positive, as otherwise the time exponential ϵ^{bt} would be increasing indefinitely. Thus it is, from (43):

$$s^2 \leqq u^2$$

Since u and b are real, s must be real. That is, by (44)

$$m^2 + \frac{a^2}{LC}$$

must be real and positive.

As m^2 is real, a^2 must be real, and must either be positive, or, if negative, $-\dfrac{a^2}{LC}$ must be less than m^2.

However, a can never be complex imaginary, without also making b complex imaginary, and impulse currents thus are characterized by the condition, that a is either real, or purely imaginary.

We thus get two cases of impulse currents:

$$b = u \mp s$$

(1) a^2 positive

$$\left.\begin{aligned} a &= \pm h \\ s &= \sqrt{m^2 + \frac{h^2}{LC}} \\ h &= \sqrt{LC(s^2 - m^2)} \\ u^2 &\geq s^2 \geq m^2 \end{aligned}\right\} \quad (45)$$

(2) a^2 negative

$$\left.\begin{aligned} a &= \pm jk \\ s &= \sqrt{m^2 - \frac{k^2}{LC}} \\ k &= \sqrt{LC(m^2 - s^2)} \\ s^2 &\leq m^2 \\ k^2 &\leq m^2 LC \end{aligned}\right\} \quad (46)$$

non-periodic in space.

a.	b:
$+h$	$u - s$
$-h$	$u - s$
$-h$	$u + s$
$+h$	$u + s$

(47)

periodic in space.

a:	b:
$+jk$	$u - s$
$-jk$	$u - s$
$+jk$	$u + s$
$-jk$	$u + s$

(48)

(1) Non-periodic Impulse Currents.

18. Substituting (47) into (1) and (2) gives

$$i = \epsilon^{-ut}\{A_1\epsilon^{-hl+st} + A_2\epsilon^{+hl+st} + A_3\epsilon^{+hl-st} + A_4\epsilon^{+hl-st}\} \quad (49)$$

$$e = \epsilon^{-ut}\left\{A_1\frac{(u-s)L-r}{h}\epsilon^{-hl+st} - A_2\frac{(u-s)L-r}{h}\epsilon^{+hl+st} - A_3\frac{(u+s)L-r}{h}\epsilon^{+hl-st} + A_4\frac{(u+s)L-r}{h}\epsilon^{-hl-sl}\right\} \quad (50)$$

It is, however,

$$\frac{(u \mp s)L - r}{h} = \frac{uL - r \mp sL}{h}$$

and, substituting for h from (45), and substituting for u

$$\frac{(u \mp s)L - r}{h} = \frac{\frac{1}{2}\left(\frac{r}{L} + \frac{g}{C}\right)L - r \mp sL}{\sqrt{LC(s^2 - m^2)}} = \frac{-mL \mp sL}{\sqrt{LC(s^2 - m^2)}},$$

$$= \mp\sqrt{\frac{L}{C}}\sqrt{\frac{s \pm m}{s \mp m}}$$

or, substituting,

$$\sqrt{\frac{s+m}{s-m}} = c \tag{51}$$

it is

$$\frac{(u-s)L-r}{h} = -c\sqrt{\frac{L}{C}}$$

$$\frac{(u+s)L-r}{h} = +\frac{1}{c}\sqrt{\frac{L}{C}}$$

and, substituting these values into (50), gives

$$e = -\sqrt{\frac{L}{C}}\,\epsilon^{-ut}\left\{c\,A_1\,\epsilon^{-hl+st} - c\,A_2\,\epsilon^{+hl+st} + \frac{1}{c}\,A_3\,\epsilon^{+hl-st} - \frac{1}{c}\,A_4\,\epsilon^{-hl-st}\right\} \tag{52}$$

The equations (49) and (52), of current and voltage respectively, of the non-periodic impulse, are the same as derived as equation (9) in Chapter III, as special case of the general circuit equation.

The constants A_1, A_2, A_3, A_4 of (60) are denoted by C_1, $-C_2$, C_3, $-C_4$ in (9) of Chapter III, and the further discussion of the equations given there.

(2) Periodic Impulse Currents.

19. Substituting (48) into (1) gives:

$$i = \epsilon^{-ut}\{\epsilon^{+st}(A_1\,\epsilon^{-jkl} + A_2\,\epsilon^{+jkl}) + \epsilon^{-st}(A_3\,\epsilon^{-jkl} + A_4\,\epsilon^{+jkl})\}$$

and, substituting the trigonometric expressions, this gives

$$i = \epsilon^{-ut}\{\epsilon^{+st}(D_1\cos kl - D_2\sin kl) + \epsilon^{-st}(D_3\cos kl - D_4\sin kl)\} \tag{53}$$

where

$$\left.\begin{aligned} D_1 &= A_1 + A_2 \\ D_2 &= j\,(A_1 - A_2) \\ D_3 &= A_3 + A_4 \\ D_4 &= j\,(A_3 - A_4) \end{aligned}\right\} \tag{54}$$

The expression

$$\frac{bL-r}{a}$$

in equation (2) assumes the form, by substituting (46) and (48):

$$\frac{bL - r}{a} = \frac{(u \mp s)\,L - r}{\pm jk} = \frac{-\,(m \pm s)\,L}{\pm j\sqrt{LC}\,(m^2 - s^2)}$$

$$= \pm j\sqrt{\frac{L}{C}}\sqrt{\frac{m \pm s}{m \mp s}}$$

and, substituting,

$$c = \sqrt{\frac{m + s}{m - s}} \tag{55}$$

gives

$$\frac{bL - r}{a} = \pm jc\sqrt{\frac{L}{C}} \text{ respectively } \pm \frac{j}{c}\sqrt{\frac{L}{C}} \tag{56}$$

Substituting now (48) and (56) into (2) gives

$$e = \sqrt{\frac{L}{C}}\,\epsilon^{-ut}\left\{jc\epsilon^{+st}(A_1\epsilon^{-jkl} - A_2\epsilon^{-jkl}) + \frac{j}{c}\,\epsilon^{-st}(A_3\epsilon^{-jkl} - A_4\epsilon^{+jkl})\right\}$$

and, substituting the trigonometric expression:

$$e = \sqrt{\frac{L}{C}}\,\epsilon^{-ut}\left\{c\epsilon^{+st}(D_2 \cos kl + D_1 \sin kl) + \frac{1}{c}\,\epsilon^{-st}\,(D_4 \cos kl + D_3 \sin kl)\right\} \tag{57}$$

The equations (53) and (57), of current and voltage repectively, of the periodic impulse, are the equations (24) of Chapter III, derived by specialization from the general circuit equations, and their further discussion is given there.

V. Special Case: Alternating Currents

$b =$ imaginary.

20. If

$$b = \pm jq$$

by equation (4), it is

$$a = \pm\sqrt{(r \mp jqL)\,(g \mp jqC)} \tag{58}$$

$$= \pm\,(h \pm jk)$$

That is, a, as square root of a complex quantity, also becomes complex imaginary.

For

$$b = +jq$$

it is

$$a^2 = (r - jqL)(g - jqC)$$
$$= (rg - q^2L) - jq\,(rC + gL)$$

that is, the sign of the imaginary term of a^2 is negative, and the sign of the imaginary term of a must be negative, that is, $a = \pm\,(h - jk)$, and inversely.

The corresponding values of b and a thus are:

b:	*a:*
$+\,jq$	$+\,h - jk$
$-\,jq$	$+\,h + jk$
$-\,jq$	$-\,h - jk$
$+\,jq$	$-\,h + jk$

Substituting these values into (1) gives:

$$i = \epsilon^{-hl}\{A_1\epsilon^{+j(kl-qt)} + A_2\epsilon^{-j(kl-qt)}\} + \epsilon^{+hl}\{A_3\epsilon^{+j(kl+qt)} + A_4\epsilon^{-j(kl+qt)}\}$$

Thus, in trigonometric expression:

$$i = \epsilon^{-hl}\{B_1 \cos(kl - qt) + B_2 \sin(kl - qt)\}$$
$$+ \epsilon^{+hl}B_3 \cos(kl + qt) + B_4 \sin(kl + qt)\} \quad (59)$$

where

$$\left.\begin{array}{ll} B_1 = A_1 + A_2 & B_3 = A_3 + A_4 \\ B_2 = j(A_1 - A_2) & B_4 = j(A_3 - A_4) \end{array}\right\}$$

Resolving in equation (59) the trigonometric function, and re-arranging, gives:

$$i = \epsilon^{-hl}\{(B_1 \cos qt - B_2 \sin qt) \cos kl + (B_2 \cos qt + B_1 \sin qt) \sin kl\}$$
$$+ \epsilon^{+hl}\{(B_3 \cos qt + B_4 \sin qt) \cos kl + (B_4 \cos qt - B_3 \sin qt) \sin kl\} \quad (60)$$

21. The equations (60) can be written in symbolic expression, by representing the terms with cos qt by the real, the terms with sin qt by the imaginary vector, that is, substituting,

$$B_1 \cos qt - B_2 \sin qt = B_1 + jB_2 = \dot{A}_1$$

hence

$$B_2 \cos qt + B_1 \sin qt = B_2 - jB_1 = -\,j\dot{A}_1$$
$$B_3 \cos qt + B_4 \sin qt = B_3 - jB_4 = -\,\dot{A}_2$$
$$B_4 \cos qt - B_3 \sin qt = B_4 + jB_3 = -\,j\dot{A}_2$$

hence

$$\dot{I} = \dot{A}_1\epsilon^{-hl}(\cos kl - j \sin kl) - \dot{A}_2\epsilon^{+hl}(\cos kl + j \sin kl) \quad (61)$$

This is the equation (23) on page 71 in Part II, derived there as the current in a general alternating-current circuit of distributed constants.

Analogous, by substituting into equation (2), the equation of the voltage is derived:

$$\dot{E} = \sqrt{\frac{Z}{Y}}\{\dot{A}_1\epsilon^{-hl}(\cos kl - j\sin kl) + \dot{A}_2\epsilon^{+hl}(\cos kl + j\sin kl)\} \quad (62)$$

Where

$$Z = r + jqL$$
$$Y = g + jqC$$

and

$$q = 2\pi f$$

Alternating currents, representing the special case, where b is purely imaginary, and impulse currents, representing the case, where b is real, thus represent two analogous special cases of the general circuit equations.

These two classes are industrially the most important types of current, though their relation to the operation of electric systems is materially different:

The alternating currents are the useful currents in our large electric power systems.

The impulse currents are the harmful currents in our large electric power systems.

The alternating currents have been extensively studied, and the most of the preceding Part II is devoted to the general alternating-current circuit.

Very little work has been done in the study of impulse current, and the next Chapter thus shall be devoted to an outline of their theory.

CHAPTER III

IMPULSE CURRENTS

22. The terms of the general equations of the electric circuit (53) and (54) contain the constants:

The eight values, C and C'.

The four values, c and c'.

And the exponential constants u, m, h, k, s, q.

Of these, the values c and c' are expressions of L and of the exponential constants, by equation (55), thus are not independent constants.

u and m are circuit constants, and as such are the same for all terms of the general equation.

Of the four terms h, k, s, q, two are dependent upon the other two by the equations (57) and (59).

Thus there remains ten independent constants, which are to be determined by the terminal conditions:

Two of the four exponential terms h, k, s, q, and the eight coefficients C and C'.

h and s are attenuation constants, in length or space and in time respectively, and k and q are wave constants, in space and in time respectively.

In a non-periodic electrical effect, k and q thus would be zero, while $k = 0$ but $q \neq 0$ gives a phenomenon periodic in time, but non-periodic in space, and inversely, $q = 0$, but $k \neq 0$ give a phenomenon, periodic in space, but non-periodic in time.

If of the four constants: s, q, h, k, one equals zero, another one also must be zero; if

$$s = 0$$

it is, by equations (58) and (57):

$$R_1^2 = q^2 + m^2$$

$$h = 0$$

$$k = \sqrt{LC\,(q^2 + m^2)}$$

if

$$q = 0$$

it is, either

$$R_1^2 = s^2 - m^2$$
$$h = \sqrt{LC\,(s^2 - m^2)}$$
$$k = 0$$

or,

$$R_1^2 = m^2 - s^2$$
$$h = 0$$
$$k = \sqrt{LC\,(m^2 - s^2)}$$

if,

$$h = 0$$

it is, either

$$R_2^2 = k^2 - LCm^2$$
$$s = 0$$
$$q = \sqrt{\frac{k^2}{LC} - m^2}$$

or

$$R_2^2 = LCm^2 - k^2$$
$$s = \sqrt{m^2 - \frac{k^2}{LC}}$$
$$q = 0$$

if

$$k = 0$$

it is

$$R_2^2 = h^2 + LCm^2$$
$$s = \sqrt{\frac{h^2}{LC} + m^2}$$
$$q = 0$$

and this gives the three sets of values:

s	q	h	k
0	$\sqrt{\frac{k^2}{LC} - m^2}$	0	$\sqrt{LC\,(q^2 + m^2)}$
$\sqrt{m^2 - \frac{k^2}{LC}}$	0	0	$\sqrt{LC\,(m^2 - s^2)}$
$\sqrt{\frac{h^2}{LC} + m^2}$	0	$\sqrt{LC\,(s^2 - m^2)}$	0

$s = 0$ and $h = 0$ gives an electrical effect which is periodic in time and in space, and is transient in time. This case leads to the equation of the stationary oscillation of the circuit of distributed resistance, inductance, capacity and conductance, such as the transmission line, more fully discussed in Part II, and in Chapters V and VII of Part III.

$q = 0$ gives a non-periodic transient.

Such a non-periodic transient is called an *impulse*, and currents, voltages and power of this character then are denoted as *impulse currents*, *impulse voltages*, and *impulse power*.

Impulse currents thus are non-periodic transients, while alternating currents are periodic non-transients or permanents.

23. An impulse thus is characterized by the condition:

$$q = 0.$$

Substituting this in equation (57) gives:

$$\left.\begin{aligned} R_1 &= s^2 - m^2 &\quad \text{or} \quad R_1 &= m^2 - s^2 \\ h &= \sqrt{LC\,(s^2 - m^2)} &\quad \text{or} \quad h &= 0 \\ k &= 0 &\quad \text{or} \quad k &= \sqrt{LC\,(m^2 - s^2)} \end{aligned}\right\} \qquad (2)$$

and inversely, by equation (59):

$$\left.\begin{aligned} R_2 &= h^2 + LCm^2 &\quad \text{or} \quad R_2 &= LCm^2 - k^2 \\ s &= \sqrt{\frac{h^2}{LC} + m^2} &\quad \text{or} \quad s &= \sqrt{m^2 - \frac{k^2}{LC}} \\ q &= 0 &\quad \text{or} \quad q &= 0 \end{aligned}\right\} \qquad (3)$$

and

$$m^2 - s^2 + \frac{h^2}{LC} = 0 \qquad m^2 - s^2 - \frac{k^2}{LC} = 0 \qquad (4)$$

That is, either $k = 0$, that is, the impulse is non-periodic in space also, or $h = 0$, that is, the impulse is periodic in space, has no attenuation with the distance.

By the character of the space distribution, we thus distinguish non-periodic and periodic impulses. One of the two constants, h or k, must always be zero, if $q = 0$, that is, the space distribution of an impulse can not be oscillatory, but is either exponential or trigonometric.

If

$s^2 > m^2$, the impulse is non-periodic; and $h^2 > LCs^2$

if

$s^2 < m^2$, the impulse is periodic, and $k^2 < LCm^2$

We thus distinguish two classes of impulses:

Non-periodic impulses and periodic impulses. The "periodic" here refers to the distribution in space, as in time the impulse always must be exponential or transient.

A. Non-periodic Impulses.

24. A non-periodic impulse is an electrical effect, in which the electrical distribution is non-periodic, or exponential, in time as well as in space. Its condition is:

$$q = 0$$

$$s^2 > m^2$$

hence:

$$k = 0$$

$$h = \sqrt{LC\,(s^2 - m^2)} \tag{5}$$

or

$$s = \sqrt{\frac{h^2}{LC} + m^2}$$

Substituting these values in equations (55) gives

$$\begin{aligned} c_1 &= L\frac{s+m}{h} = \sqrt{\frac{L}{C}}\sqrt{\frac{s+m}{s-m}} = c\sqrt{\frac{L}{C}} \\ c_2 &= L\frac{s-m}{h} = \sqrt{\frac{L}{C}}\sqrt{\frac{s-m}{s+m}} = \frac{1}{c}\sqrt{\frac{L}{C}} \\ c_1' &= 0 \\ c_2' &= 0 \end{aligned} \tag{6}$$

where

$$c = \sqrt{\frac{s+m}{s-m}} \tag{7}$$

Substituting (5), (6) and (7) into equations (53) and (54), the sin terms vanish, and with them the integration constants C', and only the integration constants C remain; the cos terms become unity, and these equations assume the form:

$$\begin{aligned} i &= \Sigma\{C_1\epsilon^{-hl-(u-s)t} - C_2\epsilon^{+hl-(u-s)t} + C_3\epsilon^{+hl-(u+s)t} - C_4\epsilon^{-hl-(u+s)t}\} \\ e &= -\Sigma\sqrt{\frac{L}{C}}\left\{cC_1\epsilon^{-hl-(u-s)t} + cC_2\epsilon^{+hl-(u-s)t} + \frac{1}{c}C_3\epsilon^{+hl-u+s)t} + \frac{1}{c}C_4\epsilon^{-hl-u+s)t}\right\} \end{aligned} \tag{8}$$

or, considering only one group of the series, and separating ϵ^{-ut}:

$$\left.\begin{aligned} i &= \epsilon^{-ut}\{C_1\epsilon^{-hl+st} - C_2\epsilon^{+hl+st} + C_3\epsilon^{+hl-st} - C_4\epsilon^{-hl-st}\} \\ e &= -\sqrt{\frac{L}{C}}\,\epsilon^{-ut}\Big\{cC_1\epsilon^{-hl+st} + cC_2\epsilon^{+hl+st} + \frac{1}{c}C_3\epsilon^{+hl-st} + \\ &\qquad\qquad \frac{1}{c}C_4\epsilon^{-hl-st}\Big\} \end{aligned}\right\} \quad (9)$$

25. The equations (9) can be simplified by shifting the zero points of time and distance, by the substitution:

$$\left.\begin{aligned} C_1 &= D_1\epsilon^{+hl_1-st_1} \\ C_3 &= \pm D_1\epsilon^{-hl_1+st_1} \end{aligned}\right\} \qquad \left.\begin{aligned} C_2 &= D_2\epsilon^{-hl_1-st_1} \\ C_4 &= \pm D_2\epsilon^{+hl_1+st_1} \end{aligned}\right\} \quad (10)$$

hence,

$$D_1 = \sqrt{\pm C_1C_3} \qquad D_2 = \sqrt{\pm C_2C_4} \quad (11)$$

$$\epsilon^{+2hl_1-2st_1} = \pm\frac{C_1}{C_3} \qquad \epsilon^{+2hl_1+2st_1} = \pm\frac{C_4}{C_2}$$

$$\left.\begin{aligned} \epsilon^{+4hl_1} &= \pm\frac{C_1C_4}{C_2C_3} \\ \epsilon^{+4st_1} &= \pm\frac{C_3C_4}{C_1C_2} \end{aligned}\right\} \quad (12)$$

or

$$\left.\begin{aligned} l_1 &= \frac{1}{4h}\log\left(\pm\frac{C_1C_4}{C_2C_3}\right) \\ t_1 &= \frac{1}{4s}\log\left(\pm\frac{C_3C_4}{C_1C_2}\right) \end{aligned}\right\} \quad (13)$$

where the $\pm$ denotes the positive value of the C product.

Making also the further substitution:

$$c = \epsilon^{+st_0}; \qquad \frac{1}{c} = \epsilon^{-st_0} \quad (14)$$

hence,

$$t_0 = \frac{\log c}{s} = \frac{1}{2s}\log\frac{s+m}{s-m} \quad (15)$$

(10) and (14) substituted in (9) gives:

$$\left.\begin{aligned} i &= \epsilon^{-ut}\{D_1(\epsilon^{-hl'+s(t'-t_0)} \pm \epsilon^{+hl'-s(t'-t_0)}) - D_2(\epsilon^{+hl'+s(t'-t_0)} \\ &\qquad\qquad \pm \epsilon^{-hl'-s(t'-t_0)})\} \\ e &= -\sqrt{\frac{L}{C}}\,\epsilon^{-ut}\{D_1(\epsilon^{-hl'+st'} \pm \epsilon^{+hl'-st'}) + D_2(\epsilon^{+hl'+st'} \pm \\ &\qquad\qquad \epsilon^{-hl'-st'})\} \end{aligned}\right\} \quad (16)$$

where

$$\left.\begin{aligned} t' &= t - t_1 + t_0 \\ l_1 &= l - l_1 \end{aligned}\right\} \tag{17}$$

By the substitution:

$$\left.\begin{aligned} \epsilon^{+x} + \epsilon^{-x} &= 2 \cosh x \\ \epsilon^{+x} - \epsilon^{-x} &= 2 \sinh x \end{aligned}\right\}$$

Equations (16) can be written in hyperbolic form thus:

$$\left.\begin{aligned} i &= \epsilon^{-ut}\{D_1' \cosh [hl' - s(t'-t_0)] - D_2' \cosh [hl' + s(t'-t_0)]\} \\ e &= -\sqrt{\frac{L}{C}}\, \epsilon^{-ut}\{D_1' \cosh [hl' - st'] + D_2' \cosh [hl' + st']\} \end{aligned}\right\} \tag{17b}$$

Or the corresponding sinh function, in case of the minus sign in equations (16).

26. The impulse thus is the combination of two single impulses of the form

$$\epsilon^{-ut}(\epsilon^{-hl+st} \pm \epsilon^{+hl-st})$$

which move in opposite direction, the D_1 impulse toward rising l: $\frac{dl}{dt} > 0$, and the D_2 impulse toward decreasing l: $\frac{dl}{dt} < 0$

The voltage impulse differs from the current impulse by the factor $\sqrt{\frac{L}{C}}$ (the "surge impedance"), and by a *time displacement* t_0. That is, in the general impulse, voltage e and current i are displaced in time.

t_0 thus may be called the time displacement, or time lag of the current impulse behind the voltage impulse.

t_0 is positive, that is, the current *lags* behind the voltage impulse, if in equation (15) the log is positive, that is, m is positive, or: $\frac{r}{L} > \frac{g}{C}$, that is, the resistance-inductance term preponderates.

Inversely, t_0 is negative, and the current *leads*, or the voltage impulse lags behind the current impulse, if m is negative, that is, $\frac{r}{L} < \frac{g}{C}$, or the capacity term preponderates.

If $g = 0$, that is, no shunted conductance, the current impulse always lags behind the voltage impulse.

If $m = 0$, that is, $\frac{r}{L} = \frac{g}{C}$, or $\frac{r}{g} = \frac{L}{C}$, $t_0 = 0$, that is, the voltage impulse and the current impulse are in phase with each other, that is, there is no time displacement, and current and voltage impulses have at any time or at any space the same shape: *distortionless circuit.* m therefore is called the *distortion constant* of the circuit.

27. If

$$s = m \tag{18}$$

it is

$$h = 0 \tag{19}$$

hence, substituting into equations (16)

$$\left.\begin{aligned} i &= A\epsilon^{-ut}(\epsilon^{+m(t'-t_0)} \pm \epsilon^{-m(t'-t_0)}) \\ e &= B\sqrt{\frac{L}{C}}\,\epsilon^{-ut}(\epsilon^{+mt'} \pm \epsilon^{-mt'}) \end{aligned}\right\} \tag{20}$$

where

$$\left.\begin{aligned} A &= D_1 - D_2 \\ B &= D_1 + D_2 \end{aligned}\right\}$$

that is, a simple impulse.

Or, substituting for u and m their values

$$\left.\begin{aligned} i &= A_1\epsilon^{-\frac{g}{C}t} \pm A_2\epsilon^{-\frac{r}{L}t} \\ e &= \sqrt{\frac{L}{C}}\left\{B_1\epsilon^{-\frac{g}{C}t} \pm B_2\epsilon^{-\frac{r}{L}t}\right\} \end{aligned}\right\} \tag{21}$$

Thus, the capacity effect, as first term, and the inductance effect, as second term, appear separate.

28. In the individual impulse:

$$\epsilon^{-ut}(\epsilon^{-hl+st} \pm \epsilon^{+hl-st}) = \epsilon^{-hl}\epsilon^{-(u-s)t} \pm \epsilon^{+hl}\epsilon^{-(u+s)t} \tag{18}$$

the term ϵ^{-ut} is the attenuation of the impulse by the energy dissipation in the circuit, that is, represents the rate at which the impulse would die out by its energy dissipation.

The first term, $\epsilon^{-(u-s)t}$, dies out at a slower rate than given by the energy dissipation, that is, in this term, at any point l, energy is supplied, is left behind by the passing impulse, and as the result, this term decreases with increasing distance l, by the factor ϵ^{-hl}; inversely, the second term, $\epsilon^{-(u+s)t}$, dies out more rapidly with the time, than corresponds to the energy losses, that is, at any point l, this term abstracts energy and shifts it along the circuit, and thereby gives an increase of energy in the direction of propagation, by ϵ^{+hl}. In other words, of the two terms of the impulse, the one drops energy while moving along the line, and the other picks it up and carries it along.

The terms $\epsilon^{\pm st}$ thus represent the dropping and picking up of energy with the time, the terms $\epsilon^{\pm hl}$ the dropping and picking up of energy in space along the line. In distinction to u, which may be called the *energy dissipation constant*, s (and its corresponding h) thus may be called the *energy transfer constant* of the impulse. The higher s is, the greater then is the rate of energy transfer, that is, the steeper the wave front, and s thus may also be called the *wave front constant* of the impulse.

The transformation from the four constants C to the two constants D obviously merely represents the shifting of the zero value of time and distance to the center of the impulse.

Also, in equations (16), in the energy dissipation term ϵ^{-ut}, obviously also the new time t' may be used: $\epsilon^{-ut'} = \epsilon^{-u(t-t_1)}$, and this would merely mean a change of the constants D_1 and D_2 by the constant factor ϵ^{-ut_1}.

Substituting in equations (16): $l = 0$, gives the equation of the impulse in a circuit with massed constants

$$\left.\begin{aligned} i &= A\epsilon^{-ut}(\epsilon^{+st} \pm \epsilon^{-st}) \\ e &= B\sqrt{\frac{L}{C}}\,\epsilon^{-ut}(\epsilon^{+s)t-t_0)} \pm \epsilon^{-s(t-t_0)}) \end{aligned}\right\}$$

where

$$\left.\begin{aligned} A &= D_1 - D_2 \\ B &= D_1 + D_2 \end{aligned}\right\}$$

29. The equations (9) can be brought into a different form by the substitution,

$$\left.\begin{aligned} C_1 &= B\epsilon^{+hl_1}\epsilon^{-st_1} \\ C_2 &= \pm B\epsilon^{-h(l_1+l_0)}\epsilon^{-st_1} \\ C_3 &= \pm B\epsilon^{-h(l_1+l_0)}\epsilon^{+st_1} \\ C_4 &= \pm B\epsilon^{+hl_1}\epsilon^{-st_1} \end{aligned}\right\} \qquad (22)$$

where the + sign applies, if C_2, etc., is of the same sign with C_1, the − sign, if it is of opposite sign.

Substituting also equations (14), this transforms equations (9) into the form

$$\left.\begin{aligned} i &= B\epsilon^{-ut}\{(\epsilon^{-hl'+s(t'-t_0)} \mp \epsilon^{-hl'-s(t'-t_0)}) - (\epsilon^{-h(l_0-l')+s(t'-t_0)} \\ &\qquad \mp \epsilon^{-h(l_0-l')-s(t'-t_0)})\} \\ e &= -\sqrt{\frac{L}{C}}\,B\epsilon^{-ut}\{(\epsilon^{-hl'+st'} \pm \epsilon^{-hl'-st'}) + (\epsilon^{-h(l_0-l')+st'} \\ &\qquad \pm \epsilon^{-h(l_0-l')-st'})\} \end{aligned}\right\} \quad (23)$$

where

$$\left.\begin{aligned} l' &= l - l_1 \\ t' &= t - t_1 + t_0 \end{aligned}\right\} \quad (24)$$

or, re-arranging equations (23)

$$\left.\begin{aligned} i &= B\epsilon^{-ut}\{\epsilon^{-hl'} - \epsilon^{-h(l_0-l')}\}\{\epsilon^{+s(t'-t_0)} \pm \epsilon^{-s(t'-t_0)}\} \\ e &= -\sqrt{\frac{L}{C}}\,B\,\epsilon^{-ut}\{\epsilon^{-hl'} + \epsilon^{-h(l_0-l')}\}\{\epsilon^{+st'} \pm \epsilon^{-st'}\} \end{aligned}\right\} \quad (25)$$

Or substituting

$$\left.\begin{aligned} B &= A\,\epsilon^{+\frac{hl_0}{2}} \\ &\text{and} \\ l &= l' - \frac{l_0}{2} \end{aligned}\right\} \quad (26)$$

gives

$$\left.\begin{aligned} i &= A\epsilon^{-ut}\{\epsilon^{-hl} - \epsilon^{+hl}\}\{\epsilon^{+st('-t_0(} \mp \epsilon^{-s(t'-t_0)}\} \\ e &= \sqrt{\frac{L}{C}}\,A\epsilon^{-ut}\{\epsilon^{-hl} + \epsilon^{+hl}\}\{\epsilon^{+st'} \pm \epsilon^{-st'}\} \end{aligned}\right\} \quad (27)$$

30. These equations show the non-periodic impulse as consisting of the product of a time impulse

$$\epsilon^{+st} \pm \epsilon^{-st}$$

and a space impulse

$$\epsilon^{-hl} \pm \epsilon^{+hl}$$

Or, the impulse of current and of voltage consists of a main impulse, decaying along the line by the factor

$$\epsilon^{-hl'}$$

and a reflected impulse, or impulse traveling in the opposite direction, from the reflection point $\frac{l_0}{2}$, and dying out at the rate

$$\epsilon^{-h(l_0-l')}$$

and, by shifting the starting point of distance l to the reflection point $\frac{l_0}{2}$, the simplest form of impulse equations (27) are derived.

As seen from equations (27), and also from (25):

In the voltage impulse, the main and the reflected impulse add, in the current impulse they subtract.

The current impulse lags behind the voltage impulse by time t_0.

t_0 may be positive, or negative, that is, the current lag or lead, depending whether the resistance—inductance term, or the capacity term preponderates in the line constants, as discussed before.

Equations (27) may also be written in the form of hyperbolic function, as

$$\left.\begin{aligned} i &= A_0\epsilon^{-ut} \sinh hl \sinh s(t' - t_0) \\ e &= \sqrt{\frac{L}{C}}\, A_0\epsilon^{-ut} \cosh hl \cosh st' \end{aligned}\right\}$$

or

$$\left.\begin{aligned} i &= A_0\epsilon^{-ut} \sinh hl \cosh s(t' - t_0) \\ e &= \sqrt{\frac{L}{C}}\, A_0\epsilon^{-ut} \cosh hl \sinh st' \end{aligned}\right\} \qquad (28)$$

31. Still another form of equations (9) is given by the substitution,

$$\left.\begin{aligned} C_1 &= B\epsilon^{+hl_1}\epsilon^{-st_1} \\ C_2 &= B\epsilon^{-hl_1}\epsilon^{-st_1} \\ C_3 &= B\epsilon^{-h(l_1+l_0)}\epsilon^{+st_1} \\ C_4 &= B\epsilon^{+h(l_1+l_0)}\epsilon^{+st_1} \\ c &= \epsilon^{+st_0} \end{aligned}\right\} \qquad (29)$$

this gives

$$\left.\begin{aligned} i &= B\epsilon^{-ut}\{\epsilon^{+s(t'-t_0)}(\epsilon^{+hl'} \mp \epsilon^{-hl'}) - \epsilon^{-s(t'-t_0)}(\epsilon^{+h(l'-l_0)} \pm \epsilon^{-h(l'-l_0)})\} \\ e &= -\sqrt{\frac{L}{C}}\, B\epsilon^{-ut}\{\epsilon^{+st'}(\epsilon^{+hl'} \pm \epsilon^{-hl'}) + \epsilon^{-st'}(\epsilon^{+h(l'-l_0)} \pm \epsilon^{-h(l'-l_0)})\} \end{aligned}\right\} \qquad (30)$$

where

$$\left.\begin{aligned} l' &= l - l_1 \\ t' &= t - t_1 + t_0 \end{aligned}\right\} \qquad (31)$$

Equations (30) can be written in hyperbolic functions in the form

$$\left.\begin{array}{l}\left.\begin{array}{l} i = B\epsilon^{-ut}\{\epsilon^{+s(t'-t_0}\cosh hl' - \epsilon^{-s(t'-t_0}\cosh h(l'-l_0)\} \\ e = -B\sqrt{\frac{L}{C}}\epsilon^{-ut}\{\epsilon^{+st'}\sinh hl' + \epsilon^{-st'}\sinh h(l'-l_0)\}\end{array}\right\} \\ \text{or} \\ \left.\begin{array}{l} i = B\epsilon^{-ut}\{\epsilon^{+s\, t'-t_0)}\sinh hl' - \epsilon^{-s(t'-t_0)}\sinh h(l'-l_0)\} \\ e = -B\sqrt{\frac{L}{C}}\,\epsilon^{-ut}\{\epsilon^{+st'}\cosh hl' + \epsilon^{-st'}\cosh h(l'-l_0)\}\end{array}\right\}\end{array}\right\} \quad (32)$$

B. Periodic Impulses

32. A periodic impulse is an electrical effect, in which the electrical distribution is periodic in space, or can be represented by a periodic function, such as a trigonometric series. In time, however, it is non-periodic.

As any function can be represented within limits by a trigonometric series, it follows, that any electrical distribution can be represented by a trigonometric series with the complete circuit as one fundamental wave, or a fraction thereof, and thus every electrical impulse can be considered as periodic within the limits of its circuit.

As seen in the preceding (page 257), the periodic impulse is characterized by

$$\left.\begin{array}{ll} & s^2 < m^2 \\ & q = 0 \\ & h = 0 \\ & k = \sqrt{LC\,(m^2 - s^2)} \\ \text{or} & s = \sqrt{m^2 - \frac{k^2}{LC}} \end{array}\right\} \quad (33)$$

Substituting these values in equations (55) gives

$$\left.\begin{array}{l} c_1 = 0 \\ c_2 = 0 \\ c_1' = L\frac{m+s}{k} = \sqrt{\frac{L}{C}}\sqrt{\frac{m+s}{m-s}} = c\sqrt{\frac{L}{C}} \\ c_2' = L\frac{m-s}{k} = \sqrt{\frac{L}{C}}\sqrt{\frac{m-s}{m+s}} = \frac{1}{c}\sqrt{\frac{L}{C}} \end{array}\right\} \quad (34)$$

where

$$c = \sqrt{\frac{m+s}{m-s}} \tag{35}$$

Substituting (33), (34) and (35) into equations (53) and (54), and substituting,

$$\left.\begin{aligned} C_1 - C_2 &= D_1 \\ C_1' + C_2' &= D_2 \\ C_3 - C_4 &= D_3 \\ C_3' + C_4' &= D_4 \end{aligned}\right\} \tag{36}$$

gives

$$\left.\begin{aligned} i &= \Sigma\{\epsilon^{-(u-s)t}(D_1 \cos kl - D_2 \sin kl) + \\ &\qquad \epsilon^{-(u+s)t}(D_3 \cos kl - D_4 \sin kl)\} \\ e &= \Sigma\sqrt{\frac{L}{C}}\left\{c\epsilon^{-(u-s)t}(D_2 \cos kl + D_1 \sin kl) + \right. \\ &\qquad \left.\frac{1}{c}\epsilon^{-(u+s)t}(D_4 \cos kl + D_3 \sin kl)\right\} \end{aligned}\right\} \tag{37}$$

or, considering only one group of the series, separating ϵ^{-ut}, and substituting,

$$c = \epsilon^{+st_0} \tag{38}$$

hence,

$$t_0 = + \frac{\log c}{s} = \frac{1}{2s} \log \frac{m+s}{m-s} \tag{39}$$

gives

$$\left.\begin{aligned} i &= \epsilon^{-ut}\{\epsilon^{+st}(D_1 \cos kl - D_2 \sin kl) + \epsilon^{-st}(D_3 \cos kl - \\ &\qquad D_4 \sin kl)\} \\ e &= \sqrt{\frac{L}{C}}\,\epsilon^{-ut}\{\epsilon^{+s(t+t_0)}(D_2 \cos kl + D_1 \sin kl) + \\ &\qquad \epsilon^{-s(t+t_0)}(D_4 \cos kl + D_3 \sin kl)\} \end{aligned}\right\} \tag{40}$$

33. Substituting,

$$\left.\begin{aligned} D_1 &= A_1 \cos kl_1 & D_3 &= A_2 \cos k\,(l_1 - l_0) \\ D_2 &= A_1 \sin kl_1 & D_4 &= A_2 \sin k\,(l_1 - l_0) \end{aligned}\right\} \tag{41}$$

and

$$\left.\begin{aligned} A_1 &= A\epsilon^{-st_1} \\ A_2 &= \pm A\epsilon^{+st_1} \end{aligned}\right\} \tag{42}$$

gives

$$\left.\begin{aligned} i &= A\epsilon^{-ut}\{\epsilon^{+s(t-t_1)} \cos k\,(l + l_1) \pm \epsilon^{-s(t-t_1)} \cos k\,(l + l_1 - l_0)\} \\ e &= A\sqrt{\frac{L}{C}}\,\epsilon^{-ut}\{\epsilon^{+s)t-t_1+t_0)} \sin kl\,(l + l_1) \pm \epsilon^{-s(t-t_1+t_0)} \\ &\qquad \sin k(l + l_1 - l_0)\} \end{aligned}\right\} \tag{43}$$

substituting the new coördinates of time and distance, that is, changing the zero point of time and of distance, by the expression:

$$\left.\begin{aligned} t' &= t - t_1 + t_0 \\ l' &= l + l_1 \end{aligned}\right\} \tag{44}$$

gives

$$\left.\begin{aligned} i &= A\epsilon^{-ut}\{\epsilon^{+s(t'-t_0)} \cos kl' \pm \epsilon^{-s(t'-t_0)} \cos k\,(l' - l_0)\} \\ e &= A\sqrt{\frac{L}{C}}\,\epsilon^{-ut}\{\epsilon^{+st'} \sin kl' \pm \epsilon^{-st'} \sin k\,(l' - l_0)\} \end{aligned}\right\} \tag{45}$$

or, substituting,

$$A = B\,\epsilon^{+u(t_1-t_0)} \tag{46}$$

gives

$$\left.\begin{aligned} i &= B\epsilon^{-ut'}\{\epsilon^{+s(t'-t_0)} \cos kl' \pm \epsilon^{-s(t'-t_0')} \cos k\,(l' - l_0)\} \\ e &= B\sqrt{\frac{L}{C}}\,\epsilon^{-ut'}\{\epsilon^{+st'} \sin kl' \pm \epsilon^{-st'} \sin k\,(l' - l_0)\} \end{aligned}\right\} \tag{47}$$

where the + sign applies, if A_1 and A_2 are of the same, the − sign, if A_1 and A_2 are of different sign.

Substituting, instead of the equations (41), the equations

$$\left.\begin{aligned} D_1 &= A_1 \sin kl_1 \qquad & D_3 &= A_2 \sin k\,(l_1 - l_0) \\ D_2 &= A_1 \cos kl_1 \qquad & D_4 &= A_2 \cos k\,(l_1 - l_0) \end{aligned}\right\} \tag{48}$$

gives

$$\left.\begin{aligned} i &= B\epsilon^{-ut'}\{\epsilon^{+s(t'-t_0)} \sin kl' \pm \epsilon^{-s(t'-t_0)} \sin k\,(l' - l_0)\} \\ e &= -B\sqrt{\frac{L}{C}}\,\epsilon^{-ut'}\{\epsilon^{+st'} \cos kl' \pm \epsilon^{-st'} \cos k\,(l' - l_0)\} \end{aligned}\right\} \tag{49}$$

The equations (47) and (49) of the periodic impulse, are of the same form as the equations (32) of the non-periodic impulse, except that in the latter the hyperbolic functions take the place of the trigonometric functions in (47) and (49).

34. From these equations (37), (40), (43), (45), (47) and (49) it follows:

In the periodic impulse, (37), current and voltage each consist of two components, which are periodic functions of distance, but exponential functions of time. The second component dies out at a faster rate than the first component. Since the attenuation due to the energy dissipation by resistance and by conductance follows the exponential term ϵ^{-ut}, it follows, that the first component of current and voltage respectively dies out with the time at a slower rate, the second component at a faster rate than corresponds to the energy dissipation in the circuit. That is, the first component receives energy, the second gives off energy, and the second component thus continuously transfers energy to the first component.

The coefficient u thus may be called the *energy dissipation constant*, the coefficient s the *energy transfer constant*, while $u - s$ and $u + s$ respectively are the *attenuation constants* of the two respective components.

By energy transfer, the first component thus increases in energy by ϵ^{+st}, the second decreases by ϵ^{-st}, while both simultaneously decrease by ϵ^{-ut}, by energy dissipation, as seen from (40).

From (43), (45) and (47) it follows, that current and voltage are in quadrature with each other in their distribution in space, in either of the two components of the periodic impulse. That is, in each of the two components maximum current coincides with zero voltage, and inversely.

From (45) and (47) it follows, that the two components of the periodic impulse differ in the phase of their space distribution by the distance l_0, the second component lagging behind the first component by the distance l_0.

In each of the two components of the periodic impulse, the current lags behind the voltage by the time t_0.

Current and voltage thus are in quadrature with each other in space, and displaced from each other in time, by the "time displacement" t_0.

From equations (39) it follows, that t_0 is positive, that is, the current lags behind the voltage by time t_0, if m is positive, and t_0 is negative, that is, the current leads the voltage, if m is negative.

Since $m = \frac{1}{2}\left(\frac{r}{L} - \frac{g}{C}\right)$, it follows:

The current lags behind the voltage, if $\frac{r}{L} > \frac{g}{C}$, that is, if the inductance effect preponderates, and

The current leads the voltage, if $\frac{g}{C} > \frac{r}{L}$, that is, if the capacity effect preponderates.

35. From (47) it follows:

The voltage equals the current times the surge impedance $z = \sqrt{\frac{L}{C}}$, but is in quadrature with it in space, and the current is lagging by t_0 in time.

By the conditions of existence of the periodic impulse, s must numerically be smaller than m.

$$s = 0 \text{ gives}$$

By (39)
$$st_0 = 0$$

By (33)
$$k = m\sqrt{LC}$$

and by (49)

$$\left.\begin{aligned} i &= B\epsilon^{-ut}\{\cos kl' \pm \cos k\,(l' - l_0)\} \\ e &= B\sqrt{\frac{L}{C}}\,\epsilon^{-ut'}\{\sin kl' \pm \sin k\,(l' - l_0)\} \end{aligned}\right\} \tag{50}$$

hence, current and voltage are in phase in time, but in quadrature in space.

$$s = m \text{ gives}$$
$$k = 0$$

Hence, from (40)

$$i = \epsilon^{-ut}(D_1\epsilon^{+mt} + D_3\epsilon^{-mt})$$

$$e = \sqrt{\frac{L}{C}}\,\epsilon^{-ut}(D_2\epsilon^{+m(t+t_0)} + D_4\epsilon^{-m(t+t_0)})$$

hence, substituting for u and m,

$$\left.\begin{aligned} i &= D_1\epsilon^{-\frac{g}{C}t} + D_3\epsilon^{-\frac{r}{L}t} \\ e &= \sqrt{\frac{L}{C}}\,\{D_2'\epsilon^{-\frac{g}{C}(t+t_0)} + D_4'\epsilon^{-\frac{r}{L}(t+t_0)}\} \end{aligned}\right\} \tag{51}$$

where

$$\left.\begin{aligned} D_2' &= D_2\epsilon^{+ut_0} \\ D_4' &= D_4\epsilon^{+ut_0} \end{aligned}\right\} \tag{52}$$

In this impulse, the capacity terms and the inductance terms are separate, and current and voltage are uniform throughout the entire circuit.

36. The constants D or A or B are determined, as integration constants, by the terminal conditions of the problem.

For instance, if at the starting moment of the impulse, that is, at time $t = 0$, the distribution of current and of voltage throughout the circuit are given, it is, by (37), for $t = 0$,

$$\left.\begin{aligned} i &= \Sigma\{(D_1 + D_3)\cos kl - (D_2 + D_4)\sin kl\} \\ e &= \sqrt{\frac{L}{C}}\,\Sigma\left\{\left(cD_2 + \frac{D_4}{c}\right)\cos kl + \left(cD_1 + \frac{D_3}{c}\right)\sin kl\right\} \end{aligned}\right\} \qquad (53)$$

The development of the given distribution of current and voltage into a Fourier series thus gives in the coefficients of this series the equations determining the constants D_1, D_2, D_3, D_4.

CHAPTER IV

DISCUSSION OF GENERAL EQUATIONS

37. In Chapter I the general equations of current and voltage were derived for a circuit or section of a circuit having uniformly distributed and constant values of r, L, g, C. These equations appear as a sum of groups of four terms each, characterized by the feature that the four terms of each group have the same values of s, q, h, k.

Of the four terms of each group, i_1, i_2, i_3, i_4 or e_1, e_2, e_3, e_4 respectively (equations (53) and (54) of Chapter I), two contain the angles $(qt - kl)$: i_1, e_1 and i_3, e_3; and two contain the angles $(qt + kl)$: i_2, e_2 and i_4, e_4.

In the terms i_1, e_1 and i_3, e_3, the speed of propagation of the phenomena follows from the equation

$$qt - kl = \text{constant},$$

thus:

$$\frac{dl}{dt} = +\frac{q}{k},$$

hence is positive, that is, the propagation is from lower to higher values of l, or towards increasing l.

In the terms i_2, e_2 and i_4, e_4, the speed of propagation from

$$qt + kl = \text{constant}$$

is

$$\frac{dl}{dt} = -\frac{q}{k}$$

hence is negative, that is, the propagation is from higher to lower values of l, or towards decreasing l.

Considering therefore i_1, e_1 and i_3, e_3 as direct or main waves, i_2, e_2 and i_4, e_4 are their return waves, or reflected waves, and i_2, e_2 is the reflected wave of i_1, e_1; i_4, e_4 is the reflected wave of i_3, e_3.

Obviously, i_2, e_2 and i_4, e_4 may be considered as main waves, and then i_1, e_1 and i_3, e_3 are reflected waves. Substituting $(-l)$ for $(+l)$ in equations (53) and (54) of Chapter I, that is, looking at the circuit in the opposite direction, terms i_2, e_2 and i_1, e_1 and terms i_4, e_4 and i_3, e_3 merely change places, but otherwise the equations remain the same, except that the sign of i is reversed, that is, the current is now considered in the opposite direction.

Each group thus consists of two waves and their reflected waves: $i_1 - i_2$ and $e_1 + e_2$ is the first wave and its reflected wave, and $i_3 - i_4$ and $e_3 + e_4$ is the second wave and its reflected wave.

In general, each wave and its reflected wave may be considered as one unit, that is, we can say: $i' = i_1 - i_2$ and $e' = e_1 + e_2$ is the first wave, and $i'' = i_3 - i_4$ and $e'' = e_3 + e_4$ is the second wave.

In the first wave, i', e', the amplitude decreases in the direction of propagation, ε^{-hl} for rising, ε^{+hl} for decreasing l, and the wave dies out with increasing time t by $\varepsilon^{-(u-s)t} = \varepsilon^{-ut}\,\varepsilon^{+st}$.

In the second wave, i'', e'', the amplitude increases in the direction of propagation, ε^{+hl} for rising, ε^{-hl} for decreasing l, but the wave dies out with the increasing time t by $\varepsilon^{-(u+s)t} = \varepsilon^{-ut}\,\varepsilon^{-st}$, that is, faster than the first wave.

If the amplitude of the wave remained constant throughout the circuit — as would be the case in a free oscillation of the circuit, in which the stored energy of the circuit is dissipated, but no power supplied one way or the other — that is, if $h = 0$, from equation (59) of Chapter I, $s = 0$; that is, both waves coincide and form one, which dies out with the time by the decrement ϵ^{-ut}.

It thus follows: In general, two waves, with their reflected waves, traverse the circuit, of which the one, i'', e'', increases in amplitude in the direction of propagation, but dies out correspondingly more rapidly in time, that is, faster than a wave of constant amplitude, while the other, i', e', decreases in amplitude but lasts a longer time, that is, dies out slower than a wave of constant amplitude. In the one wave, i'', e'', an increase of amplitude takes place at a sacrifice of duration in time, while in the other wave, i', e', a slower dying out of the wave with the time is produced at the expense of a decrease of amplitude during its propagation, or, in i'', e'' duration in time is sacrificed to duration in distance, and inversely in i', e'.

It is interesting to note that in a circuit having resistance, inductance, and capacity, the mathematical expressions of the two cases of energy flow; that is, the gradual or exponential and the oscillatory or trigonometric, are both special cases of the equations (63) and (64) of Chapter I, corresponding respectively to $q = 0$, $k = 0$ and to $h = 0$, $s = 0$.

38. In the equations (53) and (54) of Chapter I

$$qt = 2\pi$$

gives the time of a complete cycle, that is, the *period of the wave*,

$$t_0 = \frac{2\pi}{q},$$

and *the frequency of the wave* is

$$f = \frac{q}{2\pi}.$$

$$kl = 2\pi$$

gives the distance of a complete cycle, that is, the *wave length*,

$$l_w = \frac{2\pi}{k},$$

$$(u - s)\,t = 1 \quad \text{and} \quad (u + s)\,t = 1$$

give the time,

$$t_1' = \frac{1}{u - s} \quad \text{and} \quad t_1'' = \frac{1}{u + s},$$

during which the wave decreases to $\frac{1}{\varepsilon} = 0.3679$ of its value, and

$$hl = 1$$

gives the distance,

$$l_1 = \frac{1}{h},$$

over which the wave decreases to $\frac{1}{\varepsilon} = 0.3679$ of its value;

that is, q is the *frequency constant* of the wave,

$$f = \frac{q}{2\pi}, \qquad t_0 = \frac{2\pi}{q}; \tag{1}$$

k is the *wave length constant*,

$$l_w = \frac{2\pi}{k}; \tag{2}$$

$(u - s)$ and $(u + s)$ are the *time attenuation constants* ot the wave,

$$\left.\begin{aligned} t_1' &= \frac{1}{u - s}, \\ t_1'' &= \frac{1}{u + s}, \end{aligned}\right\} \tag{3}$$

and h is the *distance attenuation constant* of the wave,

$$l_1 = \frac{1}{h}. \tag{4}$$

39. If the frequency of the current and e.m.f. is very high, thousands of cycles and more, as with traveling waves, lightning disturbances, high-frequency oscillations, etc., q is a very large quantity compared with s, u, m, h, k, and k is a large quantity compared with h, then by dropping in equations (53) to (64) of Chapter I the terms of secondary order the equations can be simplified.

From (57) of Chapter I,

$$\begin{aligned} R_1^2 &= \sqrt{(s^2 + q^2 - m^2)^2 + 4\,q^2m^2} = \sqrt{(q^2 + m^2)^2 + 2\,s^2\,(q^2 - m^2)} \\ &= (q^2 + m^2)\left\{1 + \frac{2\,s^2\,(q^2 - m^2)}{(q^2 + m^2)^2}\right\}^{\frac{1}{2}} \\ &= q^2 + m^2 + s^2\,\frac{q^2 - m^2}{q^2 + m^2} \\ &= q^2 + m^2 + s^2 \\ &= q^2, \end{aligned}$$

$$\left.\begin{aligned} h &= \sqrt{LC}\sqrt{\tfrac{1}{2}\left\{R_1^2 + s^2 - q^2 - m^2\right\}} = s\sqrt{LC}, \\ k &= \sqrt{LC}\sqrt{\tfrac{1}{2}\left\{R_1^2 - s^2 + q^2 + m^2\right\}} = \sqrt{LC\,(q^2 + m^2)} = q\sqrt{LC}, \end{aligned}\right\} \tag{5}$$

$$h^2 + k^2 = (s^2 + q^2)\,LC = q^2LC,$$

and

$$c_1 = \frac{qk + h\,(m + s)}{h^2 + k^2}\,L = \frac{qL}{k} = \sqrt{\frac{L}{C}},$$

$$c_2 = \frac{qk - h(m - s)}{h^2 + k^2} L = \frac{qL}{k} = \sqrt{\frac{L}{C}},$$

$$c_1' = \frac{k(m + s) - qh}{h^2 + k^2} L = \frac{q\sqrt{LC}\,(m + s) - qs\sqrt{LC}}{q^2 LC} L = \frac{m}{q}\sqrt{\frac{L}{C}},$$

$$c_2' = \frac{k(m - s) + qh}{h^2 + k^2} L = \frac{q\sqrt{LC}\,(m - s) + qs\sqrt{LC}}{q^2 LC} L = \frac{m}{q}\sqrt{\frac{L}{C}};$$

that is,

and

$$\left.\begin{aligned} c_1 = c_2 &= \sqrt{\frac{L}{C}} \\ c_1' = c_2' &= \frac{m}{q}\sqrt{\frac{L}{C}}. \end{aligned}\right\} \tag{9}$$

Writing

$$\left.\begin{aligned} \sigma &= \sqrt{LC}, \\ c &= \sqrt{\frac{L}{C}}, \end{aligned}\right\} \tag{7}$$

where σ is the reciprocal of the speed of propagation (velocity of light), we have

$$\left.\begin{aligned} h &= \sigma s, \\ k &= \sigma q, \end{aligned}\right\} \tag{8}$$

and

$$\left.\begin{aligned} c_1 = c_2 &= c \\ c_1' = c_2' &= \frac{m}{q} c \end{aligned}\right\} \tag{9}$$

and introducing the new independent variable, as distance,

$$\lambda = \sigma l, \tag{10}$$

we have

and

$$\left.\begin{aligned} kl &= q\lambda \\ hl &= s\lambda; \end{aligned}\right\} \tag{11}$$

hence, the wave length is given by

$$q\lambda = 2\pi$$

as

$$\lambda_0 = \frac{2\pi}{q}, \tag{12}$$

and since the period is

$$t_0 = \frac{2\pi}{q},$$

it follows that *by the introduction of the denotation* (10) *distances are measured with the velocity of propagation as unit length,* and wave length l_w and period t_0 thus have the same numerical values.

The use of the velocity of propagation as unit of length of electric circuits such as transmission lines offers many advantages in dealing with transients, and therefore is generally advisable.

Substituting now in equations (63) and (64) of Chapter I gives

$$i = \varepsilon^{-ut}\sum\{\varepsilon^{+s(t-\lambda)} D_1[q(t-\lambda)] - \varepsilon^{+s(t+\lambda)} D_2[q(t+\lambda)]\ (i')$$
$$+ \varepsilon^{-s(t-\lambda)} D_3[q(t-\lambda)] - \varepsilon^{-s(t+\lambda)} D_4[q(t+\lambda)]\}\ (i'') \tag{13}$$

and

$$e = \varepsilon^{-ut}\sum\{\varepsilon^{+s(t-\lambda)} H_1[q(t-\lambda)] + \varepsilon^{+s(t+\lambda)} H_2[q(t+\lambda)]\ (e')$$
$$+ \varepsilon^{-s(t-\lambda)} H_3[q(t-\lambda)] + \varepsilon^{-s(t+\lambda)} H_4[q(t+\lambda)]\},\ (e'') \tag{14}$$

where

$$D[q(t \pm \lambda)] = C\cos q(t \pm \lambda) + C'\sin q(t \pm \lambda)$$

and

$$H[q(t \pm \lambda)] = \sqrt{\frac{L}{C}}\left\{\left(\frac{m}{q}C' - C\right)\cos q(t \pm \lambda)\right.$$
$$\left. - \left(\frac{m}{q}C + C'\right)\sin q(t \pm \lambda)\right\}. \tag{15}$$

40. As seen from equations (13) and (14), the waves are products of ϵ^{-ut} and a function of $(t - \lambda)$ for the main wave,

$(t + \lambda)$ for the reflected wave, thus:

$$\left.\begin{aligned} i_1 + i_3 &= \varepsilon^{-ut} f_1 (t - \lambda) \\ \text{and} \qquad i_2 + i_4 &= \varepsilon^{-ut} f_2 (t + \lambda); \end{aligned}\right\} \tag{16}$$

hence, for constant $(t - \lambda)$ on the main waves, and for constant $(t + \lambda)$ on the reflected waves, we have

$$\left.\begin{aligned} i_1 + i_3 &= B\varepsilon^{-ut} \\ \text{and} \qquad i_2 + i_4 &= B'\varepsilon^{-ut}; \end{aligned}\right\} \tag{17}$$

that is, during its passage along the circuit the wave decreases by the decrement ε^{-ut}, or at a constant rate, independent of frequency, wave length, etc., and depending merely on the circuit constants r, L, g, C. The decrement of the traveling wave in the direction of its motion is

$$\varepsilon^{-ut} = \varepsilon^{-\frac{1}{2}\left(\frac{r}{L} + \frac{g}{C}\right)t},$$

and therefore is independent of the character of the wave, for instance its frequency, etc.

41. The physical meaning of the two waves i' and e' can best be appreciated by observing the effect of the wave when traversing a fixed point λ of the circuit.

Consider as example the main wave only, $i' = i_1 + i_3$, and neglect the reflected waves, for which the same applies.

From equation (74),

$$i = \varepsilon^{-s\lambda - (u-s)t} D_1 [q (t - \lambda)] + \varepsilon^{+s\lambda - (u+s)t} D_3 [q (t - \lambda)]; \tag{18}$$

or the absolute value is

$$I = D_1 \varepsilon^{-s\lambda - (u-s)t} + D_3 \varepsilon^{+s\lambda - (u+s)t}, \tag{19}$$

where D_1 and D_3 have to be combined vectorially.

Assuming then that at the time $t = 0$, $I = 0$, for constant λ we have

$$I = D (\varepsilon^{-(u-s)t} - \varepsilon^{-(u+s)t}), \tag{20}$$

the amplitude of I at point λ.

Since (81) is the difference of two exponential functions of different decrement, it follows that as function of the time t, I

rises from 0 to a maximum and then decreases again to zero, as shown in Fig. 1, where

$$I_1 = D\varepsilon^{-(u-s)t},$$
$$I_3 = D\varepsilon^{-(u+s)t},$$
$$I = I_1 - I_3,$$

and the actual current i is the oscillatory wave with I as envelope.

The combination of two waves thus represents the passage of a wave across a given point, the amplitude rising during the arrival and decreasing again after the passage of the wave.

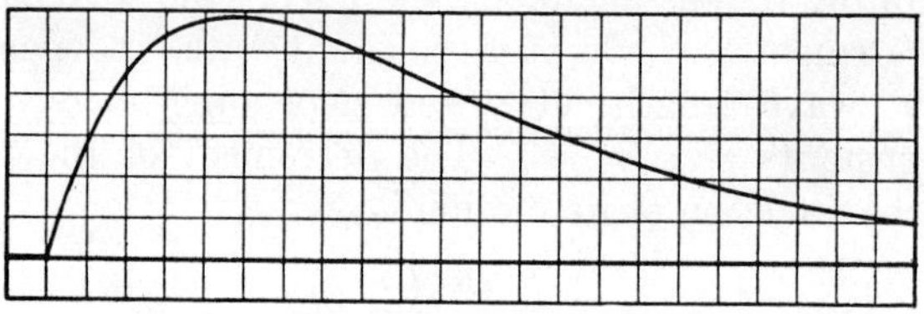

FIG. 1. Amplitude of electric traveling wave.

42. If h and so also s equal zero, i', e' and i'', e'' coincide in equations (13) and (14), and C_1 and C_3 thus can be combined into one constant B_1, C_2 and C_4 into one constant B_2, thus:

$$\left.\begin{aligned} C_1 + C_3 &= B_1, \\ C_2 + C_4 &= B_2, \\ C_1' + C_3' &= B_1', \\ C_2' + C_4' &= B_2', \end{aligned}\right\} \tag{21}$$

and (13), (14) then assume the form

$$\begin{aligned} i = \varepsilon^{-ut} \sum \{ &[B_1 \cos q\,(t - \lambda) + B_1' \sin q\,(t - \lambda)] \\ &- [B_2 \cos q\,(t + \lambda) + B_2' \sin q\,(t + \lambda)]\}, \end{aligned} \tag{22}$$

$$e = \sqrt{\frac{L}{C}}\,\varepsilon^{-ut} \sum \left\{ \left[\left(\frac{m}{q} B_1' - B_1\right) \cos q\,(t-\lambda) - \left(\frac{m}{q} B_1 + B_1'\right) \sin q\,(t-\lambda) \right] + \left[\left(\frac{m}{q} B_2' - B_2\right) \cos q\,(t+\lambda) - \left(\frac{m}{q} B_2 + B_2'\right) \sin q\,(t+\lambda) \right] \right\}. \tag{23}$$

These equations contain the distance λ only in the trigonometric but not in the exponential function; that is, i and e vary in phase throughout the circuit, but not in amplitude; or, in other words, the oscillation is of uniform intensity throughout the circuit, dying out uniformly with the time from an initial maximum value; however, the wave does not travel along the circuit, but is a stationary or standing wave. It is an oscillatory discharge of a circuit containing a distributed r, L, g, C, and therefore is analogous to the oscillating condenser discharge through an inductive circuit, except that, due to the distributed capacity, the phase changes along the circuit. The free oscillations of a circuit such as a transmission line are of this character.

For $\lambda = 0$, that is, assuming the wave length of the oscillation as so great, hence the circuit as such a small fraction of the wave length, that the phase of i and e can be assumed as uniform throughout the circuit, the equations (22) and (23) assume the form

$$i = \varepsilon^{-ut}\{B_0 \cos qt + B_0' \sin qt\}$$

and (24)

$$e = \sqrt{\frac{L}{C}}\varepsilon^{-ut}\left\{\left(\frac{m}{q}B' - B\right)\cos qt - \left(\frac{m}{q}B + B'\right)\sin qt\right\};$$

these are the usual equations of the condenser discharge through an inductive circuit, which here appear as a special case of a special case of the general circuit equations.

If q equals zero, the functions D and H in equations (13) and (14) become constant, and these equations so assume the form

$$\left.\begin{aligned} i &= \varepsilon^{-ut}\sum\{[C_1\varepsilon^{+s(t-\lambda)} + C_3\varepsilon^{-s(t-\lambda)}] \\ &\qquad - [C_2\varepsilon^{+s(t+\lambda)} + C_4\varepsilon^{-s(t+\lambda)}]\} \\ &\text{and} \\ e &= \sqrt{\frac{L}{C}}\varepsilon^{-ut}\quad\{[B_1\varepsilon^{+s(t-\lambda)} + B_3\varepsilon^{-s(t-\lambda)}] \\ &\qquad + [B_2\varepsilon^{+s(t+\lambda)} + B_4\varepsilon^{-s(t+\lambda)}]\}, \end{aligned}\right\} \qquad (25)$$

where

$$B = \frac{m}{q} C' - C. \tag{26}$$

This gives expressions of current and e.m.f. which are no longer oscillatory but exponential, thus representing a gradual change of i and e as functions of time and distance, corresponding to the gradual or logarithmic condenser discharge. For $\lambda = 0$, these equations change to the equations of the logarithmic condenser discharge.

These equations (25) are only approximate, however, since in them the quantities s, u, h have been neglected compared with q, assuming the latter as very large, while now it is assumed as zero.

43. If, however,

$$m = \frac{1}{2}\left(\frac{r}{L} - \frac{g}{C}\right) = 0, \tag{27}$$

that is,

$$\frac{r}{L} = \frac{g}{C}, \tag{28}$$

or

$$r \div g = L \div C,$$

or, in words, the power coefficients of the circuit are proportional to the energy storage coefficients, or the time constant of the electromagnetic field of the circuit, $\frac{r}{L}$, equals the time constant of the electrostatic field of the circuit, $\frac{g}{C}$, then

$$u = \frac{r}{L} = \frac{g}{C} = \text{time constant of the circuit}, \tag{29}$$

and from equation (57) of Chapter I

$$\left.\begin{aligned} R_1{}^2 &= s^2 + q^2, \\ h &= s\sqrt{LC} = \sigma s, \\ k &= q\sqrt{LC} = \sigma q, \end{aligned}\right\} \tag{30}$$

and from equation (55) of Chapter I

$$\left.\begin{aligned} c_1 &= \frac{L}{\sigma} = \sqrt{\frac{L}{C}} = c, \\ c_1' &= 0, \\ c_2 &= \frac{L}{\sigma} = \sqrt{\frac{L}{C}} = c, \\ c_2' &= 0; \end{aligned}\right\} \quad (31)$$

and

hence, substituting in equations (53) and (54) of Chapter I,

$$i = \varepsilon^{-\frac{r}{L}t} \left\{ \varepsilon^{+s(t-\lambda)} D_1 [q(t-\lambda)] - \varepsilon^{+s(t+\lambda)} D_2 [q(t+\lambda)] + \varepsilon^{-s(t-\lambda)} D_3 [q(t-\lambda)] - \varepsilon^{-s(t+\lambda)} D_4 [q(t+\lambda)] \right\} \quad (32)$$

and

$$e = -\sqrt{\frac{L}{C}} \varepsilon^{-\frac{r}{L}t} \left\{ \varepsilon^{+s(t-\lambda)} D_1 [q(t-\lambda)] + \varepsilon^{+s(t+\lambda)} D_2 [q(t+\lambda)] + \varepsilon^{-s(t-\lambda)} D_3 [q(t-\lambda)] + \varepsilon^{-s(t+\lambda)} D_4 [q(t+\lambda)] \right\}. \quad (33)$$

These equations are similar to (13) and (14), but are derived here for the case $m = 0$, without assumptions regarding the relative magnitude of q and the other quantities: "distortionless circuit."

These equations (32) and (33) therefore also apply for $q = 0$, and then assume the form

$$i = \varepsilon^{-\frac{r}{L}t} \left\{ [C_1 \varepsilon^{+s(t-\lambda)} + C_3 \varepsilon^{-s(t-\lambda)}] - [C_2 \varepsilon^{+s(t+\lambda)} + C_4 \varepsilon^{-s(t+\lambda)}] \right\}, \quad (34)$$

$$e = -\sqrt{\frac{L}{C}} \varepsilon^{-\frac{r}{L}t} \left\{ [C_1 \varepsilon^{+s(t-\lambda)} + C_3 \varepsilon^{-s(t-\lambda)}] + [C_2 \varepsilon^{+s(t+\lambda)} + C_4 \varepsilon^{-s(t+\lambda)}] \right\}. \quad (35)$$

These equations (34) and (35) are the same as (25), but in the present case, where $m = 0$, apply irrespective of the relative values of the quantities s, etc.

Therefore in a circuit in which $m = 0$ a transient term may appear which is not oscillatory in time nor in space, but changing gradually.

If the constant h in equations (53) and (54) differs from zero, the oscillation (using the term *oscillation* here in the most general sense, that is, including also *alternation*, as an oscillation of zero attenuation) travels along the circuit, but it becomes stationary, as a standing wave, for $h = 0$; that is, the distance attenuation constant h may also be called the *propagation constant* of the wave.

$h = 0$ thus represents a wave which does not propagate or move along the circuit, but stands still, that is, a stationary or standing wave.

If the constant h in equations (53) and (54) of Chapter I differs from zero, the oscillation (using the term *oscillation* here in the most general sense, that is, including also *alternation*, as an oscillation of zero attenuation) travels along the circuit, but it becomes stationary, as a standing wave, for $h = 0$; that is, the distance attenuation constant h may also be called the *propagation constant* of the wave.

$h = 0$ thus represents a wave which does not propagate or move along the circuit, but stands still, that is, a stationary or standing wave.

CHAPTER V

STANDING WAVES

44. If the propagation constant of the wave vanishes,

$$h = 0,$$

the wave becomes a stationary or standing wave, and the equations of the standing wave are thus derived from the general equations (53) to (64) of Chapter I, by substituting therein $h = 0$, which gives

$$R_2^2 = \sqrt{(k^2 - LCm^2)^2}; \tag{1}$$

hence, if

$$k^2 > LCm^2,$$
$$R_2^2 = k^2 - LCm^2;$$

and if

$$k^2 < LCm^2,$$
$$R_2^2 = LCm^2 - k^2.$$

Therefore, two different cases exist, depending upon the relative values of k^2 and LCm^2, and in addition thereto the intermediary or critical case, in which $k^2 = LCm^2$.

These three cases require separate consideration.

$$LCm^2 = LC\left\{\frac{1}{2}\left(\frac{r}{L} - \frac{g}{C}\right)\right\}^2 = \left\{\frac{1}{2}\left(r\sqrt{\frac{C}{L}} - g\sqrt{\frac{L}{C}}\right)\right\}^2 \tag{2}$$

is a circuit constant, while k is the wave length constant, that is, the higher k the shorter the wave length.

A. Short waves,

$$k^2 > LCm^2, \tag{3}$$

hence,

$$R_2^2 = k^2 - LCm^2 \tag{4}$$

and

$$s = 0,$$

$$q = \sqrt{\frac{k^2}{LC} - m^2}, \tag{5}$$

or approximately, for very large k,

$$q = \frac{k}{\sqrt{LC}}. \tag{6}$$

Herefrom then follows

$$\left.\begin{aligned} c_1 &= \frac{qL}{k} = c, \\ c_1' &= \frac{mL}{k} = c', \\ c_2 &= \frac{qL}{k} = c, \\ \text{and} \qquad c_2' &= \frac{mL}{k} = c'. \end{aligned}\right\} \tag{7}$$

Substituting now $h = 0$ and (5), (6) in equations (53), (54), of Chapter I, the two waves i', e' and i'', e'' coincide, and all the exponential terms reduce to ϵ^{-ut}; hence, substituting

$$\left.\begin{aligned} B_1 &= C_1 + C_3, \\ B_2 &= C_2 + C_4, \\ B_1' &= C_1' + C_3', \\ \text{and} \qquad B_2' &= C_2' + C_4', \end{aligned}\right\} \tag{8}$$

gives

$$\begin{aligned} i = \varepsilon^{-ut} \{&[B_1 \cos (qt - kl) + B_1' \sin (qt - kl)] \\ &- [B_2 \cos (qt + kl) + B_2' \sin (qt + kl)]\} \end{aligned} \tag{9}$$

and

$$\begin{aligned} e = \frac{L}{k} \varepsilon^{-ut} \{&[(mB_1' - qB_1) \cos (qt - kl) - (mB_1 + qB_1') \sin (qt - kl)] \\ &+ [(mB_2' - qB_2) \cos (qt + kl) - (mB_2 + qB_2') \sin (qt + kl)]\}. \end{aligned} \tag{10}$$

Equations (9) and (10) represent a *stationary electrical oscillation* or standing wave on the circuit.

B. Long waves,

$$k^2 < LCm^2; \tag{11}$$

hence,

$$R_2^2 = LCm^2 - k^2. \tag{12}$$

and

$$\left.\begin{aligned} s &= \sqrt{m^2 - \frac{k^2}{LC}}, \\ q &= 0, \end{aligned}\right\} \tag{13}$$

or approximately, for very small values of k,

$$s = m = \frac{1}{2}\left(\frac{r}{L} - \frac{g}{C}\right); \tag{14}$$

herefrom then follows

$$\left.\begin{aligned} c_1 &= c_2 = 0, \\ c_1' &= \frac{(m+s)\,L}{k} \\ \text{and} \qquad c_2' &= \frac{(m-s)\,L}{k}. \end{aligned}\right\} \tag{15}$$

Substituting now $h = 0$ and (13), (15) into (53) and (54) of Chapter I, the two waves i', e' and i'', e'' remain separate, having different exponential terms, $\epsilon^{-(u-s)t}$ and $\epsilon^{-(u+s)t}$, but in each of the two waves the main wave and the reflected wave coincide, due to the vanishing of q.

Substituting then

$$\left.\begin{aligned} B_1 &= C_1 - C_2, \\ B_1' &= C_1' + C_2', \\ B_2 &= C_3 - C_4, \\ \text{and} \qquad B_2' &= C_3' + C_4', \end{aligned}\right\} \tag{16}$$

gives

$$i = \varepsilon^{-ut}\{(B_1\,\varepsilon^{+st} + B_2\varepsilon^{-st}) \cos kl - (B_1'\varepsilon^{+st} + B_2'\,\varepsilon^{-st}) \sin kl\} \tag{17}$$

and

$$e = \frac{L}{k} \varepsilon^{-ut} \{ [(m + s) B_1' \varepsilon^{+st} + (m - s) B_2' \varepsilon^{-st}] \cos kl$$

$$+ [(m + s) B_1 \varepsilon^{+st} + (m - s) B_2 \varepsilon^{-st}] \sin kl \}$$

$$\left. \begin{aligned} &= \frac{L}{k} \varepsilon^{-ut} \{ m[(B_1' \varepsilon^{+st} + B_2' \varepsilon^{-st}) \cos kl \\ &\qquad + (B_1 \varepsilon^{+st} + B_2 \varepsilon^{-st}) \sin kl] \\ &\quad + s[(B_1' \varepsilon^{+st} - B_2' \varepsilon^{-st}) \cos kl \\ &\qquad + (B_1 \varepsilon^{+st} - B_2 \varepsilon^{-st}) \sin kl] \} \end{aligned} \right\} \quad (18)$$

Equations (17) and (18) represent a *gradual or exponential circuit discharge*, and the distribution still is a trigonometric function of the distance, that is, a wave distribution, but dies out gradually with the time, without oscillation.

C. Critical case,

$$k^2 = LCm^2; \quad (19)$$

hence,

$$\left. \begin{aligned} R_2^2 &= 0, \\ s &= 0, \\ q &= 0, \end{aligned} \right\} \quad (20)$$

and

$$\left. \begin{aligned} c_1 &= c_2 = 0, \\ c_1' &= c_2' = \frac{mL}{k} = \sqrt{\frac{L}{C}} = c', \end{aligned} \right\} \quad (21)$$

and all the main waves and their reflected waves coincide when substituting $h = 0$, (20), (21) in (53) and (54) of Chapter I.

Hence, writing

$$\left. \begin{aligned} B &= C_1 - C_2 + C_3 - C_4 \\ \text{and} \quad B' &= C_1' + C_2' + C_3' + C_4' \end{aligned} \right\} \quad (22)$$

gives

$$i = \varepsilon^{-ut} \{ B \cos kl - B' \sin kl \} \quad (23)$$

and

$$e = \sqrt{\frac{L}{C}} \epsilon^{-ut} \left\{ B' \cos kl + B \sin kl \right\}. \tag{24}$$

In the critical case (23) and (24), the wave is distributed as a trigonometric function of the distance, but dies out as a simple exponential function of the time.

45. An electrical standing wave thus can have two different forms: it can be either oscillatory in time or exponential in time, that is, gradually changing. It is interesting to investigate the conditions under which these two different cases occur.

The transition from gradual to oscillatory takes place at

$$k_0^2 = m^2 LC; \tag{25}$$

for larger values of k the phenomenon is oscillatory; for smaller, exponential or gradual.

Since k is the wave length constant, the wave length, at which the phenomenon ceases to be oscillatory in time and becomes a gradual dying out, is given by (2) of Chapter IV as

$$\left. \begin{aligned} l_{w0} &= \frac{2\pi}{k_0} \\ &= \frac{2\pi}{m\sqrt{LC}}. \end{aligned} \right\} \tag{26}$$

In an undamped wave, that is, in a circuit of zero r and zero g, in which no energy losses occur, the speed of propagation is

$$S = \frac{q}{k} = \frac{1}{\sqrt{LC}}, \tag{27}$$

and if the medium has unit permeability and unit inductivity, it is the speed of light,

$$S_0 = 3 \times 10^{10}. \tag{28}$$

In an undamped circuit, this wave length l_{w_0} would correspond to the frequency,

$$q_0 = \frac{k_0}{\sqrt{LC}} = m;$$

hence, from (1) of Chapter IV,

$$f_0 = \frac{q_0}{2\pi} = \frac{m}{2\pi}. \tag{29}$$

The frequency at the wave length l_{w_0} is zero, since at this wave length the phenomenon ceases to be oscillatory; that is, due to the energy losses in the circuit, by the effective resistance r and effective conductance g, the frequency f of the wave is reduced below the value corresponding to the wave length l_w, the more, the greater the wave length, until at the wave length l_{w_0} the frequency becomes zero and the phenomenon thereby non-oscillatory. This means that with increasing wave length the velocity of propagation of the phenomenon decreases, and becomes zero at wave length l_{w_0}.

If

$$m^2LC = 0,$$

$$k_0 = 0 \text{ and } l_{w_0} = \infty;$$

that is, the standing wave is always oscillatory.

If

$$m^2LC = \infty,$$

$$k_0 = \infty \text{ and } l_{w_0} = 0;$$

that is, the standing wave is always non-oscillatory, or gradually dying out.

In the former case, $m^2LC = 0$, or oscillatory phenomenon, substituting for m^2, we have

$$\frac{1}{4}\left(\frac{r}{L} - \frac{g}{C}\right)^2 LC = 0,$$

$$r\sqrt{\frac{C}{L}} - g\sqrt{\frac{L}{C}} = 0,$$

and

$$\frac{r}{g} = \frac{L}{C};$$

or

$$rC - gL = 0 \text{ (distortionless circuit).}$$

In the latter case, $m^2LC = \infty$, or non-oscillatory or exponential standing wave, we have

$$r\sqrt{\frac{C}{L}} - g\sqrt{\frac{L}{C}} = \infty,$$

and since neither r, g, L, nor C can be equal infinity it follows that either $L = 0$ or $C = 0$.

Therefore, the standing wave in a circuit is always oscillatory, regardless of its wave length, if

$$rC - gL = 0, \tag{30}$$

or

$$\frac{r}{g} = \frac{L}{C}; \tag{31}$$

that is, the ratio of the energy coefficients is equal to the ratio of the reactive coefficients of the circuit.

The standing wave can never be oscillatory, but is always exponential, or gradually dying out, if either the inductance L or the capacity C vanishes; that is, the circuit contains no capacity or contains no inductance.

In all other cases the standing wave is oscillatory for waves shorter than the critical value $l_{w_0} = \frac{2\pi}{k_0}$, where

$$k_0^2 = m^2LC = \frac{1}{4}\left\{r\sqrt{\frac{C}{L}} - g\sqrt{\frac{L}{C}}\right\}^2, \tag{32}$$

and is exponential or gradual for standing waves longer than the critical wave length l_{w_0}; or for $k < k_0$ the standing wave is exponential, for $k > k_0$ it is oscillatory.

The value $k_0 = m\sqrt{LC}$ thus takes a similar part in the theory of standing waves as the value $r_0^2 = 4\,L_0C_0$ in the condenser discharge through an inductive circuit; that is, it separates the exponential or gradual from trigonometric or oscillatory conditions.

The difference is that the condenser discharge through an inductive circuit is gradual, or oscillatory, depending on the circuit constants, while in a general circuit, with the same circuit constants, usually gradual as well as oscillatory standing waves exist, the former with greater wave length, or

$$m\sqrt{LC} > k, \tag{33}$$

the latter with shorter wave length, or

$$m\sqrt{LC} < k. \tag{34}$$

An idea of the quantity k_0, and therewith the wave length l_{w_0}, at which the frequency of the standing wave becomes zero, or the wave non-oscillatory, and of the frequency f_0, which, in an undamped circuit, will correspond to this critical wave length l_{w_0}, can best be derived by considering some representative numerical examples.

As such may be considered:

(1) A high-power high-potential overhead transmission line.
(2) A high-potential underground power cable.
(3) A submarine telegraph cable.
(4) A long-distance overhead telephone circuit.

(1) *High-power high-potential overhead transmission line.*

46. Assume energy to be transmitted 120 miles, at 40,000 volts between line and ground, by a three-phase system with grounded neutral. The line consists of copper conductors, wire No. 00 B. and S. gage, with 5 feet between conductors.

Choosing the mile as unit length,

$$r = 0.41 \text{ ohm per mile.}$$

The inductance of a conductor is given by

$$L = l\left(2 \log_\epsilon \frac{l_d}{l_r} + \frac{\mu}{2}\right) 10^{-9}, \text{ in henrys,} \tag{35}$$

where l = the length of conductor, in cm.; l_r = the radius of conductor; l_d = the distance from return conductor, and μ = the permeability of conductor material. For copper, $\mu = 1$.

As one mile equals 1.61×10^5 cm., substituting this, and reducing the natural logarithm to the common logarithm, by the factor 2.3026, gives

$$L = \left(0.7415 \log \frac{l_d}{l_r} + 0.0805\right), \text{ in mh. per mile.} \tag{36}$$

For
$$l_r = 0.1825 \text{ inch and } l_d = 60 \text{ inches,}$$
$$L = 1.95 \text{ mh. per mile.}$$

The capacity of a conductor is given by

$$C = l\left(\frac{1 + \delta}{2\, S_0^2 \log_\epsilon \frac{l_d}{l_r}}\right) 10^9, \text{ in farads,} \tag{37}$$

where $S_0 = 3 \times 10^{10}$ = the speed of light, and δ = the allowance for capacity of insulation, tie wires, supports, etc., assumed as 5 per cent.

Substituting S_0, and reducing to one mile and common logarithm, gives

$$C = \frac{0.0408}{\log \frac{l_d}{l_r}}, \text{in mf.}; \tag{38}$$

hence, in this instance,

$$C = 0.0162 \text{ mf.}$$

Estimating the loss in the static field of the line as 400 watts per mile of conductor gives an effective conductance,

$$g = \frac{400}{40{,}000^2} = 0.25 \times 10^{-6} \text{ mho},$$

which gives the line constants per mile as $r = 0.41$ ohm; $L = 1.95 \times 10^{-3}$ henry; $g = 0.25 \times 10^{-6}$ mho, and $C = 0.0162 \times 10^{-6}$ farad.

Herefrom then follows

$$u = \frac{1}{2}\left(\frac{r}{L} + \frac{g}{C}\right) = \frac{1}{2}(210 + 15.5) = 113,$$

$$m = \frac{1}{2}\left(\frac{r}{L} - \frac{g}{C}\right) = 97,$$

$$\sigma = \sqrt{LC} = \sqrt{31.6} \times 10^{-6} = 5.62 \times 10^{-6},$$

$$k_0 = m\sqrt{LC} = 545 \times 10^{-6};$$

hence, the critical wave length is

$$l_{w_0} = \frac{2\pi}{k_0} = 11{,}500 \text{ miles},$$

and in an undamped circuit this wave length would correspond to the frequency of oscillation,

$$f_0 = \frac{m}{2\pi} = 15.7 \text{ cycles per sec.}$$

Since the shortest wave at which the phenomenon ceases to be oscillatory is 11,500 miles in length, and the longest wave which can originate in the circuit is four times the length of the circuit, or 480 miles, it follows that whatever waves may originate in this circuit are by necessity oscillatory, and non-oscillatory currents or voltages can exist in this circuit only when impressed upon it by some outside source, and then are of such great wave length that the circuit is only an insignificant fraction of the wave, and great differences of voltage and current of non-oscillatory nature cannot exist, as *standing waves*.

Since the difference in length between the shortest non-oscillatory wave and the longest wave which can originate in the circuit is so very great, it follows that in high-potential long-distance transmission circuits all phenomena which may result in considerable potential differences and differences of current throughout the circuit are oscillatory in nature, and the solution case (A) is the one the study of which is of the greatest importance in long-distance transmissions.

With a length of circuit of 120 miles, the longest standing wave which can originate in the circuit has the wave length

$$l_w = 480 \text{ miles},$$

and herefrom follows

$$k = \frac{2\pi}{l_w} = 0.0134$$

and

$$\frac{k^2}{LC} = \frac{0.0134^2}{31.6 \times 10^{-12}} = 5.7 \times 10^6;$$

hence, in the expression of q in equation (101),

$$q = \sqrt{\frac{k^2}{LC} - m^2}$$

$$= \sqrt{5.7 \times 10^6 - 0.00941 \times 10^6},$$

m^2 is negligible compared with $\frac{k^2}{LC}$; that is,

$$q = \frac{k}{\sqrt{LC}} = \frac{0.0134}{5.62 \times 10^{-6}} = 2380; \tag{39}$$

or

$$f = \frac{q}{2\pi} = 380 \text{ cycles per sec.}$$

Hence, even for the longest standing wave which may originate in this transmission line, $q = 2380$ is such a large quantity compared with $m = 97$ that m can be neglected compared with q, and for shorter waves, the overtones of the fundamental wave, this is still more the case; that is, in equation (9) and (10) the terms with m may be dropped. In equation (10) $\frac{Lq}{k}$ thus become common factors, and since from equation (39)

$$\frac{Lq}{k} = \sqrt{\frac{L}{C}}, \tag{40}$$

by substituting $m = 0$ and (40) in (9) and (10) we get the *general equations of standing waves in long-distance transmission lines*, thus:

$$i = \varepsilon^{-ut}\{[B_1 \cos (qt - kl) + B_1' \sin (qt - kl)] \\ - [B_2 \cos (qt + kl) + B_2' \sin (qt + kl)]\}, \tag{41}$$

$$e = -\sqrt{\frac{L}{C}}\varepsilon^{-ut}\{[B_1 \cos (qt - kl) + B_1' \sin (qt - kl)] \\ + [B_2 \cos (qt + kl) + B_2' \sin (qt + kl)]\}, \tag{42}$$

or

$$e = \varepsilon^{-ut}\{[A_1 \cos (qt + kl) + A_1' \sin (qt + kl)] \\ + [A_2 \cos (qt - kl) + A_2' \sin (qt - kl)]\}, \tag{43}$$

$$i = \sqrt{\frac{C}{L}}\varepsilon^{-ut}\{[A_1 \cos (qt + kl) + A_1' \sin (qt + kl)] \\ - [A_2 \cos (qt - kl) + A_2' \sin (qt - kl)]\}, \tag{44}$$

where

$$A_1 = -\sqrt{\frac{L}{C}}B_2, \quad A_1' = -\sqrt{\frac{L}{C}}B_2', \text{ etc.}$$

(2) *High-potential underground power cable.*

47. Choose as example an underground power cable of 20 miles length, transmitting energy at 7000 volts between conductor and ground or cable armor, that is, a three-phase three-conductor 12,000-volt cable.

Assume the conductor as stranded and of a section equivalent to No. 00 B. and S. G.

the expression for the capacity, equation (23), multiplies with the expression for the capacity, equation (119), multiplies with the dielectric constant or specific capacity of the cable insulation, and that $\frac{l_d}{l_r}$ is very small, about three or less; or taking the values of the circuit constants from tests of the cable, we get values of the magnitude, per mile of single conductor, $r = 0.41$ ohm; $L = 0.4 \times 10^{-3}$ henry; $g = 10^{-6}$ mho, corresponding to a power factor of the cable-charging current, at 25 cycles, of 1 per cent; $C = .6 \times 10^{-6}$ farad.

Herefrom the following values are obtained: $u = 513, m = 512$, $\sigma = \sqrt{LC} = 15.5 \times 10^{-6}$, $k_0 = m\sqrt{LC} = 7.95 \times 10^{-3}$, and the critical wave length is $l_{w_0} = 790$ miles, and the frequency of an undamped oscillation, corresponding to l_{w_0}, is $f_0 = 81.5$ cycles per second.

As seen, in an underground high-potential cable the critical wave length is very much shorter than in the overhead long-distance transmission line. At the same time, however, the length of an underground cable circuit is very much shorter than that of a long-distance transmission line, so that the critical wave length still is very large compared with the greatest wave length of an oscillation originating in the cable, at least ten times as great. Which means that the discussion of the possible phenomena in any overhead line, under (1), applies also to the underground high-potential cable circuit.

In the present example the longest standing wave which may originate in the cable has the wave length

$$l_w = 80 \text{ miles},$$

which gives

$$k = 0.0785$$

and

$$\frac{k}{\sqrt{LC}} = 5070,$$

or about ten times as large as m, so that m can still be neglected in equation (26) of Chapter IV, and we have

$$q = \frac{k}{\sqrt{LC}} = 5070,$$

or

$$f = 810 \text{ cycles per second,}$$

and the general equations of the phenomenon in long-distance transmission lines, (27) to (29), also apply as the general equations of standing waves in high-potential underground cable circuits.

(3) *Submarine telegraph cable.*

48. Choosing the following values: length of cable, single stranded-conductor, ground return, = 4000 miles; constants per mile of conductor: $r = 3$ ohms, $L = 10^{-3}$ henry, $g = 10^{-6}$ mho, and $C = 0.1 \times 10^{-6}$ farad, we get $u = 1500$; $m = 1500$; $\sigma = \sqrt{LC} = 10 \times 10^{-6}$, and $k_0 = m\sqrt{LC} = 15 \times 10^{-3}$, from which the critical wave length is $l_{w_0} = 418$ miles, and the corresponding frequency $f_0 = 239$ cycles per second.

From the above it is seen that in a submarine cable the critical wave length l_{w_0} is relatively short, so that in long submarine cables standing waves may appear which are not oscillatory in time but die out gradually, that is, are shown by the equation of case B. In such cables, due to their relatively high resistance, the damping effect is very great; $u = 1500$, and standing waves, therefore, rapidly die out.

In the investigation of the submarine cable, the complete equations must therefore be used, and q cannot always be assumed as large compared with m and u, except when dealing with local oscillations.

(4) *Long-distance overhead telephone circuit.*

49. Consider a telephone circuit of 1000 miles length, metallic return, consisting of two wires No. 4 B. and S. G., 24 inches distant from each other.

Calculating in the same way as discussed under (1), the following constants per mile of conductor are obtained: $r = 1.31$ ohms, $L = 1.84 \times 10^{-3}$ henry, and $C = .0172 \times 10^{-6}$ farad.

As conductance, g, we may assume

(*a*) $g = 0$; that is, very perfect insulation, as in dry weather.

(*b*) $g = 2.5 \times 10^{-6}$; that is, slightly leaky line.

(c) $g = 12 \times 10^{-6}$; that is, poor insulation, or a leaky line.

(d) $g = 40 \times 10^{-6}$; that is, extremely poor insulation, as during heavy rain.

The condition may also be investigated where the line is loaded with inductance coils spaced so close together that in their effect we can consider this additional inductance as uniformly distributed. Let the total inductance per unit length be increased by the loading coils to

$$L_1 = 9 \times 10^{-3}\, h,$$

or about five times the normal value.

Denoting then the constants of the loaded line by the index 1, we have:

Quantity	(a)	(b)	(c)	(d)
$u =$	356	429	706	1,518
$u_1 =$	73	146	423	1,236
$m =$	356	283	6	-806
$m_1 =$	73	0	-277	$-1{,}090$
$\sigma = \sqrt{LC} =$	5.63×10^{-6}			
$\sigma_1 = \sqrt{L_1 C} =$	12.45×10^{-6}			
$k_0 = m\sqrt{LC} =$	2×10^{-3}	1.6×10^{-3}	33.7×10^{-6}	4.56×10^{-3}
$k_{01} = m_1\sqrt{L_1 C} =$	910×10^{-6}	0	3.45×10^{-3}	13.5×10^{-3}
$l_{w_0} = \frac{2\pi}{k_0} =$	3,140	3,920	187,000	1,380
$l_{w_{01}} = \frac{2\pi}{k_{01}} =$	6,900	∞	1,820	464
$f_0 = \frac{m}{2\pi} =$	55.6	45	0.96	128
$f_{01} = \frac{m_1}{2\pi} =$	11.6	0	44	173

In a long-distance telephone line, distributed leakage up to a certain amount increases the critical wave length and thus makes even the long wave oscillatory. Beyond this amount leakage again decreases the wave length. Distributed inductance, as by loading the line, increases the critical wave length if the leakage is small, but in a very leaky line it decreases the critical wave length, and the amount of leakage up to which an increase of the critical wave length occurs is less in a loaded line, that is, in a line of higher inductance.

In other words, a moderate amount of distributed leakage improves a long-distance telephone line, an excessive amount of leakage spoils it. An increase of inductance, by loading the line, improves the line if the leakage is small, but may spoil the line if the leakage is considerable. The amount of leakage up to which improvement in the telephone line occurs is less in a loaded than in an unloaded line; that is, a loaded telephone line requires a far better insulation than an unloaded line.

CHAPTER VI

TRAVELING WAVES

50. As seen in Chapter V, especially in electric power circuits, overhead or underground, the longest existing standing wave has a wave length which is so small compared with the critical wave length — where the frequency becomes zero — that the effect of the damping constant on the frequency and the wave length is negligible. The same obviously applies also to traveling waves, generally to a still greater extent, since the lengths of traveling waves are commonly only a small part of the length of the circuit. Usually, therefore, in the discussion of traveling waves, the effect of the damping constants on the frequency constant q and the wave length constant k can be neglected, that is, frequency and wave length assumed as independent of the energy loss in the circuit.

Usually, therefore, the equations (13) and (14) of Chapter IV can be applied in dealing with the traveling wave.

In these equations the distance traveled by the wave per second is used as unit length by the substitution

$$\lambda = \sigma l,$$

where

$$\sigma = \sqrt{LC},$$

as this brings t and λ into direct comparison and eliminates h and k from the equations by the equation (11) of Chapter IV.

With this unit length the critical value of k, $k_0 = m\sqrt{LC}$, by substituting (8) and (7) of Chapter IV, gives $q_0 = m$, and the condition of the applicability of equations (13) and (14) of Chapter IV, therefore, is that q be a large quantity compared with $q_0 = m$.

In this case $\frac{m}{q}$ is a small quantity, and thus can usually be neglected in equations (15) and (14) of Chapter IV, except when C and C' are very different in magnitude.

This gives, under the limiting conditions discussed above, the *general equations of the traveling wave*, thus:

$$\begin{aligned} i = \varepsilon^{-ut} \{ & \varepsilon^{+s(t-\lambda)} [C_1 \cos q (t - \lambda) + C_1' \sin q (t - \lambda)] \\ & - \varepsilon^{+s(t+\lambda)} [C_2 \cos q (t + \lambda) + C_2' \sin q (t + \lambda)] \\ & + \varepsilon^{-s(t-\lambda)} [C_3 \cos q (t - \lambda) + C_3' \sin q (t - \lambda)] \\ & - \varepsilon^{-s(t+\lambda)} [C_4 \cos q (t + \lambda) + C_4' \sin q (t + \lambda)] \} \end{aligned} \tag{1}$$

and

$$\begin{aligned} e = -\sqrt{\frac{L}{C}}\, \varepsilon^{-ut} \{ & \varepsilon^{+s(t-\lambda)} [C_1 \cos q (t - \lambda) + C_1' \sin q (t - \lambda)] \\ & + \varepsilon^{+s(t+\lambda)} [C_2 \cos q (t + \lambda) + C_2' \sin q (t + \lambda)] \\ & + \varepsilon^{-s(t-\lambda)} [C_3 \cos q (t - \lambda) + C_3' \sin q (t - \lambda)] \\ & + \varepsilon^{-s(t+\lambda)} [C_4 \cos q (t + \lambda) + C_4' \sin q (t + \lambda)] \}, \end{aligned} \tag{2}$$

or

$$\begin{aligned} e = \varepsilon^{-ut)} \{ & \varepsilon^{+s(t+\lambda)} [A_1 \cos q (t + \lambda) + A_1' \sin q (t + \lambda)] \\ & + \varepsilon^{+s(t-\lambda)} [A_2 \cos q (t - \lambda) + A_2' \sin q (t - \lambda)] \\ & + \varepsilon^{-s(t+\lambda)} [A_3 \cos q (t + \lambda) + A_3' \sin q (t + \lambda)] \\ & + \varepsilon^{-s(t-\lambda)} [A_4 \cos q (t - \lambda) + A_4' \sin q (t - \lambda)] \} \end{aligned} \tag{3}$$

and

$$\begin{aligned} i = \sqrt{\frac{C}{L}}\, \varepsilon^{-ut} \{ & \varepsilon^{+s(t+\lambda)} [A_1 \cos q (t + \lambda) + A_1' \sin q (t + \lambda)] \\ & - \varepsilon^{+s(t-\lambda)} [A_2 \cos q (t - \lambda) + A_2' \sin q (t - \lambda)] \\ & + \varepsilon^{-s(t+\lambda)} [A_3 \cos q (t + \lambda) + A_3' \sin q (t + \lambda)] \\ & - \varepsilon^{-s(t-\lambda)} [A_4 \cos q (t - \lambda) + A_4' \sin q (t - \lambda)] \}, \end{aligned} \tag{4}$$

where

$$\left. \begin{aligned} u &= \frac{1}{2}\left(\frac{r}{L} + \frac{g}{C}\right), \\ \lambda &= \sigma l, \\ \sigma &= \sqrt{LC}. \end{aligned} \right\} \tag{5}$$

and

In these equations (1) to (4) the sign of λ may be reversed, which merely means counting the distance in opposite direction.

This gives the following equations:

$$\begin{aligned} i = \epsilon^{-ut}\{&\epsilon^{+s(t-\lambda)}[B_1 \cos q(t-\lambda) + B_1' \sin q(t-\lambda)] \\ &- \epsilon^{+s(t+\lambda)}[B_2 \cos q(t+\lambda) + B_2' \sin q(t+\lambda)] \\ &+ \epsilon^{-s(t-\lambda)}[B_3 \cos q(t-\lambda) + B_3' \sin q(t-\lambda)] \\ &- \epsilon^{-s(t+\lambda)}[B_4 \cos q(t+\lambda) + B_4' \sin q(t+\lambda)]\}, \end{aligned} \qquad (6)$$

and

$$\begin{aligned} e = \sqrt{\frac{L}{C}}\epsilon^{-ut}\{&\epsilon^{+s(t-\lambda)}[B_1 \cos q(t-\lambda) + B_1' \sin q(t-\lambda)] \\ &+ \epsilon^{+s(t+\lambda)}[B_2 \cos q(t+\lambda) + B_2' \sin q(t+\lambda)] \\ &+ \epsilon^{-s(t-\lambda)}[B_3 \cos q(t-\lambda) + B_3' \sin q(t-\lambda)] \\ &+ \epsilon^{-s(t+\lambda)}[B_4 \cos q(t+\lambda) + B_4' \sin q(t+\lambda)]\}, \end{aligned} \qquad (7)$$

or

$$\begin{aligned} e = \epsilon^{-ut}\{&\epsilon^{+s(t-\lambda)}[A_1 \cos q(t-\lambda) + A_1' \sin q(t-\lambda)] \\ &+ \epsilon^{+s(t+\lambda)}[A_2 \cos q(t+\lambda) + A_2' \sin q(t+\lambda)] \\ &+ \epsilon^{-s(t-\lambda)}[A_3 \cos q(t-\lambda) + A_3' \sin q(t-\lambda)] \\ &+ \epsilon^{-s(t+\lambda)}[A_4 \cos q(t+\lambda) + A_4' \sin q(t+\lambda)]\} \end{aligned} \qquad (8)$$

and

$$\begin{aligned} i = \sqrt{\frac{C}{L}}\epsilon^{-ut}\{&\epsilon^{+s(t-\lambda)}[A_1 \cos q(t-\lambda) + A_1' \sin q(t-\lambda)] \\ &- \epsilon^{+s(t+\lambda)}[A_2 \cos q(t+\lambda) + A_2' \sin q(t+\lambda)] \\ &+ \epsilon^{-s(t-\lambda)}[A_3 \cos q(t-\lambda) + A_3' \sin q(t-\lambda)] \\ &- \epsilon^{+s(t+\lambda)}[A_4 \cos q(t+\lambda) + A_4' \sin q(t+\lambda)]\}. \end{aligned} \qquad (9)$$

In these equations (1) to (9) the values A, B, C, etc., are integration constants, which are determined by the terminal conditions of the problem.

The terms with $(t - \lambda)$ may be considered as the main wave, the terms with $(t + \lambda)$ as the reflected wave, or inversely, depending on the direction of propagation of the wave.

51. As the traveling wave, equations (1) to (9), consists of a main wave with variable $(t - \lambda)$ and a reflected wave of the same character but moving in opposite direction, thus with the variable $(t + \lambda)$, these waves may be studied separately, and afterwards the effect of their combination investigated.

Thus, considering at first one of the waves only, that with the variable $(t - \lambda)$, from equations (8) and (9) we have

$$\begin{aligned} e &= \varepsilon^{-ut}\left\{\varepsilon^{+s(t-\lambda)}\left[A_1 \cos q(t-\lambda) + A_1' \sin q(t-\lambda)\right]\right. \\ &\quad \left. + \varepsilon^{-s(t-\lambda)}\left[A_3 \cos q(t-\lambda) + A_3' \sin q(t-\lambda)\right]\right\} \\ &= \varepsilon^{-ut}\left\{\left(A_1\varepsilon^{+s(t-\lambda)} + A_3\varepsilon^{-s(t-\lambda)}\right)\cos q(t-\lambda)\right. \\ &\quad \left. + \left(A_1'\varepsilon^{+s(t-\lambda)} + A_3'\varepsilon^{-s(t-\lambda)}\right)\sin q(t-\lambda)\right\} \end{aligned} \tag{10}$$

and

$$i = \sqrt{\frac{C}{L}}\, e; \tag{11}$$

that is, in a single traveling wave current and voltage are in phase with each other, and proportional to each other with an *effective impedance*, the *surge impedance* or *natural impedance* of the circuit

$$z = \frac{e}{i} = \sqrt{\frac{L}{C}}. \tag{12}$$

This proportionality between e and i and coincidence of phase obviously no longer exist in the combination of main waves and reflected waves, since in reflection the current reverses with the reversal of the direction of propagation, while the voltage remains in the same direction, as seen by (8) and (9).

In equation (10) the time t appears only in the term $(t - \lambda)$ except in the factor ε^{-ut}, while the distance λ appears only in the term $(t - \lambda)$. Substituting therefore

$$t_l = t - \lambda,$$

hence

$$t = t_l + \lambda;$$

that is, counting the time differently at any point λ, and counting it at every point of the circuit from the same point in the phase of the wave from which the time t is counted at the starting point of the wave, $\lambda = 0$, or, in other words, shifting the starting point of the counting of time with the distance λ, and substituting in (150), we have

$$
\left.\begin{aligned}
e &= \varepsilon^{-ut}\left\{\varepsilon^{+st_l}\left(A_1 \cos qt_l + A_1' \sin qt_l\right)\right. \\
&\qquad \left. + \varepsilon^{-st_l}\left(A_3 \cos qt_l + A_3' \sin qt_l\right)\right\} \\
&= \varepsilon^{-u\lambda}\,\varepsilon^{-ut_l}\left\{\varepsilon^{+st_l}\left(A_1 \cos qt_l + A_1' \sin qt_l\right)\right. \\
&\qquad \left. + \varepsilon^{-st_l}\left(A_3 \cos qt_l + A_3 \sin qt_l\right)\right\} \\
&= \varepsilon^{-u\lambda}\,\varepsilon^{-ut_l}\left\{\left(A_1\,\varepsilon^{+st_l} + A_3\,\varepsilon^{-st_l}\right)\cos qt_l\right. \\
&\qquad \left. + \left(A_1'\,\varepsilon^{+st_l} + A_3'\,\varepsilon^{-st_l}\right)\sin qt_l\right\}.
\end{aligned}\right\} \tag{13}
$$

The latter form of the equation is best suited to represent the variation of the wave, at a fixed point λ in space, as function of the local time t_l.

Thus the wave is the product of a term $\varepsilon^{-u\lambda}$ which decreases with increasing distance λ, and a term

$$
\begin{aligned}
e_0 &= \varepsilon^{-ut_l}\left\{\varepsilon^{+st_l}\left(A_1 \cos qt_l + A_1' \sin qt_l\right)\right. \\
&\qquad \left. + \varepsilon^{-st_l}\left(A_3 \cos qt_l + A_3' \sin qt_l\right)\right\} \\
&= \varepsilon^{-ut_l}\left\{\left(A_1\varepsilon^{+st_l} + A_3\,\varepsilon^{-st_l}\right)\cos qt_l\right. \\
&\qquad \left. + \left(A_1'\,\varepsilon^{+st_l} + A_3'\,\varepsilon^{-st_l}\right)\sin qt_l\right\},
\end{aligned} \tag{14}
$$

which latter term is independent of the distance, but merely a function of the time t_l when counting the time at any point of the line from the moment of the passage of the same phase of the wave.

Since the coefficient in the exponent of the distance decrement $\varepsilon^{-u\lambda}$ contains only the circuit constant,

$$u = \frac{1}{2}\left(\frac{r}{L} + \frac{g}{C}\right),$$

but does not contain s and q or the other integration constants, resubstituting from equations (10) to (7) of Chapter IV,

$$\lambda = \sigma l = l\sqrt{LC},$$

we have

$$
\begin{aligned}
u\lambda &= u\sqrt{LC}\,l \\
&= \frac{1}{2}\left\{r\sqrt{\frac{C}{L}} + g\sqrt{\frac{L}{C}}\right\}l,
\end{aligned}
$$

where l is measured in any desired length.

Therefore the attenuation constant of a traveling wave is

$$u_0 = u\sqrt{LC} = \frac{1}{2}\left\{r\sqrt{\frac{C}{L}} + g\sqrt{\frac{L}{C}}\right\}, \tag{15}$$

and hence the distance decrement of the wave,

$$\varepsilon^{-u\lambda} = \varepsilon^{-ul\sqrt{LC}},$$

depends upon the circuit constants r, L, g, C only, but does not depend upon the wave length, frequency, voltage, or current; hence, all traveling waves in the same circuit die out at the same rate, regardless of their frequency and therefore of their wave shape, or, in other words, a complex traveling wave retains its wave shape when traversing a circuit, and merely decreases in amplitude by the distance decrement $\varepsilon^{-u\lambda}$. The wave attenuation thus is a constant of the circuit.

The above statement obviously applies only for waves of constant velocity, that is, such waves in which q is large compared with s, u, and m, and therefore does not strictly apply to extremely long waves, as discussed in 13.

52. By changing the line constants, as by inserting inductance L in such a manner as to give the effect of uniform distribution (loading the line), the attenuation of the wave can be reduced, that is, the wave caused to travel a greater distance l with the same decrease of amplitude.

As function of the inductance L, the attenuation constant (155) is a minimum for

$$\frac{du_0}{dL} = 0;$$

hence,

$$rC - gL = 0,$$

or

$$\frac{L}{C} = \frac{r}{g}, \tag{16}$$

and if the conductance $g = 0$ we have $L = \infty$; hence, in a perfectly insulated circuit, or rather a circuit having no energy losses depending on the voltage, the attenuation decreases with increase of the inductance, that is, by "loading the line," and the more inductance is inserted the better the telephonic transmission.

In a leaky telephone line increase of inductance decreases the attenuation, and thus improves the telephonic transmission, up to the value of inductance,

$$L = \frac{rC}{g}, \tag{17}$$

and beyond this value inductance is harmful by again increasing the attenuation.

For instance, if a long-distance telephone circuit has the following constants per mile: $r = 1.31$ ohms, $L = 1.84 \times 10^{-3}$ henry, $g = 1.0 \times 10^{-6}$ mho, and $C = 0.0172 \times 10^{-6}$ farad, the attenuation of a traveling wave or impulse is

$$u_0 = 0.00217;$$

hence, for a distance or length of line of $l_0 = 2000$ miles,

$$\varepsilon^{-u_0 l_0} = \varepsilon^{-4.34} = 0.0129;$$

that is, the wave is reduced to 1.29 per cent of its original value.

The best value of inductance, according to (17), is

$$L = \frac{r}{g} C = 0.0225 \text{ henry},$$

and in this case the attenuation constant becomes

$$u_0 = 0.00114,$$

and thus

$$\varepsilon^{-u_0 l_0} = \varepsilon^{-2.24} = 0.1055,$$

or 10.55 per cent of the original value of the wave; which means that in this telephone circuit, by adding an additional inductance of 22.5 − 1.84 = 20.7 mh. per mile, the intensity of the arriving wave is increased from 1.29 per cent to 10.55 per cent, or more than eight times.

If, however, in wet weather the leakage increases to the value $g = 5 \times 10^{-6}$, we have in the unloaded line

$$u_0 = 0.00282 \text{ and } \varepsilon^{-u_0 l} = 0.0035,$$

while in the loaded line we have

$$u_0 = 0.00341 \text{ and } \varepsilon^{-u_0 l} = 0.0011,$$

and while with the unloaded line the arriving wave is still 0.35 per cent of the outgoing wave, in the loaded line it is only 0.11 per cent; that is, in this case, loading the line with inductance has badly spoiled telephonic communication, increasing the decay of the wave more than threefold. A loaded telephone line, therefore, is much more sensitive to changes of leakage g, that is, to meteorological conditions, than an unloaded line.

53. The equation of the traveling wave (13),

$$e = \varepsilon^{-u\lambda}\,\varepsilon^{-ut_l}\left\{\varepsilon^{+st_l}(A_1 \cos qt_l + A_1' \sin qt_l) + \varepsilon^{-st_l}(A_3 \cos qt_l + A_3' \sin qt_l)\right\},$$

can be reduced to the form

$$e = \varepsilon^{-u\lambda}\left\{E_1\varepsilon^{-ut_{l_1}}(\varepsilon^{+st_{l_1}} - \varepsilon^{-st_{l_1}}) \sin qt_{l_1} + E_2\varepsilon^{-ut_{l_2}}(\varepsilon^{+st_{l_2}} - \varepsilon^{-st_{l_2}}) \cos qt_{l_2}\right\}, \tag{18}$$

where

and

$$\left.\begin{aligned} t_{l_1} &= t_l - \gamma_1 = t - \lambda - \gamma_1 \\ t_{l_2} &= t_l - \gamma_2 = t - \lambda - \gamma_2. \end{aligned}\right\} \tag{19}$$

By substituting (19) in (18), expanding, and equating (18) with (13), we get the identities

$$\left.\begin{aligned} E_1\varepsilon^{-s\gamma_1} \cos q\gamma_1 - E_2\varepsilon^{-s\gamma_2} \sin q\gamma_2 &= A_1, \\ E_1\varepsilon^{-s\gamma_1} \sin q\gamma_1 + E_2\varepsilon^{-s\gamma_2} \cos q\gamma_2 &= A_1', \\ E_1\varepsilon^{+s\gamma_1} \cos q\gamma_1 - E_2\varepsilon^{+s\gamma_2} \sin q\gamma_2 &= -A_3, \\ E_1\varepsilon^{+s\gamma_1} \sin q\gamma_1 + E_2\varepsilon^{+s\gamma_2} \cos q\gamma_2 &= -A_3', \end{aligned}\right\} \tag{20}$$

and these four equations determine the four constants E_1, E_2, γ_1, γ_2.

Any traveling wave can be resolved into, and considered as consisting of, a combination of two waves:
the *traveling sine wave*,

$$e_1 = E_1\varepsilon^{-u\lambda}\,\varepsilon^{-ut_{l_1}}(\varepsilon^{+st_{l_1}} - \varepsilon^{-st_{l_1}}) \sin qt_{l_1}, \tag{21}$$

and the *traveling cosine wave*,

$$e_2 = E_2\varepsilon^{-u\lambda}\,\varepsilon^{-ut_{l_2}}(\varepsilon^{+st_{l_2}} - \varepsilon^{-st_{l_2}}) \cos qt_{l_2}. \tag{22}$$

Since q is a large quantity compared with u and s, the two component traveling waves, (21) and (22), differ appreciably from each other in appearance only for very small values of t_l, that is, near $t_{l_1} = 0$ and $t_{l_2} = 0$. The traveling sine wave rises in the first half cycle very slightly, while the traveling cosine wave rises rapidly; that is, the tangent of the angle which the wave makes with the horizontal, or $\frac{de}{dt}$, equals 0 with the sine wave and has a definite value with the cosine wave.

All traveling waves in an electric circuit can be resolved into constituent elements, traveling sine waves and traveling cosine waves, and the general traveling wave consists of four component waves, a sine wave, its reflected wave, a cosine wave and its reflected wave.

The elements of the traveling wave, the traveling sine wave e_1, and the traveling cosine wave e_2 contain four constants: the intensity constant, E; the attenuation constant, u, and u_0 respectively; the frequency constant, q, and the constant, s.

The wave starts from zero, builds up to a maximum, and then gradually dies out to zero at infinite time.

The absolute term of the wave, that is, the term which represents the values between which the wave oscillates, is

$$e_0 = E\varepsilon^{-u\lambda}\varepsilon^{-ut_l}\left(\varepsilon^{+st_l} - \varepsilon^{-st_l}\right). \tag{23}$$

The term e_0 may be called the *amplitude* of the wave. It is a maximum for the value of t_l, given by

$$\frac{de_0}{dt_l} = 0,$$

which gives

$$-(u - s)\,\varepsilon^{-(u-s)\,t_{l_0}} + (u + s)\,\varepsilon^{-(u+s)\,t_{l_0}} = 0;$$

hence,

$$\varepsilon^{-2st_{l_0}} = \frac{u + s}{u - s}$$

and

$$t_{l_0} = \frac{1}{2s}\log\frac{u + s}{u - s}, \tag{24}$$

and substituting this value into the equation of the absolute term of the wave, (163), gives

$$e_m = E\varepsilon^{-u\lambda} \frac{2\,s}{\sqrt{u^2 - s^2}} \left(\frac{u + s}{u - s}\right)^{-\frac{u}{2s}} \qquad ((25)$$

The rate of building up of the wave, or the steepness of the wave front, is given by

$$G_0 = \left[\frac{de_0}{dt_l}\right]_{t_l = 0}$$

as

$$G_0 = E\varepsilon^{-u\lambda}\left[-(u - s)\,\varepsilon^{-(u-s)\,t_l} + (u + s)\,\varepsilon^{-(u+s)\,t_l}\right]_0$$
$$= 2\,sE\varepsilon^{-u\lambda}; \qquad ((26)$$

that is, the constant s, which above had no interpretation, represents the rapidity of the rise of the wave.

Referring, however, the rise of the wave to the maximum value e_m of the wave, and combining (165) with (166), we have

$$G_0 = e_m \frac{(u + s)^{\frac{u+s}{2s}}}{(u - s)^{\frac{u-s}{2s}}}. \qquad ((27)$$

The rapidity of the rise of the wave is a maximum, that is, t_l a minimum, for the value of s, in equation (164), given by

$$\frac{dt_{l_0}}{ds} = 0,$$

which gives

$$\log \frac{u + s}{u - s} = \frac{2\,us}{u^2 - s^2};$$

hence, $s = 0$, or the standing wave, which rises infinitely fast, that is, appears instantly.

The smaller therefore s is, the more rapidly is the rise of the traveling wave, and therefore s may be called the *acceleration constant* of the traveling wave.

54. In the components of the traveling wave, equations (21) and (22), the traveling sine wave,

$$e_1 = E\epsilon^{-u\lambda}\epsilon^{-ut_l}\left(\epsilon^{+st_l} - \epsilon^{-st_l}\right) \sin qt_l \text{ (21)},$$

and the traveling cosine wave (22),

$$e_2 = E\varepsilon^{-u\lambda}\varepsilon^{-ut_l}(\varepsilon^{+st_l} - \varepsilon^{-st_l}) \cos qt_l$$

with the amplitude,

$$e_0 = E\varepsilon^{-u\lambda}\varepsilon^{-ut_l}(\varepsilon^{+st_l} - \varepsilon^{-st_l}), \tag{28}$$

we have

$$\left.\begin{aligned} e_1 &= e_0 \sin qt_l \\ e_2 &= e_0 \cos qt_l. \end{aligned}\right\} \tag{29}$$

and

If $t_l = 0$, $e_0 = 0$; that is, t_l is the time counted from the beginning of the wave.

It is

$$t_l = t - \lambda - \gamma,$$

or, if we change the zero point of distance, that is, count the distance λ from that point of the line at which the wave starts at time $t = 0$, or, in other words, count time t and distance λ from the origin of the wave,

$$t_l = t - \lambda,$$

and the traveling wave thus may be represented by the amplitude,

$$e_0 = E\varepsilon^{-ut}(\varepsilon^{+st_l} - \varepsilon^{-st_l});$$

the sine wave,

$$e_1 = E\varepsilon^{-ut}(\varepsilon^{+st_l} - \varepsilon^{-st_l}) \sin qt_l = e_0 \sin qt_l; \tag{30}$$

the cosine wave,

$$e_2 = E\varepsilon^{-ut}(\varepsilon^{+st_l} - \varepsilon^{-st_l}) \cos qt_l = e_0 \cos qt_l;$$

and $t_l = t - \lambda$ can be considered as the distance, counting backwards from the wave front, or the temporary distance; that is, distance counted with the point λ, which the wave has just reached, as zero point, and in opposite direction to λ.

Equation (30) represents the distribution of the wave along the line at the moment t.

As seen, the wave maintains its shape, but progresses along the line, and at the same time dies out, by the time decrement ε^{-ut}.

Resubstituting,

$$t_l = t - \lambda,$$

the equation of the amplitude of the wave is

$$e_0 = E\varepsilon^{-ut}\left(\varepsilon^{+s(t-\lambda)} - \varepsilon^{-s(t-\lambda)}\right). \tag{31}$$

As function of the distance λ, the amplitude of the traveling wave, (171), is a maximum for

$$\frac{de_0}{d\lambda} = 0,$$

which gives

$$\lambda = 0;$$

that is, the amplitude of the traveling wave is a maximum at all times at its origin, and from there decreases with the distance.

This obviously applies only to the single wave, but not to a combination of several waves, as a complex traveling wave.

For

$$\lambda = 0,$$

$$e_0 = E\varepsilon^{-ut}\left(\varepsilon^{+st} - \varepsilon^{-st}\right),$$

and as function of the time t this amplitude is a maximum, according to equations (23) to (25), at

$$\left.\begin{aligned} t_0 &= \frac{1}{2s}\log\frac{u+s}{u-s}, \\ \text{and is}\qquad E_0 &= E\frac{2s}{\sqrt{u^2-s^2}}\left(\frac{u+s}{u-s}\right)^{-\frac{u}{2s}}. \end{aligned}\right\} \tag{32}$$

At any other point λ of the circuit, the amplitude therefore is a maximum, according to equation (24), at the time

$$\left.\begin{aligned} t_m &= t_0 + \lambda, \\ \text{and is}\qquad e_m &= E\frac{2s\varepsilon^{-u\lambda}}{\sqrt{u^2-s^2}}\left(\frac{u+s}{u-s}\right)^{-\frac{u}{2s}}. \end{aligned}\right\} \tag{33}$$

55. As an example may be considered a traveling wave having the constants $u = 115$, $s = 45$, $q = 2620$, and $E = 100$, hence,

$$e_0 = 100\, \varepsilon^{-115t} (\varepsilon^{+45t_l} - \varepsilon^{-45t_l}),$$
$$= 100\, \varepsilon^{-115\lambda} (\varepsilon^{-70t_l} - \varepsilon^{-160t_l}),$$

where $t_l = t - \lambda$.

In Fig. 2 is shown the amplitude e_0 as function of the distance λ, for the different values of time,

$$t = 2, 4, 8, 12, 16, 20, 24, \text{ and } 32 \times 10^{-3},$$

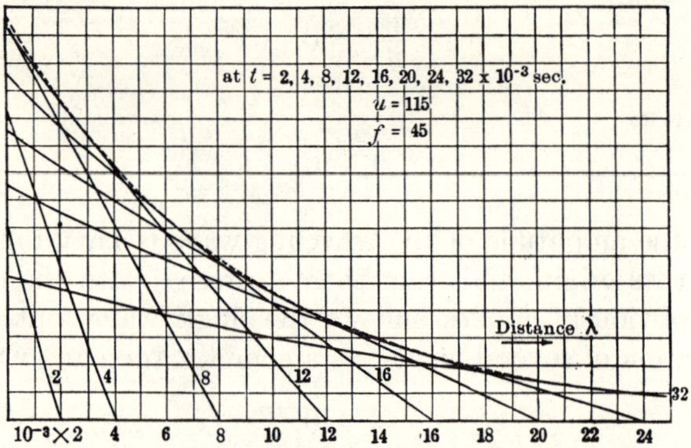

FIG. 2. **Spread of amplitude of electric traveling wave.**

with the maximum amplitude e_m, in dotted line, as envelope of the curve of e_0.

As seen, the amplitude of the wave gradually rises, and at the same time spreads over the line, reaching the greatest value at the starting point $\lambda = 0$ at the time $t_0 = 9.2 \times 10^{-3}$ sec., and then decreases again while continuing to spread over the line, until it gradually dies out.

It is interesting to note that the distribution curves of the amplitude are nearly straight lines, but also that in the present instance even in the longest power transmission line the wave has reached the end of the line, and reflection occurs before the maximum of the curve is reached. The unit of length λ is the distance traveled by the wave per second, or 188,000 miles, and during the rise of the wave, at the origin, from its start to the maximum, or 9.2×10^{-3} sec., the wave thus has traveled 1760 miles, and the reflected wave would have returned to the origin before the maximum of the wave is reached, providing the circuit is shorter than 880 miles.

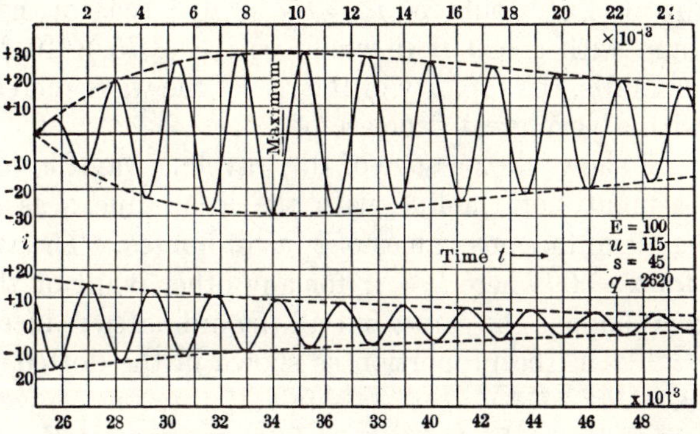

FIG. 3. Passage of traveling wave at a given point of a transmission line.

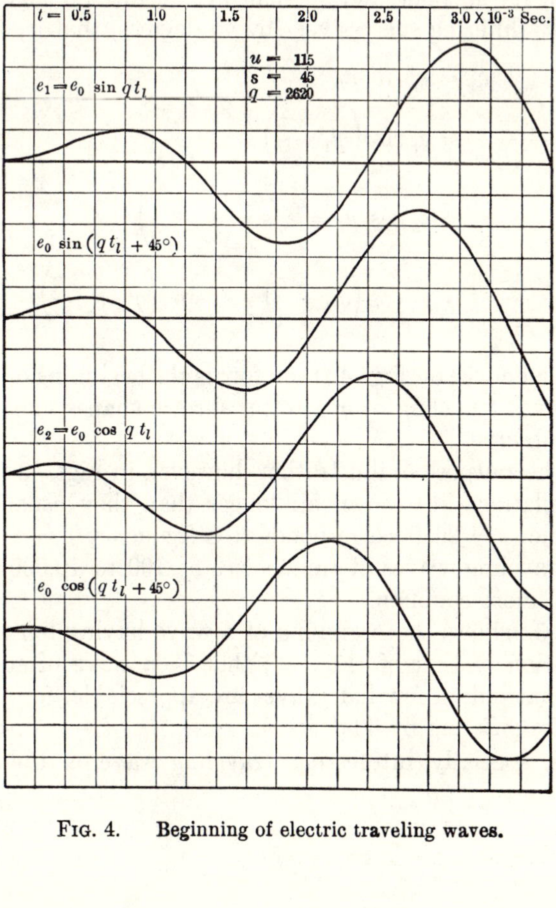

FIG. 4. Beginning of electric traveling waves.

With $s = 1$ it would be $t_0 = 8.7 \times 10^{-3}$ sec., or nearly the same, and with $s = 0.01$ it would be $t_0 = 3.75 \times 10^{-3}$ sec., or, in other words, the rapidity of the rise of the wave increases very little with a very great decrease of s.

Fig. 3 shows the passage of the traveling wave, $e_1 = e_0 \sin qt_l$, across a point λ of the line, with the local time t_l as abscissas and the instantaneous values of e_1 as ordinates. The values are given for $\lambda = 0$, where $t_l = t$; for any other point of the line λ the wave shape is the same, but all the ordinates reduced by the factor $\varepsilon^{-115\lambda}$ in the proportion as shown in the dotted curve in Fig. 99.

Fig. 4 shows the beginning of the passage of the traveling wave across a point $\lambda = 0$ of the line, that is, the starting of a wave, or its first one and one-half cycles, for the trigonometric functions differing successively by 45 degrees, that is,

$$e_1 = e_0 \sin qt_l,$$

$$\frac{e_1 + e_2}{\sqrt{2}} = e_0 \sin \left(qt_l + \frac{\pi}{4}\right),$$

$$e_2 = e_0 \cos qt_l = e_0 \sin \left(qt_l + \frac{\pi}{2}\right),$$

$$\frac{e_2 - e_1}{\sqrt{2}} = e_0 \cos \left(qt_l + \frac{\pi}{4}\right) = e_0 \sin \left(qt_l + \frac{3\pi}{4}\right).$$

The first curve of Fig. 4 therefore is the beginning of Fig. 3.

In waves traveling over a water surface shapes like Fig. 101 can be observed.

For the purpose of illustration, however, in Figs. 3 and 4 the oscillations are shown far longer than they usually occur; the value $q = 2620$ corresponds to a frequency $f = 418$ cycles, while traveling waves of frequencies of 100 to 10,000 times as high are more common.

Fig. 5 shows the beginning of a wave having ten times the attenuation of that of Fig. 4, that is, a wave of such rapid decay that only a few half waves are appreciable, for values of the phase differing by 30 degrees.

56. A specially interesting traveling wave is the wave in which

$$s = u, \tag{34}$$

since in this wave the time decrement of the first main wave and its reflected wave vanishes,

$$\varepsilon^{-(u-s)t} = 1; \tag{35}$$

that is, the first main wave and its reflected wave are not transient but permanent or alternating waves, and the equations of

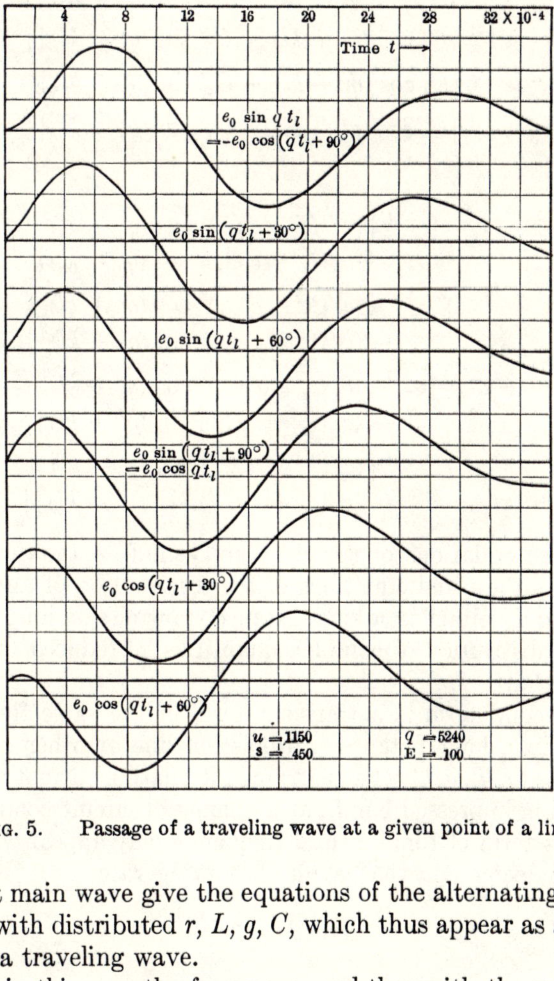

FIG. 5. **Passage of a traveling wave at a given point of a line.**

the first main wave give the equations of the alternating-current circuit with distributed r, L, g, C, which thus appear as a special case of a traveling wave.

Since in this case the frequency, and therewith the value of q, are low and comparable with u and s, the approximations made in the previous discussion of the traveling wave are not permissible, but the general equations (53) and (54) of Chapter I have to be used.

Substituting therefore in (53) and (54) of Chapter I,

$$s = u,$$

gives

$$\begin{aligned} i = [\varepsilon^{-hl} \{C_1 \cos (qt - kl) + C_1' \sin (qt - kl)\} \\ - \varepsilon^{+hl} \{C_2 \cos (qt + kl) + C_2' \sin (qt + kl)\}] \\ - \varepsilon^{-2ut} [\varepsilon^{-hl} \{C_4 \cos (qt + kl) + C_4' \sin (qt + kl)\} \\ - \varepsilon^{+hl} \{C_3 \cos (qt - kl) + C_3' \sin (qt - kl)\}] \end{aligned} \tag{36}$$

and

$$\begin{aligned} e = [\varepsilon^{-hl} \{(c_1'C_1' - c_1C_1) \cos (qt - kl) \\ - (c_1'C_1 + c_1C_1') \sin (qt - kl)\} \\ + \varepsilon^{+hl} \{(c_1'C_2' - c_1C_2) \cos (qt + kl) \\ - (c_1'C_2 + c_1C_2') \sin (qt + kl)\}] \\ + \varepsilon^{-2ut} [\varepsilon^{-hl} \{(c_2'C_4 - c_2C_4) \cos (qt + kl) \\ - (c_2'C_4 + c_2C_4') \sin (qt + kl)\} \\ + \varepsilon^{+hl} \{(c_2'C_3' - c_2C_3) \cos (qt - kl) \\ - (c_2'C_3 + c_2C_3') \sin (qt - kl)\}]. \end{aligned} \tag{37}$$

In these equations of current i and e.m.f. e the first term represents the usual equations of the distribution of alternating current and voltage in a long-distance transmission line, and can by the substitution of complex quantities be reduced to a form given in Part II.

The second term is a transient term of the same frequency; that is, in a long-distance transmission line or other circuit of distributed r, L, g, C, when carrying alternating current under an alternating impressed e.m.f., at a change of circuit conditions, a transient term of fundamental frequency may appear which has the time decrement, that is, dies out at the rate

$$\varepsilon^{-2ut} = \varepsilon^{-\left(\frac{r}{L} + \frac{g}{C}\right) t}.$$

In this decrement the factor

$$\delta_1 = \varepsilon^{-\frac{r}{L} t}$$

is the usual decrement of a circuit of resistance r and inductance L, while the other factor,

$$\delta_2 = \varepsilon^{-\frac{g}{C}t}$$

may be attributed to the conductance and capacity of the circuit, and the total decrement is the product,

$$\delta = \delta_1 \delta_2.$$

A further discussion of the equations (36) and (37) and the meaning of their transient term requires the consideration of the terminal conditions of the circuit.

57. The alternating components of (36) and (37),

$$\begin{aligned} i_0 = \varepsilon^{-hl}\{C_1 \cos(qt - kl) + C_1' \sin(qt - kl)\} \\ - \varepsilon^{+hl}\{C_2 \cos(qt + kl) + C_2' \sin(qt + kl)\} \end{aligned} \tag{38}$$

and

$$\begin{aligned} e_0 = \varepsilon^{-hl}\{(c_1'C_1' - c_1C_1)\cos(qt-kl) - (c_1'C_1 + c_1C_1')\sin(qt-kl)\} \\ + \varepsilon^{+hl}\{(c_1'C_2' - c_1C_2)\cos(qt+kl) - (c_1'C_2 + c_1C_2')\sin(qt+kl)\}, \end{aligned} \tag{39}$$

are reduced to their usual form in complex quantities by resolving the trigonometric function into functions of single angles, qt and kl, then dropping $\cos qt$, and replacing $\sin qt$ by the imaginary unit j. This gives

$$\begin{aligned} i_0 = \varepsilon^{-hl}\{(C_1 \cos kl - C_1' \sin kl)\cos qt \\ + (C_1' \cos kl + C_1 \sin kl)\sin qt\} \\ - \varepsilon^{+hl}\{(C_2 \cos kl + C_2' \sin kl)\cos qt \\ + (C_2' \cos kl - C_2 \sin kl)\sin qt\}; \end{aligned}$$

hence, in complex expression,

$$\begin{aligned} \dot{I} = \varepsilon^{-hl}\{(C_1 - jC_1')\cos kl - (C_1' + jC_1)\sin kl\} \\ - \varepsilon^{+hl}\{(C_2 - jC_2')\cos kl + (C_2' + jC_2)\sin kl\}, \end{aligned} \tag{40}$$

and in the same manner,

$$\begin{aligned} \dot{E} = \varepsilon^{-hl}\{[c_1'(C_1' + jC_1) - c_1(C_1 - jC_1')]\cos kl \\ + [c_1'(C_1 - jC_1') + c_1(C_1' + jC_1)]\sin kl\} \\ + \varepsilon^{+hl}\{[c_1'(C_2' + jC_2) - c_1(C_2 - jC_2')]\cos kl \\ - [c_1'(C_2 - jC_2') + c_1(C_2' + jC_2)]\sin kl\}. \end{aligned} \tag{41}$$

However, from equation (55) of Chapter I,

$$c_1 = \frac{qk + h(m + s)}{h^2 + k^2} L \quad \text{and} \quad c_1' = \frac{k(m + s) - qh}{h^2 + k^2} L,$$

since

$$\left.\begin{aligned} q &= 2\pi f, \\ s &= u = \frac{1}{2}\left(\frac{r}{L} + \frac{g}{C}\right), \\ m &= \frac{1}{2}\left(\frac{r}{L} - \frac{g}{C}\right), \\ \text{and} \quad s + m &= \frac{r}{L}, \end{aligned}\right\} \tag{42}$$

we have

$$\left.\begin{aligned} c_1 &= \frac{2\pi fLk + rh}{h^2 + k^2} = \frac{xk + rh}{h^2 + k^2} \\ \text{and} \quad c_1' &= \frac{rk - 2\pi fLh}{h^2 + k^2} = \frac{rk - xh}{h^2 + k^2}, \end{aligned}\right\} \tag{43}$$

where $x = 2\pi fL$ = reactance per unit length. (44)

From equation (54),

$$R_1^2 = \sqrt{(s^2 + q^2 - m^2)^2 + 4q^2m^2};$$

hence, substituting (42) and (44) and also

$$b = 2\pi fC, \tag{45}$$

we have

$$R_1^2 = \frac{1}{LC}\sqrt{(r^2 + x^2)(g^2 + b^2)}$$

$$= \frac{zy}{LC},$$

where

$$\left.\begin{aligned} z &= \sqrt{r^2 + x^2} = \text{impedance per unit length} \\ \text{and} \quad y &= \sqrt{g^2 + b^2} = \text{admittance per unit length.} \end{aligned}\right\} \tag{46}$$

From the above it follows that

$$\left.\begin{aligned} h &= \sqrt{LC}\sqrt{\tfrac{1}{2}\ \{R_1^2 + s^2 - q^2 - m^2\}} \\ &= \sqrt{\tfrac{1}{2}\ (zy + rg - xb)} \\ \text{and} \qquad k &= \sqrt{\tfrac{1}{2}(zy - rg + xb)}. \end{aligned}\right\} \qquad ((47)$$

If we now substitute

$$\left.\begin{aligned} C_1 - jC_1' &= B_2\sqrt{Y} \\ \text{and} \qquad C_2 - jC_2' &= -B_1\sqrt{Y}, \end{aligned}\right\} \qquad ((48)$$

or

$$\left.\begin{aligned} C_1' + jC_1 &= +jB_2\sqrt{Y} \\ \text{and} \qquad C_2' + jC_2 &= -jB_1\sqrt{Y}, \end{aligned}\right\} \qquad ((49)$$

where

$$\left.\begin{aligned} Z &= r + jx \\ \text{and} \qquad Y &= g + jb; \end{aligned}\right\} \qquad ((50)$$

in (40) and (41) we have

$$I = \sqrt{Y}\ \{B_1\varepsilon^{+hl}(\cos kl + j\sin kl) + B_2\varepsilon^{-hl}(\cos kl - j\sin kl)\} \qquad ((51)$$

and

$$E = (c_1 - jc_1')\sqrt{Y}\ \{B_1\varepsilon^{+hl}(\cos kl + j\sin kl) - B_2\varepsilon^{-hl}(\cos kl - j\sin kl)\}, \qquad ((52)$$

and substituting (43) gives

$$c_1 - jc_1' = \frac{h\,(r - jx) - k\,(x - jr)}{h^2 + k^2} = \frac{(r + jx)\,(h - jk)}{h^2 + k^2} = \frac{r + jx}{h + jk}. \qquad ((53)$$

However,

$$\begin{aligned} h + jk = \sqrt{(h + jk)^2} &= \sqrt{(h^2 - k^2) + 2\,jhk} \\ &= \sqrt{(rg - xb) + j\sqrt{(zy)^2 - (rg - xb)^2}} \\ &= \sqrt{(rg - xb) + j\sqrt{(r^2 + x^2)\,(g^2 + b^2) - (rg - xb)^2}} \\ &= \sqrt{(rg - xb) + j\sqrt{r^2b^2 + x^2g^2 + 2\,rgxb}} \\ &= \sqrt{(rg - xb) + j\,(rb + xg)} \\ &= \sqrt{(r + jx)\ (g + jb)}; \end{aligned}$$

or

$$h + jk = \sqrt{ZY} \qquad (194)$$

substituted in (33) gives

$$c_1 - jc_1' = \sqrt{\frac{Z}{Y}}, \tag{55}$$

and (55) substituted in (52) gives

$$E = \sqrt{Z}\left\{B_1\epsilon^{+hl}(\cos kl + j\sin kl) - B_2\epsilon^{-hl}(\cos kl - j\sin kl)\right\}, \tag{56}$$

where B_1 and B_2 are the complex imaginary integration constants.

Writing

$$h = \alpha \text{ and } k = \beta,$$
$$B_1 = D_1 \text{ and } B_2 = -D_2$$

the equations (51) and (56) become identical with the equations of the long-distance transmission line derived in Part II, equations (22) of paragraph 8.

It is interesting to note that here the general equations of alternating-current long-distance transmission appear as a special case of the equations of the traveling wave, and indeed can be considered as a section of a traveling wave, in which the acceleration constant s equals the exponential decrement u.

CHAPTER VII

FREE OSCILLATIONS

58. The general equations of the electric circuit, (53) and (54) of Chapter I, contain eight terms: four waves: two main waves and their reflected waves, and each wave consists of a sine term and a cosine term.

The equations contain five constants, namely: the frequency constant, q; the wave length constant, k; the time attenuation constant, u; the distance attenuation constant, h, and the time acceleration constant, s; among these, the time attenuation, u, is a constant of the circuit, independent of the character of the wave.

By the value of the acceleration constant, s, waves may be subdivided into three classes, namely: $s = 0$, standing waves, as discussed in Chapter V; $u > s > 0$, traveling waves, as discussed in Chapter VI; $s = u$, alternating-current and e.m.f. waves, as discussed in Part II.

The general equations contain eight integration constants C and C', which have to be determined by the terminal conditions of the problem.

Upon the values of these integration constants C and C' largely depends the difference between the phenomena occurring in electric circuits, as those due to direct currents or pulsating currents, alternating currents, oscillating currents, inductive discharges, etc., and the study of the *terminal conditions* thus is of the foremost importance.

59. By *free oscillations* are understood the transient phenomena occurring in an electric circuit or part of the circuit to which neither electric energy is supplied by some outside source nor from which electric energy is abstracted.

Free oscillations thus are the transient phenomena resulting from the dissipation of the energy stored in the electric field of the circuit, or inversely, the accumulation of the energy of the electric field; and their appearance therefore presupposes the possibility of energy storage in more than one form so as to allow an interchange or surge of energy between its different forms, electromagnetic and electrostatic energy. Free oscillations occur only in circuits containing both capacity C and inductance, L.

The absence of energy supply or abstraction defines the free oscillations by the condition that the power $p = ei$ at the two ends of the circuit or section of the circuit must be zero at all times, or the circuit must be closed upon itself.

The latter condition, of a circuit closed upon itself, leads to a full-wave oscillation, that is, an oscillation in which the length of the circuit is a complete wave or a multiple thereof. With a circuit of uniform constants as discussed here such a full-wave oscillation is hardly of any industrial importance. While the most important and serious case of an oscillation is that of a closed circuit, such a closed circuit never consists of a uniform conductor, but comprises sections of different constants; generating system, transmission line and load, thus is a complex circuit comprising transition points between the sections, at which partial reflection occurs.

The full-wave oscillation thus is that of a complex circuit, which will be discussed in the following chapters.

Considering then the free oscillations of a circuit having two ends at which the power is zero, and representing the two ends of the electric circuit by $l = 0$ and $l = l_0$, that is, counting the distance from one end of the circuit, the conditions of a free oscillation are

$$l = 0, \qquad p = 0.$$
$$l = l_0, \qquad p = 0.$$

Since $p = ei$, this means that at $l = 0$ and $l = l_0$ either e or i must be zero, which gives four sets of terminal conditions:

$$\left.\begin{aligned}
&(1)\ e = 0 \text{ at } l = 0; \qquad && i = 0 \text{ at } l = l_0.\\
&(2)\ i = 0 \text{ at } l = 0; \qquad && e = 0 \text{ at } l = l_0.\\
&(3)\ e = 0 \text{ at } l = 0; \qquad && e = 0 \text{ at } l = l_0.\\
&(4)\ i = 0 \text{ at } l = 0; \qquad && i = 0 \text{ at } l = l_0.
\end{aligned}\right\} \qquad (1)$$

Case (2) represents the same conditions as (1), merely with the distance l counting from the other end of the circuit — a line open at one end and grounded at the other end. Case (3) represents a circuit grounded at both ends, and case (4) a circuit open at both ends.

60. In either of the different cases, at the end of the circuit $l = 0$, either $e = 0$, or $i = 0$.

Substituting $l = 0$ into the equations (53) and (54) of Chapter I gives

$$\begin{aligned} e_0 = \varepsilon^{-(u-s)t}\{&[c_1'(C_1' + C_2') - c_1(C_1 + C_2)]\cos qt \\ &- [c_1'(C_1 + C_2) + c_1(C_1' + C_2')]\sin qt\} \\ + \varepsilon^{-(u+s)t}\{&[c_2'(C_3' + C_4') - c_2(C_3 + C_4)]\cos qt \\ &- [c_2'(C_3 + C_4) + c_2(C_3' + C_4')]\sin qt\} \end{aligned} \tag{2}$$

and

$$\begin{aligned} i_0 = \varepsilon^{-(u-s)t}&\{(C_1 - C_2)\cos qt + (C_1' - C_2')\sin qt\} \\ + \varepsilon^{-(u+s)t}&\{(C_3 - C_4)\cos qt + (C_3' - C_4')\sin qt\}. \end{aligned} \tag{3}$$

If neither q nor s equals zero,

for $e_0 = 0$,

$$\left.\begin{aligned} c_1'(C_1' + C_2') - c_1(C_1 + C_2) &= 0 \\ \text{and} \quad c_1'(C_1 + C_2) + c_1(C_1' + C_2') &= 0; \end{aligned}\right\}$$

hence,

$$\left.\begin{aligned} C_2 &= -C_1, \quad & C_4 &= -C_3, \\ C_2' &= -C_1', \quad & C_4' &= -C_3', \end{aligned}\right\} \tag{4}$$

and for $i_0 = 0$,

$$\left.\begin{aligned} C_2 &= C_1, \quad & C_4 &= C_3, \\ C_2' &= C_1', \quad & C_4' &= C_3'. \end{aligned}\right\} \tag{5}$$

Substituting in (53) and (54) of Chapter I,

$$\begin{aligned} i = \varepsilon^{-(u-s)t}&\{C_1[\varepsilon^{-hl}\cos(qt - kl) \pm \varepsilon^{+hl}\cos(qt + kl)] \\ &+ C_1'[\varepsilon^{-hl}\sin(qt - kl) \pm \varepsilon^{+hl}\sin(qt + kl)]\} \\ + \varepsilon^{-(u+s)t}&\{C_3[\varepsilon^{+hl}\cos(qt - kl) \pm \varepsilon^{-hl}\cos(qt + kl)] \\ &+ C_3'[\varepsilon^{+hl}\sin(qt - kl) \pm \varepsilon^{-hl}\sin(qt + kl)]\} \end{aligned} \tag{6}$$

and

$$
\begin{aligned}
e = \varepsilon^{-(u-s)t} \{ (c_1'C_1' - c_1C_1) [\varepsilon^{-hl} \cos (qt-kl) \\
\mp \varepsilon^{+hl} \cos (qt+kl)] \\
- (c_1'C_1 + c_1C_1') [\varepsilon^{-hl} \sin (qt-kl) \\
\mp \varepsilon^{+hl} \sin (qt+kl)] \} \\
+ \varepsilon^{-(u+s)t} \{ (c_2'C_3' - c_2C_3) [\varepsilon^{+hl} \cos (qt-kl) \\
\mp \varepsilon^{-hl} \cos (qt+kl)] \\
- (C_2'C_3 + c_2C_3') [\varepsilon^{+hl} \sin (qt-kl) \\
\mp \varepsilon^{-hl} \sin (qt+kl)] \},
\end{aligned}
\tag{7}
$$

where the upper sign refers to $e = 0$, the lower sign to $i = 0$ for $l = 0$.

61. In a free oscillation, either e or i must be zero at the other end of the oscillating circuit, or at $l = l_0$.

Substituting, therefore, $l = l_0$ in equations (6) and (7), and resolving and arranging the terms by functions of t, the respective coefficients of

$$\varepsilon^{-(u-s)t} \cos qt, \quad \varepsilon^{-(u-s)t} \sin qt, \quad \varepsilon^{-(u+s)t} \cos qt, \text{ and } \varepsilon^{-(u+s)t} \sin qt$$

must equal zero, either in equation (6), if $i = 0$ at $l = l_0$, or in equation (7), if $e = 0$ at $l = l_0$, provided that, as assumed above, neither s nor q vanishes.

This gives, for $i = 0$ at $l = l_0$, from equation (202),

$$
\left.
\begin{aligned}
C_1 (\varepsilon^{-hl_0} \pm \varepsilon^{+hl_0}) \cos kl_0 - C_1'(\varepsilon^{-hl_0} \mp \varepsilon^{+hl_0}) \sin kl_0 = 0, \\
C_1 (\varepsilon^{-hl_0} \mp \varepsilon^{+hl_0}) \sin kl_0 + C_1'(\varepsilon^{-hl_0} \pm \varepsilon^{-hl_0}) \cos kl_0 = 0,
\end{aligned}
\right\}
\tag{8}
$$

and analogously for C_3 and C_3'.

In equations (8), either C_1, C_1', C_3, C_3' vanish, and then the whole oscillation vanishes, or, by eliminating C_1 and C_1' from equations (8), we get

$$(\varepsilon^{-hl_0} \pm \varepsilon^{+hl_0})^2 \cos^2 kl_0 + (\varepsilon^{-hl_0} \mp \varepsilon^{+hl_0})^2 \sin^2 kl_0 = 0; \tag{9}$$

hence,

$$
\text{and} \quad
\left.
\begin{aligned}
(\varepsilon^{-hl_0} \pm \varepsilon^{+hl_0}) \cos kl_0 = 0 \\
(\varepsilon^{-hl_0} \mp \varepsilon^{+hl_0}) \sin kl_0 = 0;
\end{aligned}
\right\}
$$

hence, for the upper sign, or if $e = 0$ for $l = 0$,

thus:
$$\left.\begin{aligned} h = 0 \text{ and } \cos kl_0 = 0, \\ kl_0 = \frac{(2n+1)\pi}{2}, \end{aligned}\right\} \qquad (10)$$

and for the lower sign, or if $i = 0$ for $l = 0$,

thus:
$$\left.\begin{aligned} h = 0 \text{ and } \sin kl_0 = 0, \\ kl_0 = n\pi. \end{aligned}\right\} \qquad (11)$$

In the same manner it follows, for $e = 0$ at $l = l_0$, from equation (7), if $e = 0$ for $l = 0$, (6),

thus:
$$\left.\begin{aligned} h = 0, \quad \sin kl_0 = 0 \\ kl = n\pi, \end{aligned}\right\} \qquad (12)$$

and if $i = 0$ for $l = 0$,

thus:
$$\left.\begin{aligned} h = 0 \text{ and } \cos kl_0 = 0, \\ kl_0 = \frac{(2n+1)\pi}{2}. \end{aligned}\right\} \qquad (13)$$

From equations (10) to (13) it thus follows that $h = 0$, that is, *the free oscillation of a uniform circuit is a standing wave.* Also

$$kl_0 = \frac{(2n+1)\pi}{2}, \qquad (14)$$

if $e = 0$ at one, $i = 0$ at the other end of the circuit, and

$$kl_0 = n\pi \qquad (15)$$

if either $e = 0$ at both ends of the circuit or $i = 0$ at both ends of the circuit.

62. From (14) it follows that

$$kl_0 = \frac{\pi}{2},$$

or an *odd* multiple thereof; that is, the longest wave which can exist in the circuit is that which makes the circuit a quarter-wave length. Besides this fundamental wave, all its odd multiples can exist. Such an oscillation may be called a *quarter-wave oscillation.*

The oscillation of a circuit which is open at one end, grounded at the other end, is a quarter-wave oscillation, which can contain only the odd harmonics of the fundamental wave of oscillation.

From (15) it follows that

$$kl_0 = \pi,$$

or a multiple thereof; that is, the longest wave which can exist in such a circuit is that wave which makes the circuit a half-wave length. Besides this fundamental wave, all its multiples, odd as well as even, can exist. Such an oscillation may be called a *half-wave oscillation.*

The oscillation of a circuit which is open at both ends, or grounded at both ends, is a half-wave oscillation, and a half-wave oscillation can also contain the even harmonics of the fundamental wave of oscillation, and therefore also a constant term for $n = 0$ in (15).

It is interesting to note that in the half-wave oscillation of a circuit we have a case of a circuit in which higher *even* harmonics exist, and the e.m.f. and current wave, therefore, are not symmetrical.

From $h = 0$ follows, by equation (59) of Chapter I,

$$\left.\begin{aligned} s &= 0, \quad \text{if} \quad k^2 > LCm^2, \\ \text{and} \qquad q &= 0, \quad \text{if} \quad k^2 < LCm^2. \end{aligned}\right\} \tag{16}$$

The smallest value of k which can exist from equation (14) is

$$k = \frac{\pi}{2\,l_0},$$

and, as discussed in paragraph 15, this value in high-potential high-power circuits usually is very much larger than LCm^2, so that the case $q = 0$ is realized only in extremely long circuits, as long-distance telephone or submarine cable, but not in transmission lines, and the first case, $s = 0$, therefore, is of most importance.

Substituting, therefore, $h = 0$ and $s = 0$ into the equation (55) of Chapter I, gives

$$\left.\begin{aligned} c_1 &= c_2 = \frac{q}{k}L = c \\ \text{and} \qquad & \\ c_1' &= c_2' = \frac{m}{k}L = c', \end{aligned}\right\} \tag{17}$$

and substituting into equations (6) and (7) of the free oscillation gives

$$\begin{aligned} i = \varepsilon^{-ut}\{A_1 [\cos (qt - kl) \pm \cos (qt + kl)] \\ + A_2 [\sin (qt - kl) \pm \sin (qt + kl)]\} \end{aligned} \tag{18}$$

and

$$\begin{aligned} e = \frac{L}{k}\varepsilon^{-ut} \{(mA_2 - qA_1) [\cos (qt-kl) \mp \cos (qt + kl)] \\ - (mA_1 + qA_2) [\sin (qt-kl) \mp \sin (qt + kl)]\}. \end{aligned} \tag{19}$$

where: $\qquad A_1 = C_1 + C_3$ and $A_2 = C_1' + C_3'$.

Since k and therefore q are large quantities, m can be neglected compared with q, and

$$k = \sqrt{LC}q;$$

hence

$$\frac{Lq}{k} = \sqrt{\frac{L}{C}}$$

and the equation (19) assumes, with sufficient approximation, the form

$$\begin{aligned} e = -\sqrt{\frac{L}{C}}\varepsilon^{-ut} \{A_1 [\cos (qt - kl) \mp \cos (qt + kl)] \\ + A_2[\sin (qt - kl) \mp \sin (qt + kl)]\}, \end{aligned} \tag{20}$$

where the upper sign in (18) and (20) corresponds to $e = 0$ at $l = 0$, the lower sign to $i = 0$ at $l = 0$, as is obvious from the equations.

Substituting

$$A_1 = A \cos \gamma \text{ and } A_2 = A \sin \gamma \tag{21}$$

into (18) and (20) gives the equations of the free oscillation, thus:

$$\left.\begin{aligned} i &= A\varepsilon^{-ut}\{\cos(qt-kl-\gamma) \mp \cos(qt+kl-\gamma)\} \\ \text{and} \\ e &= -A\sqrt{\frac{L}{C}}\varepsilon^{-ut}\{\cos(qt-kl-\gamma) \mp \cos(qt+kl-\gamma)\}. \end{aligned}\right\} \tag{22}$$

With the upper sign, or for $e = 0$ at $l = 0$, this gives

$$\left.\begin{aligned} i &= 2\,A\varepsilon^{-ut}\cos kl \cos(qt-\gamma) \\ \text{and} \\ e &= -2\,A\sqrt{\frac{L}{C}}\varepsilon^{-ut}\sin kl \sin(qt-\gamma). \end{aligned}\right\} \tag{23}$$

With the lower sign, or for $i = 0$ at $l = 0$, this gives

$$\left.\begin{aligned} i &= 2\,A\varepsilon^{-ut}\sin kl \sin(qt-\gamma) \\ \text{and} \\ e &= -2\,A\sqrt{\frac{L}{C}}\varepsilon^{-ut}\cos kl \cos(qt-\gamma). \end{aligned}\right\} \tag{24}$$

63. While the free oscillation of a circuit is a standing wave, the general standing wave, as represented by equations (43) and (44) of Chapter V with four integration constants A_1, A_1', A_2, A_2', is not necessarily a free oscillation.

To be a free oscillation, the power ei, that is, either e or i, must be zero at two points of the circuit, the ends of the circuit or section of circuit which oscillates.

At a point l_1 of the circuit at which $e = 0$, the coefficients of $\cos qt$ and $\sin qt$ in equation (43) of Chapter V must vanish. This gives

$$\left.\begin{aligned} (A_1 + A_2)\cos kl_1 + (A_1' - A_2')\sin kl_1 &= 0 \\ \text{and} \quad -(A_1 - A_2)\sin kl_1 + (A_1' + A_2')\cos kl_1 &= 0. \end{aligned}\right\} \tag{25}$$

Eliminating $\sin kl_1$ and $\cos kl_1$ from these two equations gives

$$\left.\begin{aligned} (A_1^2 - A_2^2) + (A_1'^2 - A_2'^2) &= 0, \\ \text{or} \\ A_1^2 + A_1'^2 &= A_2^2 + A_2'^2, \end{aligned}\right\} \tag{26}$$

as the condition which must be fulfilled between the integration constants.

The value l_1 then follows from (25) as

$$\tan kl_1 = \frac{A_1' + A_2'}{A_1 - A_2} = - \frac{A_1 + A_2}{A_1' - A_2'}. \tag{27}$$

At a point l_2 of the circuit at which $i = 0$ the coefficients of $\cos qt$ and $\sin qt$ in equation (44) of Chapter V must vanish. This gives, in the same manner as above,

$$(A_1^2 - A_2^2) + (A_1'^2 - A_2'^2) = 0,$$

that is, the same conditions as (221), and gives for l_2 the value

$$\tan kl_2 = \frac{A_1' - A_2'}{A_1 + A_2} = - \frac{A_1 - A_2}{A_1' + A_2'} \cdot \tag{28}$$

From (223) and (224) it follows that

$$\tan kl_2 = - \frac{1}{\tan kl_1}; \tag{29}$$

That is, the angles kl_1 and kl_2 differ by one quarter-wave length or an odd multiple thereof.

Herefrom it then follows that if the integration constants of a standing wave fulfill the conditions

$$A_1^2 + A_1'^2 = A_2^2 + A_2'^2 = B^2, \tag{30}$$

the circuit of this wave contains points l_1, distant from each other by a half-wave length, at which $e = 0$, and points l_2, distant from each other by a half-wave length, at which $i = 0$, and the points l_2 are intermediate between the points l_1, that is, distant therefrom by one quarter-wave length. Any section of the circuit, from a point l_1 or l_2 to any other point l_1 or l_2, then is a freely oscillating circuit.

In the free oscillation of the circuit the circuit is bounded by one point l_1 and one point l_2; that is, the e.m.f. is zero at one end and the current zero at the other end of the circuit, case (1) or (2) of equation (1), and the circuit is then a quarter-wave or an odd multiple thereof, or the circuit is bounded by two points l_1 or by two points l_2, and then the voltage is zero at both ends of the circuit in the former case, number (3) in equation (1), or the current is zero at both ends of the circuit in the latter case, number (4) in equation (1), and in either case the circuit is one half-wave or a multiple thereof.

Choosing one of the points l_1 or l_2 as starting point of the distance, that is, substituting $l - l_1$ or $l - l_2$ respectively, instead of l, in the equations (43) and (44) of Chapter V, with some transformation these equations convert into the equations (23) or (24). In other words, the equation (30), as relation between the integration constants of a standing wave, is the necessary and sufficient condition that this standing wave be a free oscillation.

64. A single term of a free oscillation of a circuit, with the distance counted from one end of the circuit, that is, one point of zero power, thus is represented by equations (23) or (24), respectively.

Reversing the sign of l, that is, counting the distance in the opposite direction, and substituting $B = \pm 2\,A\sqrt{\frac{L}{C}}$, these equations assume a more convenient form, thus:

for $e = 0$ at $l = 0$,

and
$$\left.\begin{aligned} e &= B\varepsilon^{-ut} \sin kl \sin (qt - \gamma) \\ i &= B\sqrt{\frac{C}{L}}\,\varepsilon^{-ut} \cos kl \cos (qt - \gamma), \end{aligned}\right\} \qquad (31)$$

and for $i = 0$ at $l = 0$,

and
$$\left.\begin{aligned} e &= B\varepsilon^{-ut} \cos kl \cos (qt - \gamma) \\ i &= B\sqrt{\frac{C}{L}}\,\varepsilon^{-ut} \sin kl \sin (qt - \gamma). \end{aligned}\right\} \qquad (32)$$

Introducing again the velocity of propagation as unit distance,

$$\left.\begin{aligned} \lambda &= \sigma l, \\ \sigma &= \sqrt{LC}, \end{aligned}\right\} \qquad (33)$$

from equation (5) of Chapter IV and (33) we get:

$$\begin{aligned} kl &= \lambda\sqrt{q^2 + m^2} \\ &= q\lambda\sqrt{1 + \left(\frac{m}{q}\right)^2}; \end{aligned}$$

hence, if m is small compared with q,

$$kl = q\lambda, \tag{34}$$

and substituting (33) in (34) gives

$$k = \sigma q = q\sqrt{LC}, \tag{35}$$

and from (14) and (15), for a quarter-wave oscillation, we have

$$\left.\begin{aligned} k &= \frac{(2\,n+1)\,\pi}{2\,l_0} \\ \text{and}\quad q &= \frac{(2\,n+1)\,\pi}{2\,l_0\sqrt{LC}}; \end{aligned}\right\} \tag{36}$$

for a half-wave oscillation,

$$\left.\begin{aligned} k &= \frac{n\pi}{l_0}, \\ q &= \frac{n\pi}{l_0\sqrt{LC}}. \end{aligned}\right\} \tag{37}$$

Denoting the length of the circuit in a quarter-wave oscillation by

$$\lambda_1 = \sigma l_0, \tag{38}$$

and the length of the circuit in a half-wave oscillation by

$$\lambda_2 = \sigma l_0, \tag{39}$$

the wave length of the fundamental or lowest frequency of oscillation is

$$\lambda_0 = 4\,\lambda_1 = 2\,\lambda_2; \tag{40}$$

or the length of the fundamental wave, with the velocity of propagation as distance unit, in a quarter-wave oscillation is

$$\left.\begin{aligned} &\lambda_0 = 4\,l_0\sqrt{LC}, \\ &\text{and in a half-wave oscillation is} \\ &\lambda_0 = 2\,l_0\sqrt{LC}. \end{aligned}\right\} \tag{41}$$

Substituting (41) into (36) and (37) for a quarter-wave oscillation gives

$$\left.\begin{aligned} k &= (2n+1)\frac{2\pi\sqrt{LC}}{\lambda_0} \\ \text{and}\qquad q &= (2n+1)\frac{2\pi}{\lambda_0}, \end{aligned}\right\} \tag{42}$$

and for a half-wave oscillation gives

$$\left.\begin{aligned} k &= n\frac{2\pi\sqrt{LC}}{\lambda_0} \\ \text{and}\qquad q &= n\frac{2\pi}{\lambda_0}. \end{aligned}\right\} \tag{43}$$

Writing now

$$\left.\begin{aligned} \theta &= \frac{2\pi}{\lambda_0}t, \\ \tau &= \frac{2\pi}{\lambda_0}\lambda = \frac{2\pi\sqrt{LC}}{\lambda_0}l, \end{aligned}\right\} \tag{44}$$

that is, representing a complete cycle of the fundamental frequency, or complete wave in time, by $\theta = 2\pi$, and a complete wave in space by $\tau = 2\pi$, from (43) and (44) we have

$$\left.\begin{aligned} kl &= n\tau \\ \text{and}\qquad qt &= n\theta, \end{aligned}\right\} \tag{45}$$

where n may be any integer number with a half-wave oscillation, but only an odd number with a quarter-wave oscillation.

65. Substituting (45) into (31) and (32) gives as the complete expression of a free oscillation the following equation

A. Quarter-wave oscillation.

(*a*) $e = 0$ at $l = 0$ (or $\tau = 0$)

$$\left.\begin{aligned} e &= \varepsilon^{-ut}\sum_0^\infty{}^n B_n \sin(2n+1)\tau \sin[(2n+1)\vartheta - \gamma_n] \\ \text{and}\qquad i &= \sqrt{\frac{C}{L}}\,\varepsilon^{-ut}\sum_0^\infty{}^n B_n \cos(2n+1)\tau\cos[(2n+1)\theta - \gamma_n]; \end{aligned}\right\} \tag{46}$$

(b) $i = 0$ at $l = 0$ (or $\tau = 0$)

$$\left.\begin{aligned} e &= \varepsilon^{-ut} \sum_0^\infty {}^n B_n \cos(2n+1)\tau \cos[(2n+1)\theta - \gamma_n] \\ &\text{and} \\ i &= \sqrt{\frac{C}{L}}\varepsilon^{-ut} \sum_0^\infty {}^n B_n \sin(2n+1)\tau \sin[(2n+1)\theta - \gamma_n]. \end{aligned}\right\} \quad (47)$$

B. Half-wave oscillation.

(a) $e = 0$ at $l = 0$ (or $\tau = 0$)

$$\left.\begin{aligned} e &= \varepsilon^{-ut} \sum_0^\infty {}^n B_n \sin n\tau \sin(n\theta - \gamma_n) \\ &\text{and} \\ i &= \sqrt{\frac{C}{L}}\varepsilon^{-ut} \sum_0^\infty {}^n B_n \cos n\tau \cos(n\theta - \gamma_n); \end{aligned}\right\} \quad (48)$$

(b) $i = 0$ at $l = 0$ (or $\tau = 0$)

$$\left.\begin{aligned} e &= \varepsilon^{-ut} \sum_0^\infty {}^n B_n \cos n\tau \cos(n\theta - \gamma_n) \\ &\text{and} \\ i &= \sqrt{\frac{C}{L}}\varepsilon^{-ut} \sum_0^\infty {}^n B_n \sin n\tau \sin(n\theta - \gamma_n), \end{aligned}\right\} \quad (49)$$

where

$$\left.\begin{aligned} \theta &= \frac{2\pi}{\lambda_0} t, \\ \tau &= \frac{2\pi\sqrt{LC}}{\lambda_0} l, \end{aligned}\right\} \quad (44)$$

$$\left.\begin{aligned} \lambda_0 &= 4\, l_0 \sqrt{LC} \text{ in a quarter-wave} \\ &= 2\, l_0 \sqrt{LC} \text{ in a half-wave oscillation,} \end{aligned}\right\} \quad (41)$$

and

$$\varepsilon^{-ut} = \varepsilon^{-\frac{u\lambda_0\theta}{2\pi}}. \quad (50)$$

λ_0 is the wave length, and thus $\frac{1}{\lambda_0}$ the frequency, of the fundamental wave, with the velocity of propagation as distance unit.

It is interesting to note that the time decrement of the free oscillation, ε^{-ut}, is the same for all frequencies and wave lengths, and that the relative intensity of the different harmonic components of the oscillation, and thereby the wave shape of the oscillation, remains unchanged during the decay of the oscillation.

This result, analogous to that found in the chapter on traveling waves, obviously is based on the assumption that the constants of the circuit do not change with the frequency. This, however, is not perfectly true. At very high frequencies r increases, due to unequal current distribution in the conductor and the appearance of the radiation resistance, as discussed in Part II, L slightly decreases hereby, g increases by the energy losses resulting from brush discharges and from electrostatic radiation, etc., so that, in general, at very high frequency an increase of $\frac{r}{L}$, and $\frac{g}{C}$, and therewith of u, may be expected; that is, very high harmonics would die out with greater rapidity, which would result in smoothing out the wave shape with increasing decay, making it more nearly approach the fundamental and its lower harmonics, as discussed in the Chapters on "Variation of Constants."

66. The equations of a free oscillation of a circuit, as quarter-wave or half-wave, (46) to (49), still contain the pairs of integration constants B_n and γ_n, representing, respectively, the intensity and the phase of the nth harmonic.

These pairs of integration constants are determined by the terminal conditions of time; that is, they depend upon the amount and the distribution of the stored energy of the circuit at the starting moment of the oscillation, or, in other words, on the distribution of current and e.m.f. at $t = 0$.

The e.m.f., e_0, and the current, i_0, at time $t = 0$, can be expressed as an infinite series of trigonometric functions of the distance l; that is, the distance angle τ, or a Fourier series of such character as also to fulfill the terminal conditions in space, as discussed above, that is, $e = 0$, and $i = 0$, respectively, at the ends of the circuit.

The voltage and current distribution in the circuit, at the starting moment of the oscillation, $t = 0$, or, $\theta = 0$, can be represented by the Fourier series, thus:

$$\left.\begin{aligned} e_0 &= \sum_0^{\infty}{}^n (a_n \cos n\tau + a_n' \sin n\tau) \\ \text{and} \qquad & \\ i_0 &= \sum_0^{\infty}{}^n (b_n \cos n\tau + b_n' \sin n\tau), \end{aligned}\right\} \qquad (51)$$

where

$$\left.\begin{aligned} a_0 &= \frac{1}{2\pi}\int_0^{2\pi} e_0 d\tau = \text{avg}\,[e_0]_0^{2\pi}, \\ a_n &= \frac{1}{\pi}\int_0^{2\pi} e_0 \cos n\tau\, d\tau = 2\,\text{avg}\,[e_0 \cos n\tau]_0^{2\pi}, \\ a_n' &= \frac{1}{\pi}\int_0^{2\pi} e_0 \sin n\tau\, d\tau = 2\,\text{avg}\,[e_0 \sin n\tau]_0^{2\pi}, \end{aligned}\right\} \tag{52}$$

and analogously for b.

The expression avg $[F]_{a_1}^{a_2}$ denotes the average value of the function F between the limits a_1 and a_2.

Since these integrals extend over the complete wave 2π, the wave thus has to be extended by utilizing the terminal conditions regarding τ, but the wave is symmetrical with regard to $l = 0$ and with regard to $l = l_0$, and this feature in the case of a quarter-wave oscillation excludes the existence of even values of n in equations (51) and (52).

67. Substituting in equations (46) to (49),

$$t = 0, \qquad \theta = 0,$$

and then equating with (51), gives, from (46),

$$e_0 = \sum_0^\infty {}^n B_n \sin (2n+1)\tau \sin \gamma_n = \sum_0^\infty {}^n [a_n \cos (2n+1)\tau + a_n' \sin (2n+1)\tau]$$

and

$$i_0 = \sqrt{\frac{C}{L}} \sum_0^\infty {}^n B_n \cos (2n+1)\tau \cos \gamma_n = \sum_0^\infty {}^n [b_n \cos (2n+1)\tau + b_n' \sin (2n+1)\tau];$$

hence,

$$a_n = 0, \quad b_n' = 0,$$

$$B_n \sin \gamma_n = a_n' \text{ and } \sqrt{\frac{C}{L}} B_n \cos \gamma_n = b_n.$$

Equation (46) gives the constants

$$a_n = 0; \quad b_n' = 0,$$

$$\left.\begin{aligned} B_n &= \sqrt{a_n'^2 + \frac{L}{C} b_n^2}, \\ \tan \gamma_n &= \frac{a_n'}{b_n} \sqrt{\frac{L}{C}}; \end{aligned}\right\} \tag{53}$$

in the same manner equation (243) gives the constants

$$a_n' = 0; \quad b_n = 0,$$

$$\left.\begin{aligned} B_n &= \sqrt{a_n^2 + \frac{L}{C} b_n'^2}, \\ \tan \gamma_n &= \frac{b_n'}{a_n} \sqrt{\frac{C}{L}} \cdot \end{aligned}\right\} \tag{54}$$

Equation (48) gives the same values as (46), and (49) the same values as (47).

Examples.

68. As first example may be considered the discharge of a transmission line: A circuit of length l_0 is charged to a uniform voltage E, while there is no current in the circuit. This circuit then is grounded at one end, while the other end remains insulated.

Let the distance be counted from the grounded end, and the time from the moment of grounding, and introducing the denotations (39).

The terminal conditions then are:

(*a*) $\tau = 0 \quad e = 0,$

$$\tau = \frac{\pi}{2} \quad i = 0.$$

(*b*) at $\theta = 0$

$$e = 0 \text{ for } \tau = 0; \; e = E \text{ for } \tau \neq 0,$$

$$i = 0 \text{ for } \tau \neq 0; \; i = \text{indefinite for } \tau = 0.$$

The distribution of e.m.f., e_0, and current, i_0, in the circuit, at the starting moment $\theta = 0$, can be expressed by the Fourier series (51), and from (52),

$$a_n' = 2 \operatorname{avg} [E \sin (2 \pi + 1) \tau] = \frac{4 E}{(2 n + 1) \pi} \tag{55}$$

and

$$b_n = 0,$$

and from (249),

$$\left. \begin{aligned} B_n &= \frac{4 E}{(2n+1)\pi} \text{ and } \tan \gamma_n = \infty; \\ \text{hence,} \quad \gamma_n &= \frac{\pi}{2}, \end{aligned} \right\} \tag{56}$$

and substituting (56) into (46),

$$\left. \begin{aligned} e &= \frac{4 E}{\pi} \varepsilon^{-ut} \sum_1^{\infty} {}^n \frac{\sin (2 n + 1) \tau \cos (2 n + 1)\theta}{2 n + 1} \\ \text{and} \\ i &= \frac{4 E}{\pi} \sqrt{\frac{C}{L}} \varepsilon^{-ut} \sum_1^{\infty} {}^n \frac{\cos (2 n + 1) \tau \sin (2 n + 1)\theta}{2 n + 1}. \end{aligned} \right\} \tag{57}$$

From (44) it follows that

$$t = \frac{2 l_0 \sqrt{LC}}{\pi} \theta.$$

$\theta = 2 \pi$ gives the period,

$$t_1 = 4 l_0 \sqrt{LC},$$

and the frequency,

$$f_1 = \frac{1}{4 l_0 \sqrt{LC}},$$

and $\tau = 2 \pi$ gives the wave length,

$$l_w = 4 l_0,$$

of the fundamental wave, or oscillation of lowest frequency and greatest wave length.

Choosing the same line constants as in paragraph 16, namely: $l_0 = 120$ miles; $r = 0.41$ ohm per mile; $L = 1.95 \times 10^{-3}$ henry per mile; $g = .25 \times 10^{-6}$ mho per mile, and $C = 0.0162 \times 10^{-6}$ farad per mile, we have

$$u = 113,$$

$$ut = \frac{2\, l_0 \sqrt{LC}}{\pi} u\theta = 0.0485\, \theta,$$

and the fundamental frequency of oscillation is

$$f_1 = 371 \text{ cycles per second.}$$

If now the e.m.f. to which the line is charged is

$$E = 40{,}000 \text{ volts,}$$

substituting these values in equations (57) gives

$$\left.\begin{aligned} e &= 51{,}000\, \varepsilon^{-0.0485\,\theta} \left\{\sin \tau \cos \theta + \tfrac{1}{3} \sin 3\, \tau \cos 3\, \theta \right. \\ &\quad \left. + \tfrac{1}{5} \sin 5\, \tau \cos 5\, \theta + \ldots\right\}, \text{ in volts} \\ \text{and} \quad & \\ i &= 147\, \varepsilon^{-0.0485\,\theta} \left\{\cos \tau \sin \theta + \tfrac{1}{3} \cos 3\, \tau \sin 3\, \theta \right. \\ &\quad \left. + \tfrac{1}{5} \cos 5\, \tau \sin 5\, \theta + \ldots\right\}, \text{ in amp.} \end{aligned}\right\} \quad (58)$$

The maximum value of e is

$$e = E = 40{,}000 \text{ volts,}$$

and the maximum current of i is

$$i = I = 115.5 \text{ amp.}$$

Since

$$\left.\begin{aligned} \sum_1^\infty {}_n \frac{\sin (2\, n-1)\, a \cos (2\, n-1)\, b}{2\, n - 1} &= 0, \text{ if } b - \frac{\pi}{2} < a < b, \\ & \quad \text{or } b + \frac{\pi}{2} < a < b + \pi, \\ \text{and} \qquad & \\ &= +\frac{\pi}{2}, \text{ if } b < a < b + \frac{\pi}{2} \\ &= -\frac{\pi}{2}, \text{ if } b < a < b - \frac{\pi}{2}, \end{aligned}\right\} \quad (59)$$

applying (59) to (58) we have at any point τ of the line, at the time θ given by

$$0 < \theta < \tau: \qquad e = E\varepsilon^{-ut}; \qquad i = 0.$$

$$\tau < \theta < \tau + \frac{\pi}{2}: \qquad e = 0; \qquad i = I\varepsilon^{-ut}.$$

$$\tau + \frac{\pi}{2} < \theta < \tau + \pi: \qquad e = -E\varepsilon^{-ut}; \quad i = 0.$$

$$\tau + \pi < \theta < \tau + \frac{3\pi}{2}: \quad e = 0; \qquad i = -I\varepsilon^{-ut}.$$

$$\tau + \frac{3\pi}{2} < \theta < \tau + 2\pi; \quad e = E\varepsilon^{-ut}; \qquad i = 0, \text{ etc.}$$

At any moment of time θ one part of the line has voltage $e = E\varepsilon^{-ut}$ and zero current, and the other part of the line has current $i = I\varepsilon^{-ut}$ and zero voltage, and the dividing line between the two sections of the line is at $\tau = \theta \pm \frac{n\pi}{2}$, hence moves along the line at the rate $\tau = \theta$.

69. As second example may be considered the discharge of a live line into a dead line: A circuit of length l_1, charged to a uniform voltage E, but carrying no current, is connected to a circuit of the same constants, but of length l_2, and having neither voltage nor current, otherwise both circuits are insulated.

Let the total length of the circuit be denoted by

$$2\,l = l_1 + l_2,$$

and let the time be counted from the moment where the circuits l_1 and l_2 are connected together, the distance from the beginning of the live circuit l_1, whose other end is connected to the dead circuit l_2.

Introduce again the denotations (44), and represent the total length of the line $2\,l = l_1 + l_2$ by $\tau = \pi$, then write

$$\tau_1 = \frac{l_1}{l_1 + l_2}\pi.$$

As the voltage is E from $\tau = 0$ to $\tau = \tau_1$, and 0 from $\tau = \tau_1$ to $\tau = \pi$, the mean value of voltage, or the voltage which will be left on the line after the transient phenomenon has passed, is

$$e_0 = \frac{\tau_1}{\pi} E,$$

and the terminal conditions of voltage and current are

$$\begin{aligned} \theta &= 0 \\ e &= E - e_0 \quad \text{for } 0 < \tau < \tau_1, \\ e &= -e_0 \quad \text{for } \tau_1 < \tau < \pi, \\ i &= 0. \end{aligned}$$

Proceeding then in the same manner as in paragraph 34, in the present case the equations (49) and (52) apply, and

$$a_n = \frac{2}{\pi}\left\{\int_0^{\tau_1}(E - e_0)\cos n\theta\, d\theta - \int_{\tau_1}^{\pi} e_0 \cos n\theta\, d\theta\right\}$$

$$= \frac{2\,E \sin n\tau_1}{n},$$

$$a_n' = b_n = b_n' = 0;$$

hence,

$$\gamma_n = 0$$

and

$$\left.\begin{aligned} e &= \frac{2\,E}{\pi}\left\{\frac{\tau_1}{2} + \varepsilon^{-ut}\sum_1^{\infty}{}_n \frac{\sin n\tau_1}{n}\cos n\tau \cos n\theta\right\}, \\ i &= \frac{2\,E}{\pi}\sqrt{\frac{C}{L}}\,\varepsilon^{-ut}\sum_1^{\infty}{}_n \frac{\sin n\tau_1}{n}\sin n\tau \sin n\theta. \end{aligned}\right\} \quad (60)$$

Choosing the same line constants as in paragraph 35, and assuming

$$l_1 = 120 \text{ miles and } l_2 = 80 \text{ miles},$$

we have

$$l = 100 \text{ miles and } \tau_1 = 0.6\,\pi.$$

Let $E = 40{,}000$ volts, $ut = 0.0404\,\theta$, and the fundamental frequency of oscillation, f_1, = 445 cycles per second; then

$$\left.\begin{aligned} e = {} & 24{,}000 + 25{,}500\,\varepsilon^{-0.0404\,\theta}\{\sin 108^\circ \cos\tau\cos\theta + \tfrac{1}{2}\sin 216^\circ \\ & \cos 2\tau\cos 2\theta + \tfrac{1}{3}\sin 348^\circ\cos 3\tau\cos 3\theta + \cdots\}\text{ volts} \\ \text{and} & \\ i = {} & 73.5\,\varepsilon^{-0.0404\,\theta}\{\sin 108^\circ \sin\tau\sin\theta + \tfrac{1}{2}\sin 216^\circ\sin 2\tau \\ & \sin 2\theta + \tfrac{1}{3}\sin 348^\circ\sin 3\tau\sin 3\theta + \cdots\}\text{ amp.} \end{aligned}\right\} \quad (61)$$

CHAPTER VIII

TRANSITION POINTS AND THE COMPOUND CIRCUIT

70. The discussions of standing waves and free oscillations in Chapters V and VII, and traveling waves in Chapter VI, apply directly only to simple circuits, that is, circuits comprising a conductor of uniformly distributed constants r, L, g, and C. Industrial electric circuits, however, never are simple circuits, but are always complex circuits comprising sections of different constants, — generator, transformer, transmission lines, and load, — and a simple circuit is realized only by a section of a circuit, as a transmission line or a high-potential transformer coil, which is cut off at both ends from the rest of the circuit, either by open-circuiting, $i = 0$, or by short-circuiting, $e = 0$. Approximately, the simple circuit is realized by a section of a complex or compound circuit, connecting to other sections of very different constants, so that the ends of the circuit can, approximately, be considered as reflection points. For instance, an underground cable of low L and high C, when connected to a large reactive coil of high L and low C, may approximately, at its ends be considered as having reflection points $i = 0$. A high-potential transformer coil of high L and low C, when connected to a cable of low L and high C, may at its ends be considered as having reflection points $e = 0$. In other words, in the first case the reactive coil may be considered as stopping the current, in the latter case the cable considered as short-circuiting the transformer. This approximation, however, while frequently relied upon in engineering practice, and often permissible for the circuit section in which the transient phenomenon originates, is not permissible in considering the effect of the phenomenon on the adjacent sections of the circuit. For instance, in the first case above mentioned, a transient phenomenon in an underground cable connected to a high reactance, the current and e.m.f. in the cable may approximately be represented by considering the reactive coil as a reflection point, that is, an open circuit, since only a small current exists in the reactive coil. Such a small current in the reactive coil may, however, give a very high and destructive voltage in the reactive coil, due to its high L, and thus in the circuit beyond the reactive coil. In the investigation of the effect of a transient phenomenon originating in one section of a compound circuit, as an oscillating arc on an underground cable, on other sections of the circuit, as the generating station, even a very great change of circuit constants cannot be considered as a reflection point. Since this is the most important case met in industrial practice, as disturbances originating in one section of a compound circuit usually develop their destructive effects in other sections of the circuit, the investigation of the general problem of a compound circuit comprising sections of different constants thus becomes necessary. This requires the investigation of the changes occurring in an electric wave, and its equations, when passing over a transition point from one circuit or section of a circuit into another section of different constants.

71. The equations (53) to (60) of Chapter I, while most general, are less convenient for studying the transition of a wave from one circuit to another circuit of different constants, and since in industrial high-voltage circuits, at least for waves originating in the circuits, q and k are very large compared with s and h, as discussed in paragraph 16, s and h may be neglected compared with q and k. This gives, as discussed in paragraph 9,

$$\left.\begin{aligned} h &= \sigma s, \\ k &= \sigma q, \\ c_1 = c_2 &= \sqrt{\frac{L}{C}} = c, \\ c_1' = c_2' &= \frac{m}{q}\sqrt{\frac{L}{C}} = 0, \end{aligned}\right\} \qquad (1)$$

where

$$\sigma = \sqrt{LC}, \qquad (2)$$

and substituting

$$\lambda = \sigma l, \qquad (3)$$

that is

$$\left.\begin{aligned} kl &= q\lambda, \\ hl &= s\lambda, \end{aligned}\right\} \qquad (4)$$

gives

$$\begin{aligned} i = \varepsilon^{-ut} \{ & \varepsilon^{-s(\lambda - t)} [C_1 \cos q (\lambda - t) + C_1' \sin q (\lambda - t)] \\ & - \varepsilon^{+s(\lambda + t)} [C_2 \cos q (\lambda + t) + C_2' \sin q (\lambda + t)] \\ & + \varepsilon^{+s(\lambda - t)} [C_3 \cos q (\lambda - t) + C_3' \sin q (\lambda - t)] \\ & - \varepsilon^{-s(\lambda + t)} [C_4 \cos q (\lambda + t) + C_4' \sin q (\lambda + t)] \} \end{aligned} \tag{5}$$

and

$$\begin{aligned} e = \sqrt{\frac{L}{C}} \varepsilon^{-ut} \{ & \varepsilon^{-s(\lambda - t)} [C_1 \cos q (\lambda - t) + C_1' \sin q (\lambda - t)] \\ & + \varepsilon^{+s(\lambda + t)} [C_2 \cos q (\lambda + t) + C_2' \sin q (\lambda + t)] \\ & + \varepsilon^{+s(\lambda - t)} [C_3 \cos q (\lambda - t) + C_3' \sin q (\lambda - t)] \\ & + \varepsilon^{-s(\lambda + t)} [C_4 \cos q (\lambda + t) + C_4' \sin q (\lambda + t)] \}. \end{aligned} \tag{6}$$

Substituting now

$$\left.\begin{aligned} C_1^2 + C_1'^2 &= A^2, \\ C_2^2 + C_2'^2 &= B^2, \\ C_3^2 + C_3'^2 &= C^2, \\ C_4^2 + C_4'^2 &= D^2, \end{aligned}\right\} \tag{7}$$

$$\left.\begin{aligned} \frac{C_1'}{C_1} &= \tan \alpha, \\ \frac{C_2'}{C_2} &= \tan \beta, \\ \frac{C_3'}{C_3} &= \tan \gamma, \\ \frac{C_4'}{C_4} &= \tan \delta, \end{aligned}\right\} \tag{8}$$

gives

$$\begin{aligned} i = \varepsilon^{-ut} \{ & A\varepsilon^{-s(\lambda - t)} \cos [q(\lambda - t) - \alpha] - B\varepsilon^{+s(\lambda + t)} \cos [q(\lambda + t) - \beta] \\ & + C\varepsilon^{+s(\lambda - t)} \cos [q(\lambda - t) - \gamma] - D\varepsilon^{-s(\lambda + t)} \cos [q(\lambda + t) - \delta] \} \end{aligned} \tag{9}$$

and

$$e = \sqrt{\frac{L}{C}}\varepsilon^{-ut}\{A\varepsilon^{-s(\lambda-t)}\cos[q(\lambda-t)-\alpha]+B\varepsilon^{+s(\lambda+t)}\cos[q(\lambda+t)-\beta]$$
$$+C\varepsilon^{+s(\lambda-t)}\cos[q(\lambda-t)-\gamma]+D\varepsilon^{-s(\lambda+t)}\cos[q(\lambda+t)-\delta]\}. \tag{10}$$

72. In these equations (9) and (10) λ is the distance coördinate, using the velocity of propagation as unit distance, and at a transition point from one circuit to another, where the circuit constants change, the velocity of propagation also changes, and thus t for the same time constants s and q, h and k also change, and therewith kl, but transformed to the distance variable λ, $q\lambda$ remains the same; that is, by introducing the distance variable λ, the distance can be measured throughout the entire circuit, and across transition points, at which the circuit constants change, and the same equations (9) and (10) apply throughout the entire circuit. In this case, however, in any section of the circuit,

$$\left.\begin{aligned}\lambda &= \sigma_i l\\ &= l\sqrt{L_iC_i},\end{aligned}\right\} \tag{11}$$

where L_i and C_i are the inductance and the capacity, respectively, of the section i of the circuit, per unit length, for instance, per mile, in a line, per turn in a transformer coil, etc.

In a compound circuit the time variable t is the same throughout the entire circuit, or, in other words, the frequency of oscillation, as represented by q, and the rate of decay of the oscillation, as represented by the exponential function of time, must be the same throughout the entire circuit. Not so, however, with the distance variable l; the wave length of the oscillation and its rate of building up or down along the circuit need not be the same, and usually are not, but in some sections of the circuit the wave length may be far shorter, as in coiled circuits as transformers, due to the higher L, or in cables due to the higher C. To extend the same equations over the entire compound circuit, it therefore becomes necessary to substitute for the distance variable l another distance variable λ of such character that the wave length has the same value in all sections of the compound circuit. As the wave length of the section i is $\frac{l}{\sqrt{L_iC_i}}$, this is done by changing the unit distance by the factor $\sigma_i = \sqrt{L_iC_i}$. The distance unit of the new distance variable λ then is the distance traversed by the wave in unit time, hence different in linear measure for the different sections of the circuit, but offers the advantage of carrying the distance measurements across the entire circuit and over transition points by the same distance variable λ.

This means that the length l_i of any section i of the compound circuit is expressed by the length $\lambda_i = \sigma_i l_i$.

The introduction of the distance variable λ also has the advantage that in the determination of the constants r, L, g, C of the different sections of the circuit different linear distance measurements l may be used. For instance, in the transmission line, the constants may be given per mile, that is, the mile used as unit length, while in the high-potential coil of a transformer the turn, or the coil, or the total transformer may be used as unit of length l, so that the actual linear length of conductor may be unknown. For instance, choosing the total length of conductor in the high-potential transformer as unit length, then the length of the transformer winding in the velocity measure λ is $\lambda_0 = \sqrt{L_0 C_0}$, where L_0 = total inductance, C_0 = total capacity of transformer.

The introduction of the distance variable λ thus permits the representation in the circuit of apparatus as reactive coils, etc., in which one of the constants is very small compared with the other and therefore is usually neglected and the apparatus considered as "massed inductance," etc., and allows the investigation of the effect of the distributed capacity of reactive coils and similar matters, by representing the reactive coil as a finite (frequently quite long) section λ_0 of the circuit.

73. Let $\lambda_0, \lambda_1, \lambda_2, \ldots \lambda_n$ be a number of transition points at which the circuit constants change and the quantities may be denoted by index 1 in the section from λ_0 to λ_1, by index 2 in the section from λ_1 to λ_2, etc.

At $\lambda = \lambda_1$ it then must be $i_1 = i_2$, $e_1 = e_2$; thus substituting $\lambda = \lambda_1$ into equations (9) and (10) gives

$$\begin{aligned} &\varepsilon^{-(u_1-s_1)t}\{A_1\varepsilon^{-s_1\lambda_1}\cos[q_1(\lambda_1-t)-\alpha_1]-B_1\varepsilon^{+s_1\lambda_1}\cos[q_1(\lambda_1+t)-\beta_1]\} \\ &+\varepsilon^{-(u_1+s_1)t}\{C_1\varepsilon^{+s_1\lambda_1}\cos[q_1(\lambda_1-t)-\gamma_1]-D_1\varepsilon^{-s_1\lambda_1}\cos[q_1(\lambda_1+t)-\delta_1]\} \\ &=\varepsilon^{-(u_2-s_2)t}\{A_2\varepsilon^{-s_2\lambda_1}\cos[q_2(\lambda_1-t)-\alpha_2]-B_2\varepsilon^{+s_2\lambda_1}\cos[q_2(\lambda_1+t)-\beta_2]\} \\ &+\varepsilon^{-(u_2+s_2)t}\{C_2\varepsilon^{+s_2\lambda_1}\cos[q_2(\lambda_1-t)-\gamma_2]-D_2\varepsilon^{-s_2\lambda_1}\cos[q_2(\lambda_1+t)-\delta_2]\}. \end{aligned} \tag{12}$$

Herefrom it follows that

$$q_2 = q_1; \tag{13}$$

that is, the frequency must be the same throughout the entire circuit as is obvious, and

$$u_2 \pm s_2 = u_1 \pm s_1. \tag{14}$$

Since $u_2 \neq u_1$, only one of the two waves can exist, the $A\ B$, or the $C\ D$, and since these two waves differ from each other only by the sign of s, by assuming now that s may be either positive or negative we can select one of the two waves, for instance, the second wave, but use $\boldsymbol{A}, \boldsymbol{B}, \boldsymbol{\alpha}, \boldsymbol{\beta}$ as denotations of the integration constants.

74. The equations (9) and (10) now assume the form

$$\left.\begin{aligned} i &= \varepsilon^{-ut}\{A\varepsilon^{+s(\lambda-t)}\cos[q(\lambda-t)-\alpha] \\ &\qquad - B\varepsilon^{-s(\lambda+t)}\cos[q(\lambda+t)-\beta]\} \\ &\text{and} \\ e &= \sqrt{\frac{L}{C}}\varepsilon^{-ut}\{A\varepsilon^{+s(\lambda-t)}\cos[q(\lambda-t)-\alpha] \\ &\qquad + B\varepsilon^{-s(\lambda+t)}\cos[q(\lambda+t)-\beta]\}, \end{aligned}\right\} \tag{15}$$

or

$$\left.\begin{aligned} i &= \varepsilon^{-(u+s)t}\{A\varepsilon^{+s\lambda}\cos[q(\lambda-t)-\alpha] \\ &\qquad - B\varepsilon^{-s\lambda}\cos[q(\lambda+t)-\beta]\} \\ &\text{and} \\ e &= \sqrt{\frac{L}{C}}\varepsilon^{-(u+s)t}\{A\varepsilon^{+s\lambda}\cos[q(\lambda-t)-\alpha] \\ &\qquad + B\varepsilon^{-s\lambda}\cos[q(\lambda+t)-\beta]\}; \end{aligned}\right\} \tag{16}$$

or, using equations (5) and (6) instead of (9) and (10), the corresponding equations are of the form

$$\left.\begin{aligned} i &= \varepsilon^{-ut}\{\varepsilon^{+s(\lambda-t)}[A\cos q(\lambda-t) + B\sin q(\lambda-t)] \\ &\qquad - \varepsilon^{-s(\lambda+t)}[C\cos q(\lambda+t) + D\sin q(\lambda+t)]\} \\ &\text{and} \\ e &= \sqrt{\frac{L}{C}}\varepsilon^{-ut}\{\varepsilon^{+s(\lambda-t)}[A\cos q(\lambda-t) + B\sin q(\lambda-t)] \\ &\qquad + \varepsilon^{-s(\lambda+t)}[C\cos q(\lambda+t) + D\sin q(\lambda+t)]\}, \end{aligned}\right\} \tag{17}$$

or

$$\left.\begin{aligned} i &= \varepsilon^{-(u+s)t}\left\{\varepsilon^{+s\lambda}\left[A\cos q(\lambda - t) + B\sin q(\lambda - t)\right]\right. \\ &\qquad \left. - \varepsilon^{-s\lambda}\left[C\cos q(\lambda + t) + D\sin q(\lambda + t)\right]\right\} \\ &\text{and} \\ e &= \sqrt{\frac{L}{C}}\,\varepsilon^{-(u+s)t}\left\{\varepsilon^{+s\lambda}\left[A\cos q(\lambda - t) + B\sin q(\lambda - t)\right]\right. \\ &\qquad \left. + \varepsilon^{-s\lambda}\left[C\cos q(\lambda + t) + D\sin q(\lambda + t)\right]\right\}, \end{aligned}\right\} \tag{18}$$

where s may be positive or negative.

From equation (12) it then follows that

$$u_1 + s_1 = u_2 + s_2 = u_3 + s_3 = \ldots = u_n + s_n = u_0, \tag{19}$$

where u_1, u_2, u_3, etc., u_n are the time constants of the individual sections of the complex circuit, $\frac{1}{2}\left(\frac{r}{L} + \frac{g}{C}\right)$, and u_0 may be called the *resultant time decrement of the complex circuit.*

75. Equation (12), by canceling equal terms on both sides, then assumes the form

$$A_1\varepsilon^{+s_1\lambda_1}\cos\left[q(\lambda_1 - t) - \alpha_1\right] - B_1\varepsilon^{-s_1\lambda_1}\cos\left[q(\lambda_1 + t) - \beta_1\right] =$$
$$A_2\varepsilon^{+s_2\lambda_1}\cos\left[q(\lambda - t) - \alpha_2\right] - B_2\varepsilon^{-s_2\lambda_1}\cos\left[q(\lambda_1 + t) - \beta_2\right],$$

and, resolved for cos qt and sin qt, this gives the identities

$$\begin{aligned} &A_1\varepsilon^{+s_1\lambda_1}\cos(q\lambda_1 - \alpha_1) - B_1\varepsilon^{-s_1\lambda_1}\cos(q\lambda_1 - \beta_1) = \\ &A_2\varepsilon^{+s_2\lambda_1}\cos(q\lambda_1 - \alpha_2) - B_2\varepsilon^{-s_2\lambda_1}\cos(q\lambda_1 - \beta_2), \\ &A_1\varepsilon^{+s_1\lambda_1}\sin(q\lambda_1 - \alpha_1) + B_1\varepsilon^{-s_1\lambda_1}\sin(q\lambda_1 - \beta_1) = \\ &A_2\varepsilon^{+s_2\lambda_1}\sin(q\lambda_1 - \alpha_2) + B_2\varepsilon^{-s_2\lambda_1}\sin(q\lambda_1 - \beta_2). \end{aligned} \tag{20}$$

These identities resulted by equating $i_1 = i_2$ from equation (15). In the same manner, by equating e_1 and e_2 from equation (15) there result the two further identities

$$\sqrt{\frac{L_1}{C_1}}\left\{A_1\varepsilon^{+s_1\lambda_1}\cos(q\lambda_1 - \alpha_1) + B_1\varepsilon^{-s_1\lambda_1}\cos(q\lambda_1 - \beta_1)\right\} =$$

$$\sqrt{\frac{L_2}{C_2}}\left\{A_2\varepsilon^{+s_2\lambda_1}\cos(q\lambda_1 - \alpha_2) + B_2\varepsilon^{-s_2\lambda_1}\cos(q\lambda_1 - \beta_2)\right\},$$

$$\sqrt{\frac{L_1}{C_1}}\left\{A_1\varepsilon^{+s_1\lambda_1}\sin(q\lambda_1-\alpha_1)-B_1\varepsilon^{-s_1\lambda_1}\sin(q\lambda_1-\beta_1)\right\}=$$

$$\sqrt{\frac{L_2}{C_2}}\left\{A_2\varepsilon^{+s_2\lambda_1}\sin(q\lambda_1-\alpha_2)-B_2\varepsilon^{-s_2\lambda_1}\sin(q\lambda_1-\beta_2)\right\}. \qquad (21)$$

Equations (20) and (21) determine the constants of any section of the circuit, A_2, B_2, α_2, β_2, from the constants of the next section of the circuit, A_1, B_1, α_1, β_1.

Let

$$\left.\begin{aligned} A\varepsilon^{+s\lambda_1}\cos(q\lambda_1-\alpha)&=A';\\ A\varepsilon^{+s\lambda_1}\sin(q\lambda_1-\alpha)&=A'';\\ B\varepsilon^{-s\lambda_1}\cos(q\lambda_1-\beta)&=B',\\ B\varepsilon^{-s\lambda_1}\sin(q\lambda_1-\beta)&=B'';\end{aligned}\right\} \qquad (22)$$

$$\left.\begin{aligned} c_1&=\sqrt{\frac{L_1}{C_1}},\\ c_2&=\sqrt{\frac{L_2}{C_2}}.\end{aligned}\right\} \qquad (23)$$

Then

$$\left.\begin{aligned} 2\,c_2A_2'&=(c_1+c_2)\,A_1'+(c_1-c_2)\,B_1',\\ 2\,c_2B_2'&=(c_1+c_2)\,B_1'+(c_1-c_2)\,A_1',\\ 2\,c_2A_2''&=(c_1+c_2)\,A_1''-(c_1-c_2)\,B_1'',\\ 2\,c_2B_2''&=(c_1+c_2)\,B_1''-(c_1-c_2)\,A_1'',\end{aligned}\right\} \qquad (24)$$

and since by (22):

$$A'^2+A''^2=A^2\varepsilon^{+2s\lambda_1}\ \text{etc.}, \qquad (25)$$

substituting herein (24),

$$\begin{aligned} 4\,c_2^2A_2^2\varepsilon^{+2s_2\lambda_1}=(c_1+c_2)^2A_1^2\varepsilon^{+2s_1\lambda_1}&+(c_1-c_2)^2B_1^2\varepsilon^{-2s_1\lambda_1}. \qquad (26)\\ &+2\,(A_1'B_1'-A_1''B_1'')\,(c_1^1-c_2^2)\end{aligned}$$

and

$$\left.\begin{aligned}\tan(q\lambda_1-\alpha_2)&=\frac{1-\dfrac{c_1-c_2}{c_1+c_2}\dfrac{B_1}{A_1}\varepsilon^{-2s_1\lambda_1}\dfrac{\sin(q\lambda_1-\beta_1)}{\sin(q\lambda_1-\alpha_1)}}{1+\dfrac{c_1-c_2}{c_1+c_2}\dfrac{B_1}{A_1}\varepsilon^{-2s_1\lambda_1}\dfrac{\cos(q\lambda_1-\beta_1)}{\cos(q\lambda_1-\alpha_1)}}\tan(q\lambda_1-\alpha_1)\\ \tan(q\lambda_1-\beta_2)&=\frac{1-\dfrac{c_1-c_2}{c_1+c_2}\dfrac{A_1}{B_1}\varepsilon^{+2s_1\lambda_1}\dfrac{\sin(q\lambda_1-\alpha_1)}{\sin(q\lambda_1-\beta_1)}}{1+\dfrac{c_1-c_2}{c_1+c_2}\dfrac{A_1}{B_1}\varepsilon^{+2s_2\lambda_1}\dfrac{\cos(q\lambda_1-\alpha_1)}{\cos(q\lambda_1-\beta_1)}}\tan(q\lambda_1-\beta_1).\end{aligned}\right\} \tag{27}$$

In the same manner, equating, for $\lambda = \lambda_1$, in equations (18) the current i_1, corresponding to the section from λ_0 to λ_1, with the current i_2, corresponding to the section from λ_1 to λ_2, and also the e.m.fs., $e_2 = e_1$, gives the constants in equations (18) and (17), of one section, λ_1 to λ_2, expressed by those of the next adjoining section, λ_0 to λ_1, as

$$\left.\begin{aligned}A_2&=\varepsilon^{-s_2\lambda_1}\{a_1\varepsilon^{+s_1\lambda_1}A_1+b_1\varepsilon^{-s_1\lambda_1}(C_1\cos 2q\lambda_1+D_1\sin 2q\lambda_1)\}\\ B_2&=\varepsilon^{-s_2\lambda_1}\{a_1\varepsilon^{+s_1\lambda_1}B_1+b_1\varepsilon^{-s_1\lambda_1}(C_1\sin 2q\lambda_1-D_1\cos 2q\lambda_1)\}\\ C_2&=\varepsilon^{+s_2\lambda_1}\{a_1\varepsilon^{-s_1\lambda_1}C_1+b_1\varepsilon^{+s_1\lambda_1}(A_1\cos 2q\lambda_1-B_1\sin 2q\lambda_1)\}\\ D_2&=\varepsilon^{+s_2\lambda_1}\{a_1\varepsilon^{-s_1\lambda_1}D_1+b_1\varepsilon^{+s_1\lambda_1}(A_1\sin 2q\lambda_1-B_1\cos 2q\lambda_1)\}\end{aligned}\right\} \tag{28}$$

where

$$\left.\begin{aligned}a_1&=\frac{c_1+c_2}{2c_2},\\ b_1&=\frac{c_1-c_2}{2c_2},\end{aligned}\right\} \tag{29}$$

$$\left.\begin{aligned}c_1&=\sqrt{\frac{L_1}{C_1}},\\ c_2&=\sqrt{\frac{L_2}{C_2}}.\end{aligned}\right\} \tag{30}$$

76. The general equation of current and e.m.f. in a complex circuit thus also consists of two terms, the main wave A in equations (15), (16), and its reflected wave B.

The factor $\epsilon^{-(u+s)t} = \epsilon^{-u_0 t}$ in equations (16) and (18) represents the time decrement, or the decrease of the intensity of the wave with the time, and as such is the same throughout the entire circuit. In an isolated section, of time constant u, the time decrement, from Chapters V and VII, is, however, ϵ^{-ut}; that is, with the decrement ε^{-ut} the wave dies out in the isolated section at the rate at which its stored energy is dissipated by the power lost in resistance and conductance. In a section of the circuit connected to other sections the time decrement $\varepsilon^{-u_0 t}$ does not correspond to the power dissipation in the section; that is, the wave does not die out in each section at the rate as given by the power consumed in this section, or, in other words, power transfer occurs from section to section during the oscillation of a compound circuit

If s is negative, u_0 is less than u, and the wave dies out in that particular section at a lesser rate than corresponds to the power consumed in the section, or, in other words, in this section of the compound circuit more power is consumed by r and g than is supplied by the decrease of the stored energy, and this section, therefore, must receive energy from adjoining sections. Inversely, if s is positive, $u_0 > u$, the wave dies out more rapidly in that section than its stored energy is consumed by r and g; that is, a part of the stored energy of this section is transferred to the adjoining sections, and only a part — occasionally a very small part — dissipated in the section, and this section acts as a store of energy for supplying the other sections of the system.

The constant s of the circuit, therefore, may be called *energy transfer constant*, and positive s means transfer of energy from the section to the rest of the circuit, and negative s means reception of energy from other sections. This explains the vanishing of s in a standing wave of a uniform circuit, due to the absence of energy transfer, and the presence of s in the equations of the traveling wave, due to the transfer of energy along the circuit, and in the general equations of alternating-current circuits.

It immediately follows herefrom that in a compound circuit some of the s of the different sections must always be positive, some negative.

In addition to the time decrement $\epsilon^{-(u+s)t} = \epsilon^{-u_0 t}$ the waves in equations (16) and (18) also contain the distance decrement $\epsilon^{+s\lambda}$ for the main wave, $\epsilon^{-s\lambda}$ for the reflected wave. Negative s therefore means a decrease of the main wave for increasing λ, or in the direction of propagation, and a decrease of the reflected wave for decreasing λ, that is, also in the direction of propagation; while positive s means increase of main wave as well as reflected wave in the direction of propagation along the circuit. In other words, if s is negative and the section consumes more power than is given by its stored energy, and therefore receives power from the adjoining sections, the electric wave decreases in the direction of its propagation, or builds down, showing the gradual dissipation of the power received from adjoining sections. Inversely, if s is positive and the section thus supplies power to adjoining sections, the electric wave increases in this section in the direction of its propagation, or builds up.

In other words, in a compound circuit, in sections of low power dissipation, the wave increases and transfers power to sections of high power dissipation, in which the wave decreases.

This can still better be seen from equations (15) and (17). Here the time decrement ϵ^{-ut} represents the dissipation of stored energy by the power consumed in the section by r and g. The time distance decrement, $\epsilon^{+s(\lambda-t)}$ for the main wave, $\epsilon^{-s(\lambda+t)}$ for the reflected wave, represents the decrement of the wave for constant $(\lambda - t)$ or $(\lambda + t)$ respectively; that is, shows the change of wave intensity during its propagation. Thus for instance, following a wave crest, the wave decreases for negative s and increases for positive s, in addition to the uniform decrease by the time constant ϵ^{-ut}; or, in other words, for positive s the wave gathers intensity during its progress, for negative s it loses intensity in addition to the loss of intensity by the time constant of this particular section of the circuit.

77. Introducing the resultant time decrement u_0 of the compound circuit, the equations of any section, (16) and (18), can also be expressed by the resultant time decrement of the entire compound circuit, u_0, and the energy transfer constant of the individual section; thus

$$s = u_0 - u, \tag{31}$$

$$\left.\begin{aligned} i &= \varepsilon^{-u_0 t}\{A\varepsilon^{+s\lambda}\cos[q(\lambda-t)-\alpha]-B\varepsilon^{-s\lambda}\cos[q(\lambda+t)-\beta]\} \\ &\text{and} \\ e &= \sqrt{\frac{L}{C}}\varepsilon^{-u_0 t}\{A\varepsilon^{+s\lambda}\cos[q(\lambda-t)-\alpha]+B\varepsilon^{-s\lambda}\cos[q(\lambda+t)-\beta]\}, \end{aligned}\right\} \quad (32)$$

or

$$\left.\begin{aligned} i = \varepsilon^{-u_0 t}\{&\varepsilon^{+s\lambda}[A\cos q(\lambda-t)+B\sin q(\lambda-t)] \\ &-\varepsilon^{-s\lambda}[C\cos q(\lambda+t)+D\sin q(\lambda+t)]\} \\ &\text{and} \\ e = \sqrt{\frac{L}{C}}\varepsilon^{-u_0 t}\{&\varepsilon^{+s\lambda}[A\cos q(\lambda-t)+B\sin q(\lambda-t)] \\ &+\varepsilon^{-s\lambda}[C\cos q(\lambda+t)+D\sin q(\lambda+t)]\}. \end{aligned}\right\} \quad (33)$$

The constants A, B, C, D are the integration constants, and are such as given by the terminal conditions of the problem, as by the distribution of current and e.m.f. in the circuit at the starting moment, for $t = 0$, or at one particular point, as $\lambda = 0$.

78. The constants u_0 and q depend upon the circuit conditions. If the circuit is closed upon itself — as usually is the case with an electrical transmission or distribution circuit — and Λ is the total length of the closed circuit, the equations must give for $\lambda = \Lambda$ the same values as for $\lambda = 0$, and therefore q must be a complete cycle or a multiple thereof, $2\,n\pi$; that is,

$$q = \frac{2\,n\pi}{\Lambda}, \quad (34)$$

and the least value of q, or the fundamental frequency of oscillation, is

$$q_0 = \frac{2\,\pi}{\Lambda} \quad (35)$$

and

$$q = nq_0. \quad (36)$$

If the compound circuit is open at both ends, or grounded at both ends, and thus performs a half-wave oscillation, and Λ_1 = total length of the circuit,

$$q_0 = \frac{\pi}{\Lambda_1} \text{ and } q = nq_0, \quad (37)$$

and if the circuit is open at one end, grounded at the other end, thus performing a quarter-wave oscillation, and $\Lambda_2 =$ total length of circuit, it is

$$q_0 = \frac{\pi}{2\,\Lambda_2} \quad \text{and} \quad q = (2\,n - 1)\,q_0, \tag{38}$$

while, if the length of the compound circuit is very great compared with the frequency of the oscillation, q_0 may have any value; that is, if the wave length of the oscillation is very short compared with the length of the circuit, any wave length, and therefore any frequency, may occur. With uniform circuits, as transmission lines, this latter case, that is, the response of the line to any frequency, can occur only in the range of very high frequencies. Even in a transmission line of several hundred miles' length the lowest frequency of free oscillation is fairly high, and frequencies which are so high compared with the fundamental frequency of the circuit that, considered as higher harmonics thereof, they overlap (as discussed in the above), must be extremely high—of the magnitude of million cycles. In a compound circuit, however, the fundamental frequency may be very much lower, and below machine frequencies, as the velocity of propagation $\frac{1}{\sqrt{LC}}$ may be quite low in some sections of the circuit, as in the high-potential coils of large transformers, and the presence of iron increases the inconstancy of L for high frequencies, so that in such a compound circuit, even at fairly moderate frequencies, of the magnitude of 10,000 cycles, the circuit may respond to any frequency.

79. The constant u_0 is also determined by the circuit constants. Upon u_0 depends the energy transfer constant of the circuit section, and therewith the rate of building up in a section of low power consumption, or building down in a section of high power consumption. In a closed circuit, however, passing around the entire circuit, the same values of e and i must again be reached, and the rates of building up and building down of the wave in the different sections must therefore be such as to neutralize each other when carried through the entire circuit; that is, the total building up through the entire compound circuit must be zero. This gives an equation from which u_0 is determined.

In a complex circuit having n sections of different constants and therefore n transition points, at the distances

$$\lambda_1, \lambda_2 \ldots \lambda_n \tag{(39)}$$

where $\lambda_{n+1} = \lambda_1 + \Lambda$, and $\Lambda =$ the total length of the circuit, the equations of i and e of any section i are given by equations (33) containing the constants A_i, B_i, C_i, D_i.

The constants A, B, C, D of any section are determined by the constants of the preceding section by equations (28) to (30). The constants of the second section thus are determined by those of the first section, the constants of the third section by those of the second section, and thereby, by substituting for the latter, by the constants of the first section, and in this manner, by successive substitutions, the constants of any section i can be expressed by the constants of the first section as linear functions thereof.

Ultimately thereby the constants of section $(n + 1)$ are expressed as linear functions of the constants of the first section:

$$\left.\begin{aligned} A_{n+1} &= a'A_1 + a''B_1 + a'''C_1 + a''''D_1, \\ B_{n+1} &= b'A_1 + b''B_1 + b'''C_1 + b''''D_1, \\ C_{n+1} &= c'A_1 + c''B_1 + c'''C_1 + c''''D_1, \\ D_{n+1} &= d'A_1 + d''B_1 + d'''C_1 + d''''D_1, \end{aligned}\right\} \tag{(40)}$$

where $a', a'', a''', a'''', b', b''$, etc., are functions of s_i and λ_i.

The $(n + 1)$st section, however, is again the first section, and it is thereby, by equations (33) and (39),

$$\left.\begin{aligned} A_{n+1} &= A_1\varepsilon^{-s_1\Lambda}, \\ B_{n+1} &= B_1\varepsilon^{-s_1\Lambda}, \\ C_{n+1} &= C_1\varepsilon^{+s_1\Lambda}, \\ D_{n+1} &= D_1\varepsilon^{+s_1\Lambda}, \end{aligned}\right\} \tag{(41)}$$

and substituting (41) into (40) gives four symmetrical linear equations in A_1, B_1, C_1, D_1, from which these four constants can be eliminated, as n symmetrical linear equations with

n variables are dependent equations, containing an identity, thus:

$$\left.\begin{aligned}
(a' - \varepsilon^{-s_1\Lambda_1})\, A_1 + a''B_1 + a'''C_1 + a''''D_1 &= 0;\\
b'A_1 + (b'' - \varepsilon^{-s_1\Lambda})\, B_1 + a'''C_1 + b''''D_1 &= 0;\\
c'A_1 + c''B_1 + (c''' - \varepsilon^{+s_1\Lambda})\, C_1 + c''''D_1 &= 0;\\
d'A_1 + d''B_1 + d'''C_1 + (d'''' - \varepsilon^{+s_1\Lambda})\, D_1 &= 0,
\end{aligned}\right\} \qquad ((42)$$

and herefrom

$$\begin{vmatrix}
(a' - \varepsilon^{-s_1\Lambda}) & a'' & a''' & a'''' \\
b' & (b'' - \varepsilon^{-s_1\Lambda}) & b''' & b'''' \\
c' & c'' & (c''' - \varepsilon^{+s_1\Lambda}) & c'''' \\
d' & d'' & d''' & (d'''' - \varepsilon^{+s_1\Lambda})
\end{vmatrix} = 0. \qquad ((43)$$

Substituting in this determinant equation for s_i the values from (19))

$$s_i = u_0 - u_i \qquad ((44)$$

gives an exponential equation in u_0, thus:

$$F\,(u_0, u_i, \lambda_i, c_i) = 0, \qquad ((45)$$

from which the value u_0, or the resultant time decrement of the circuit, is determined.

In general, this equation (45) can be solved only by approximation, except in special cases.

CHAPTER IX

POWER AND ENERGY OF THE COMPOUND CIRCUIT

80. The free oscillation of a compound circuit differs from that of the uniform circuit in that the former contains exponential functions of the distance λ which represent the shifting or transfer of power between the sections of the circuit.

Thus the general expression of one term or frequency of current and voltage in a section of a compound circuit is given by equations (33) of Chapter VIII;

$$\left.\begin{aligned} i &= \varepsilon^{-u_0 t} \left\{\varepsilon^{+s\lambda} [A \cos q (\lambda - t) + B \sin q (\lambda - t)] \right. \\ &\qquad \left. - \varepsilon^{-s\lambda} [C \cos q (\lambda + t) + D \sin q (\lambda + t)]\right\} \end{aligned}\right\}$$

and

$$\left.\begin{aligned} e &= \sqrt{\frac{L}{C}} \varepsilon^{-u_0 t} \left\{\varepsilon^{+s\lambda} [A \cos q (\lambda - t) + B \sin q (\lambda - t)] \right. \\ &\qquad \left. + \varepsilon^{-s\lambda} [C \cos q (\lambda + t) + D \sin q (\lambda + t)]\right\}, \end{aligned}\right\}$$

where $q = nq_0$, $q_0 = \frac{2\pi}{\Lambda}$, Λ = total length of circuit, expressed in the distance coördinate $\lambda = \sigma l$, l being the distance coördinate of the circuit section in any measure, as miles, turns, etc., and r, L, g, C the circuit constants per unit length of l,

$$\sigma = \sqrt{LC},$$

$$c = \sqrt{\frac{L}{C}},$$

$$u = \frac{1}{2}\left(\frac{r}{L} + \frac{g}{C}\right) = \text{time constant of circuit section,}$$

$$u_0 = u + s = \text{resultant time decrement of compound circuit,}$$

$$s = u_0 - u = \text{energy transfer constant of circuit section.}$$

The instantaneous value of power at any point λ of the circuit at any time t is

$$p = ei$$

$$= \sqrt{\frac{L}{C}}\varepsilon^{-2u_0t}\left\{\varepsilon^{+2s\lambda}\left[A\cos q(\lambda - t) + B\sin q(\lambda - t)\right]^2 - \varepsilon^{-2s\lambda}\left[C\cos q(\lambda + t) + D\sin q(\lambda + t)\right]^2\right\}$$

$$= \frac{1}{2}\sqrt{\frac{L}{C}}\varepsilon^{-2u_0t}\left\{\left[\varepsilon^{+2s\lambda}(A^2 + B^2) - \varepsilon^{-2s\lambda}(C^2 + D^2)\right] + \left[\varepsilon^{+2s\lambda}(A^2 - B^2)\cos 2q(\lambda - t) - \varepsilon^{-2s\lambda}(C^2 - D^2)\cos 2q(\lambda + t)\right] + 2\left[AB\varepsilon^{+2s\lambda}\sin 2q(\lambda - t) - CD\varepsilon^{-2s\lambda}\sin 2q(\lambda + t)\right]\right\}; \quad (1)$$

that is, the instantaneous value of power consists of a constant term and terms of double frequency in $(\lambda - t)$ and $(\lambda + t)$ or in distance λ and time t.

Integrating (1) over a complete period in time gives the *effective* or *mean value of power at any point* λ as

$$P = \frac{1}{2}\sqrt{\frac{L}{C}}\varepsilon^{-2u_0t}\left\{\varepsilon^{+2s\lambda}(A^2 + B^2) - \varepsilon^{-2s\lambda}(C^2 + D^2)\right\}; \quad (2)$$

that is, the effective power at any point of the circuit is the difference between the effective power of the main wave and that of the reflected wave, and also, the instantaneous power at any time and any point of the circuit is the difference between the instantaneous power of the main wave and that of the reflected wave.

The effective power at any point of the circuit gradually decreases in any section with the resultant time decrement of the total circuit, ε^{-2u_0t}, and varies gradually or exponentially with the distance λ, the one wave increasing, the other decreasing, so that at one point of the circuit or circuit section the effective power is zero; which point of the circuit is a power node, or point across which no energy flows. It is given by

$$\varepsilon^{+2s\lambda}(A^2 + B^2) = \varepsilon^{-2s\lambda}(C^2 + D^2).$$

or

$$\left.\begin{aligned} \varepsilon^{4s\lambda} &= \frac{C^2 + D^2}{A^2 + B^2}, \\ \lambda &= \frac{1}{4s}\log\frac{C^2 + D^2}{A^2 + B^2}. \end{aligned}\right\} \quad (3)$$

The difference of power between two points of the circuit, λ_1 and λ_2, that is, the power which is supplied or received (depending upon its sign) by a section $\lambda' = \lambda_2 - \lambda_1$ of the circuit, is given by equation (2) as

$$P_0 = \frac{1}{2}\sqrt{\frac{L}{C}}\varepsilon^{-2u_0t}\left\{(\varepsilon^{+2s\lambda_2} - \varepsilon^{+2s\lambda_1})(A^2 + B^2) - (\varepsilon^{-2s\lambda_2} - \varepsilon^{-2s\lambda_1})(C^2 + D^2)\right\}. \quad (4)$$

If P_0 is > 0, this represents the power which is supplied by the section λ' to the adjoining section of the circuit; if $P_0 < 0$, this is the power received by the section from the rest of the complex circuit.

If $s\lambda_2$ and $s\lambda_1$ are small quantities, the exponential function can be resolved into an infinite series, and all but the first term dropped, as of higher orders, or negligible, and this gives the approximate value

$$\varepsilon^{\pm 2s\lambda_2} - \varepsilon^{\pm 2s\lambda_1} = \pm 2s(\lambda_2 - \lambda_1) = \pm 2s\lambda'; \quad (5)$$

hence,

$$P_0 = sl'\sqrt{\frac{L}{C}}\varepsilon^{-2u_0t}\left\{A^2 + B^2 + C^2 + D^2\right\}; \quad (6)$$

that is, the power transferred from a section of length λ' to the rest of the circuit, or received by the section from the rest of the circuit, is proportional to the length of the section, λ', to its transfer constant, s, and to the sum of the power of main wave and reflected wave.

81. The energy stored by the inductance L of a circuit element $d\lambda$, that is, in the magnetic field of the circuit, is

$$dw_1 = \frac{L'i^2}{2}d\lambda,$$

where L' = inductance per unit length of circuit expressed by the distance coördinate λ.

Since L = the inductance per unit length of circuit, of distance coördinate l, and $\lambda = \sigma l$,

$$L' = \frac{L}{\sigma} = \frac{L}{\sqrt{LC}} = \sqrt{\frac{L}{C}};$$

hence,

$$dw_1 = \frac{1}{2}\sqrt{\frac{L}{C}}\, i^2 d\lambda. \tag{7}$$

In general, the circuit constants r, L, g, C, per unit length, $l = 1$ give, per unit length, $\lambda = 1$, the circuit constants

$$\left.\begin{aligned} &\frac{r}{\sigma};\ \frac{L}{\sigma};\ \frac{g}{\sigma};\ \frac{C}{\sigma}, \\ \text{or}\quad & \\ &\frac{r}{\sqrt{LC}};\ \frac{L}{\sqrt{LC}} = \sqrt{\frac{L}{C}};\ \frac{g}{\sqrt{LC}};\ \frac{C}{\sqrt{LC}} = \sqrt{\frac{C}{L}}. \end{aligned}\right\} \tag{8}$$

Substituting (290) in equation (309) gives

$$\begin{aligned} \frac{dw_1}{d\lambda} = \frac{1}{2}\sqrt{\frac{L}{C}}\,\varepsilon^{-2u_0t}\Big\{&\varepsilon^{+2s\lambda}\,[A\cos q(\lambda - t) + B\sin q(\lambda - t)]^2 \\ &+ \varepsilon^{-2s\lambda}\,[C\cos q(\lambda + t) + D\sin q(\lambda + t)]^2 \\ &- 2\,[A\cos q(\lambda - t) + B\sin q(\lambda - t)] \\ &\quad [C\cos q(\lambda + t) + D\sin q(\lambda + t)]\Big\} \\ = \frac{1}{4}\sqrt{\frac{L}{C}}\,\varepsilon^{-2u_0t}\,\Big\{&[\varepsilon^{+2s\lambda}(A^2 + B^2) + \varepsilon^{-2s\lambda}(C^2 + D^2)] \\ &+ [\varepsilon^{+2s\lambda}(A^2 - B^2)\cos 2q(\lambda - t) \\ &+ \varepsilon^{-2s\lambda}(C^2 - D^2)\cos 2q(\lambda + t)] \\ &+ 2\,[AB\varepsilon^{+2s\lambda}\sin 2q(\lambda - t) + CD\varepsilon^{-2s\lambda}\sin 2q(\lambda + t)] \\ &- 2\,[(AC - BD)\cos 2q\lambda + (AD + BC)\sin 2q\lambda] \\ &- 2\,[(AC + BD)\cos 2qt + (AD - BC)\sin 2qt]\Big\}. \end{aligned} \tag{9}$$

Integrating over a complete period in time gives the effective energy stored in the magnetic field at point λ as

$$\begin{aligned} \frac{dW_1}{d\lambda} &= \frac{1}{2\pi}\int_0^{2\pi}\frac{dw_1}{d\lambda}\,dt \\ &= \frac{1}{4}\sqrt{\frac{L}{C}}\,\varepsilon^{-2u_0t}\,\Big\{[\varepsilon^{+2s\lambda}(A^2 + B^2) + \varepsilon^{-2s\lambda}(C^2 + D^2)] \\ &\quad - 2\,[(AC - BD)\cos 2q\lambda + (AD + BC)\sin 2q\lambda]\Big\}, \end{aligned} \tag{10}$$

and integrating over one complete period of distance λ, or one complete wave length, this gives

$$\frac{dW_1{}^0}{d\lambda} = \frac{1}{2\pi}\int_0^{2\pi}\frac{dW_1}{d\lambda}\,d\lambda \doteq \frac{1}{4}\sqrt{\frac{L}{C}}\,\varepsilon^{-2u_0t}\left\{[\varepsilon^{+2s\lambda}(A^2+B^2)\right.$$
$$\left. + \varepsilon^{-2s\lambda}(C^2+D^2)\right\}. \tag{11}$$

The energy stored by the inductance L, or in the magnetic field of the conductor, thus consists of a constant part,

$$\frac{dw_0}{d\lambda} = \frac{1}{4}\sqrt{\frac{L}{C}}\,\varepsilon^{-2u_0t}\left\{\varepsilon^{+2s\lambda}(A^2+B^2) + \varepsilon^{-2s\lambda}(C^2+D^2),\right. \tag{12}$$

a part which is a function of $(\lambda - t)$ and $(\lambda + t)$,

$$\frac{dw'}{d\lambda} = \frac{1}{4}\sqrt{\frac{L}{C}}\,\varepsilon^{-2u_0t}\left\{[\varepsilon^{+2s\lambda}(A^2-B^2)\cos 2q(\lambda - t)\right.$$
$$+ \varepsilon^{-2s\lambda}(C^2-D^2)\cos 2q(\lambda + t)]$$
$$+ 2[AB\varepsilon^{+2s\lambda}\sin 2q(\lambda - t)$$
$$\left. + CD\varepsilon^{-2s\lambda}\sin 2q(\lambda + t)]\right\}, \tag{13}$$

a part which is a function of the distance λ only but not of time t,

$$-\frac{dw''}{d\lambda} = \frac{1}{2}\sqrt{\frac{L}{C}}\,\varepsilon^{-2u_0t}\left\{(AC - BD)\cos 2q\lambda + (AD + BC)\sin 2q\lambda,\right. \tag{14}$$

and a part which is a function of time t only but not of the distance λ,

$$-\frac{dw'''}{d\lambda} = \frac{1}{2}\sqrt{\frac{L}{C}}\,\varepsilon^{-2u_0t}\left\{(AC+BD)\cos 2qt + (AD-BC)\sin 2qt\right\}, \tag{15}$$

and the total energy of the electromagnetic field of circuit element $d\lambda$ at time t is

$$\frac{dw_1}{d\lambda} = \frac{dw_0}{d\lambda} + \frac{dw'}{d\lambda} - \frac{dw''}{d\lambda} - \frac{dw'''}{d\lambda}. \tag{16}$$

82. The energy stored in the electrostatic field of the conductor or by the capacity C is given by

$$dw_2 = \frac{C'e^2}{2}\,d\lambda;$$

or, substituting (8),

$$\frac{dw_2}{d\lambda} = \frac{1}{2}\sqrt{\frac{C}{L}}\, e^2, \tag{17}$$

and substituting in (17) the value of e from equation (33) of Chapter VIII gives the same expression as (9) except that the sign of the last two terms is reversed; that is, the total energy of the electrostatic field of circuit element $d\lambda$ at time t is

$$\frac{dw^2}{d\lambda} = \frac{dw_0}{d\lambda} + \frac{dw'}{d\lambda} + \frac{dw''}{d\lambda} + \frac{dw'''}{d\lambda} \tag{18}$$

and adding (16) and (18) gives the total stored energy of the electric field of the conductor,

$$\frac{dw}{d\lambda} = \frac{dw_1}{d\lambda} + \frac{dw_2}{d\lambda} = 2\frac{dw_0}{d\lambda} + 2\frac{dw'}{d\lambda}, \tag{19}$$

and integrated over a complete period of time this gives

$$\frac{dW}{d\lambda} = 2\,\frac{dw_0}{d\lambda} = \frac{1}{2}\sqrt{\frac{L}{C}}\,\varepsilon^{-2u_0t}\left\{\varepsilon^{+2s\lambda}\,(A^2+B^2) + \varepsilon^{-2s\lambda}\,(C^2+D^2)\right\}. \tag{20}$$

The last two terms, $\frac{dw''}{d\lambda}$ and $\frac{dw'''}{d\lambda}$, thus represent the energy which is transferred, or pulsates, between the electromagnetic and the electrostatic field of the circuit; and the term $\frac{dw'}{dt}$ represents the alternating (or rather oscillating) component of stored energy.

83. The energy stored by the electric field in a circuit section λ', between λ_1 and λ_2, is given by integrating $\frac{dW}{d\lambda}$ between λ_2 and λ_1, as

$$W = \frac{1}{4\,s}\sqrt{\frac{L}{C}}\,\varepsilon^{-2u_0t}\left\{(\varepsilon^{+2s\lambda_2} - \varepsilon^{+2s\lambda_1})\,(A^2 + B^2) - (\varepsilon^{-2s\lambda_2} - \varepsilon^{-2s\lambda_1})\,(C^2 + D^2)\right\}; \tag{21}$$

or, substituting herein the approximation (307),

$$W = \frac{1}{2}\,\lambda'\sqrt{\frac{L}{C}}\,\varepsilon^{-2u_0t}\left\{A^2 + B^2 + C^2 + D^2\right\}. \tag{22}$$

Differentiating (22) with respect to t gives the power supplied by the electric field of the circuit as

$$P = -\frac{dW}{dt} = u_0 \lambda' \sqrt{\frac{L}{C}} \varepsilon^{-2u_0 t} \{A^2 + B^2 + C^2 + D^2\}, \quad (23)$$

or, more generally,

$$P = \frac{u_0}{2s} \sqrt{\frac{L}{C}} \varepsilon^{-2u_0 t} \{(\varepsilon^{+2s\lambda_2} - \varepsilon^{+2s\lambda_1})(A^2 + B^2) - (\varepsilon^{-2s\lambda_2} - \varepsilon^{-2s\lambda_1})(C^2 + D^2)\}. \quad (24)$$

84. The power dissipated in the resistance $r'd\lambda = \frac{rd\lambda}{\sqrt{LC}}$ of a conductor element $d\lambda$ is

$$dp' = r'i^2 d\lambda \quad (25)$$

$$= \frac{2r}{L} dw_1;$$

hence, substituting herein equation (16) gives the power consumed by resistance of the circuit element $d\lambda$ as

$$\frac{dp'}{d\lambda} = \frac{2r}{L} \left\{ \frac{dw_0}{d\lambda} + \frac{dw'}{d\lambda} - \frac{dw''}{d\lambda} - \frac{dw'''}{d\lambda} \right\}, \quad (26)$$

and the power consumed by the conductance $g'd\lambda = \frac{g}{\sqrt{LC}} d\lambda$ of a conductor element $d\lambda$ is

$$dp'' = g'e^2 d\lambda \quad (27)$$

$$= \frac{2g}{C} dw_2;$$

hence the power consumed by conductance of circuit element $d\lambda$ is

$$\frac{dp''}{d\lambda} = \frac{2g}{C} \left\{ \frac{dw_0}{d\lambda} + \frac{dw'}{d\lambda} + \frac{dw''}{d\lambda} + \frac{dw'''}{d\lambda} \right\}, \quad (28)$$

and the *total power dissipated* in the circuit element $d\lambda$ is

$$\frac{dp_1}{d\lambda} = \frac{dp'}{d\lambda} + \frac{dp''}{d\lambda} = 4u\left(\frac{dw_0}{d\lambda} + \frac{dw'}{d\lambda}\right) - 4m\left(\frac{dw''}{d\lambda} + \frac{dw'''}{d\lambda}\right), \quad (29)$$

where, as before,

$$u = \frac{1}{2}\left(\frac{r}{L} + \frac{g}{C}\right)$$

and

$$m = \frac{1}{2}\left(\frac{r}{L} - \frac{g}{C}\right), \tag{30}$$

and integrating over a complete period

$$\frac{dP_1}{d\lambda} = 4\,u\frac{dw_0}{d\lambda} - 4\,m\frac{dw''}{d\lambda}, \tag{31}$$

the power dissipated in the circuit thus contains a constant term, $4\,u\frac{dw_0}{d\lambda}$, and a term which is a periodic function of the distance λ, $4\,m\frac{dw''}{d\lambda}$, of double frequency.

Averaged over a half-wave of the circuit, or a multiple thereof, the second term disappears, and

$$\frac{dP_1^0}{d\lambda} = 4\,u\frac{dw_0}{d\lambda};$$

or, substituting (12),

$$\frac{dP_1^0}{d\lambda} = u\sqrt{\frac{L}{C}}\,\varepsilon^{-2u_0t}\left\{\varepsilon^{+2s\lambda}(A^2 + B^2) + \varepsilon^{-2s\lambda}(C^2 + D^2)\right\}, \tag{32}$$

thus the power dissipated in a section $\lambda' = \lambda_2 - \lambda_1$ of the circuit is, by integrating between limits λ_1 and λ_2,

$$P_1^0 = \frac{u}{2\,s}\sqrt{\frac{L}{C}}\,\varepsilon^{-2u_0t}\left\{(\varepsilon^{+2s\lambda_2} - \varepsilon^{+2s\lambda_1})(A^2 + B^2) + (\varepsilon^{-2s\lambda_2} - \varepsilon^{2s\lambda_1})(C^2 + D^2)\right\}, \tag{33}$$

or, approximately,

$$P_1^0 = u\lambda'\sqrt{\frac{L}{C}}\,\varepsilon^{-2u_0t}\left\{A^2 + B^2 + C^2 + D^2\right\}. \tag{34}$$

85. Writing, therefore,

$$H^2 = (A^2 + B^2 + C^2 + D^2)\sqrt{\frac{L}{C}}, \tag{35}$$

the *energy stored in the electric field of the circuit section of length* λ' is

$$W = \frac{1}{2} \lambda' H^2 \varepsilon^{-2 u_0 t}; \tag{36}$$

the *power supply to the conductor by the decay of the electric field of the circuit* is

$$P = u_0 \lambda' H^2 \varepsilon^{-2 u_0 t}; \tag{37}$$

the *power dissipated in the circuit section* λ' by its effective resistance and conductance is

$$P_1^0 = u \lambda' H^2 \varepsilon^{-2 u_0 t}, \tag{38}$$

and the *power transferred from the circuit section* λ' to the rest of the circuit is

$$P_0 = s \lambda' H^2 \varepsilon^{-2 u_0 t}; \tag{39}$$

that is, $\dfrac{u}{u_0}$ = ratio of power dissipated in the section to that supplied to the section by its stored energy of the electric field.

$\dfrac{s}{u_0}$ = fraction of power supplied to the section by its electric field, which is transferred from the section to adjoining sections (or, if $s < 0$, received from them).

$\dfrac{s}{u}$ = ratio of power transferred to other sections to power dissipated in the section.

$u_0 \div u \div s$ thus is the ratio of the power supplied to the section by its electric field, dissipated in the section, and transferred from the section to adjoining sections.

These relations obviously are approximate only, and applicable to the case where the wave length is short.

86. Equation (4), of the power transferred from a section to the adjoining section, can be arranged in the form

$$\begin{aligned} P_0 = \frac{1}{2}\sqrt{\frac{L}{C}}\varepsilon^{-2u_0 t}\{[\varepsilon^{+2s\lambda_2}(A^2 + B^2) - \varepsilon^{-2s\lambda_2}(C^2 + D^2)] \\ - [\varepsilon^{+2s\lambda_1}(A^2 + B^2) - \varepsilon^{-2s\lambda_1}(C^2 + D^2)]\}; \end{aligned} \tag{40}$$

that is, it consists of two parts, thus:

$$P_0' = \frac{1}{2}\sqrt{\frac{L}{C}}\varepsilon^{-2u_0 t}\{\varepsilon^{+2s\lambda_2}(A^2 + B^2) - \varepsilon^{-2s\lambda_2}(C^2 + D^2)\}, \tag{41}$$

which is the power transferred from the section to the next following section, and

$$P_0'' = +\frac{1}{2}\sqrt{\frac{L}{C}}\varepsilon^{-2u_0t}\left\{\varepsilon^{+2s\lambda_1}(A^2+B^2)-\varepsilon^{-2s\lambda_1}(C^2+D^2)\right\}, \quad (42)$$

which is the power received from the preceding section, and the difference between the two values,

$$P_0 = P_0' - P_0'', \quad (43)$$

therefore, is the excess of the power given out over that received, or the resultant power supplied by the section to the rest of the circuit.

An approximate idea of the value of the power transfer constant can now be derived by assuming H^2 as constant throughout the entire compound circuit, which is approximately the case.

In this case, as the total power transferred between the sections must be zero, thus:

$$\sum P_0 = 0;$$

hence, substituting (341),

$$\sum s_i\lambda_i' = 0, \quad (44)$$

and, since

$$s_i = u_0 - u_i,$$

$$u_0\Lambda = \sum w_i\lambda_i'; \quad (45)$$

that is, the resultant circuit decrement multiplied by the total length of the circuit equals the sum of the time constants of the sections multiplied with the respective length of the section, or, if

$\zeta_1, \zeta_2 \ldots \zeta_1$ = length of the circuit section, as fraction of the total circuit length Λ,

$$u_0 = \sum \zeta_i u_i. \quad (46)$$

Whether this expression (46) is more general is still unknown.

87. As an example assume a transmission line having the following constants per wire: $r_1 = 52$; $L_1 = 0.21$; $g_1 = 40 \times 10^{-6}$, and $C_1 = 1.6 \times 10^{-6}$.

Further assume this line to be connected to step-up and step-down transformers having the following constants per transformer high-potential circuit: $r_2 = 5$, $L_2 = 3$; $g_2 = 0.1 \times 10^{-6}$, and $C_2 = 0.3 \times 10^{-6}$; then

$$\lambda_1' = \sigma_1 = \sqrt{L_1 C_1} = 0.58 \times 10^{-3}, \quad \lambda_2' = \sigma_2 = 0.95 \times 10^{-3},$$
$$u_1 = 136, \qquad u_2 = 1.$$

The circuit consists of four sections of the lengths

$$\lambda_1' = 0.58 \times 10^{-3}, \lambda_2' = 0.95 \times 10^{-3}, \lambda_3' = 0.58 \times 10^{-3}, \lambda_4' = 0.95 \times 10^{-3};$$

hence a total length

$$\Lambda = 3.06 \times 10^{-3},$$

and the resultant circuit decrement is

$$u_0 = 2\frac{\lambda_1'}{\Lambda} u_1 + 2\frac{\lambda_2'}{\Lambda} u_2 = 51.6 + 0.59 = 52.2;$$

hence, $$s_1 = -83.8 \text{ and } s_2 = +51.2.$$

If now the current in the circuit is $i_0 = 100$ amperes, the e.m.f. $e_0 = 40{,}000$ volts, the total stored energy is

$$W = i_0^2 (L_1 + L_2) + e_0^2 (C_1 + C_2)$$
$$= 32{,}000 + 3000 = 35{,}000 \text{ joules},$$

and from equation (36) then follows, for $t = 0$,

$$\tfrac{1}{2} \Lambda H^2 = 35{,}000,$$
$$H^2 = \frac{70{,}000}{\Lambda} = 22.8 \times 10^6,$$

which gives
$$u_0 = 52.2,$$
$$H^2 = 22.8 \times 10^6,$$
$$W = 35{,}000.$$

	Line.	Step-up Transformer.	Line.	Step-down Transformer.
Length of section, $\lambda' =$	0.58×10^{-3}	0.95×10^{-3}	0.58×10^{-3}	0.95×10^{-3}
Time constant, $u =$	136	1	136	1
Transfer constant, $s =$	−83.8	+51.2	−83.8	+51.2
Energy of electric field, $W =$	6.650	10.850	6.650	10.850 kilojoules
Power supplied by electric field, $P =$	690	1132	690	1132 kilowatts
Power dissipated, $P_1^0 =$	1800	22	1800	22 kilowatts
Power transferred, $P_0 =$	−1110	1110	−1110	1110 kilowatts

Thus, of the total power produced in the transformers by the decrease of their electric field, only 22 kw. are dissipated as heat in the transformer, and 1110 kw. transferred to the transmission line. While the power available by the decrease of the electric field of the transmission line is only 690 kw., the line dissipates energy at the rate of 1800 kw., receiving 1110 kw. from the transformers.

CHAPTER X

REFLECTION AND REFRACTION AT TRANSITION POINT

88. The general equation of the current and voltage in a section of a compound circuit, from equations (33) of Chapter VIII, is

$$\left.\begin{aligned} i &= \varepsilon^{-u_0 t}\left\{\varepsilon^{+s\lambda}\left[A\cos q\,(\lambda-t)+B\sin q\,(\lambda-t)\right]\right.\\ &\qquad \left.-\,\varepsilon^{-s\lambda}\left[C\cos q\,(\lambda+t)+D\sin q\,(\lambda+t)\right]\right\}\\ e &= c\varepsilon^{-u_0 t}\left\{\varepsilon^{+s\lambda}\left[A\cos q\,(\lambda-t)+B\sin q\,(\lambda-t)\right]\right.\\ &\qquad \left.+\,\varepsilon^{-s\lambda}\left[C\cos q\,(\lambda+t)+D\sin q\,(\lambda+t)\right]\right\}, \end{aligned}\right\} \quad (1)$$

where

$\lambda = \sigma l$ = distance variable with velocity as unit;

$\sigma = \sqrt{LC}$;

$c = \sqrt{\dfrac{L}{C}}$;

$u_0 = u + s$ = resultant time decrement;

$u = \dfrac{1}{2}\left(\dfrac{r}{L}+\dfrac{g}{C}\right)$ = time constant, and

s = energy transfer constant of section.

At a transition point λ_1 between section 1 and section 2 the constants change by equations (28) and (29) of Chapter VIII

$$\left.\begin{aligned} A_2 &= \varepsilon^{-s_2\lambda_1}\left\{a_1\varepsilon^{+s_1\lambda_1}A_1+b_1\varepsilon^{-s_1\lambda_1}\,(C_1\cos 2\,q\lambda_1+D_1\sin 2\,q\lambda_1)\right\}\\ B_2 &= \varepsilon^{-s_2\lambda_1}\left\{a_1\varepsilon^{+s_1\lambda_1}B_1+b_1\varepsilon^{-s_1\lambda_1}\,(C_1\sin 2\,q\lambda_1-D_1\cos 2\,q\lambda_1)\right\}\\ C_2 &= \varepsilon^{+s_2\lambda_1}\left\{a_1\varepsilon^{-s_1\lambda_1}C_1+b_1\varepsilon^{+s_1\lambda_1}\,(A_1\cos 2\,q\lambda_1+B_1\sin 2\,q\lambda_1)\right\}\\ D_2 &= \varepsilon^{+s_2\lambda_1}\left\{a_1\varepsilon^{-s_1\lambda_1}D_1+b_1\varepsilon^{+s_1\lambda_1}\,(A_1\sin 2\,q\lambda_1-B_1\cos 2\,q\lambda_1)\right\}, \end{aligned}\right\} \quad (2)$$

where

$$a_1 = \frac{c_1+c_2}{2\,c_2} \quad\text{and}\quad b_1 = \frac{c_1-c_2}{2\,c_2}\,. \quad (3)$$

Choosing now the transition point as zero point of λ, so that $\lambda < 0$ is section 1, $\lambda > 0$ is section 2, equations (2) assume the form

$$\left.\begin{aligned} A_2 &= a_1 A_1 + b_1 C_1, \\ B_2 &= a_1 B_1 - b_1 D_1, \\ C_2 &= a_1 C_1 + b_1 A_1, \\ D_2 &= a_1 D_1 - b_1 B_1. \end{aligned}\right\} \qquad (4)$$

From equations (4) and (3) it follows that

$$\left.\begin{aligned} &c_2 (A_2^2 - C_2^2) = c_1 (A_1^2 - C_1^2) \\ \text{and} \quad &c_2 (B_2^2 - D_2^2) = c_1 (B_1^2 - D_1^2). \end{aligned}\right\} \qquad (5)$$

If now a wave in section 1, A B, travels towards transition point $\lambda = 0$, at this point a part is reflected, giving rise to the reflected wave C D in section 1, while a part is transmitted and appears as main wave A B in section 2. The wave C D in section 2 thus would not exist, as it would be a wave coming towards $\lambda = 0$ from section 2, so not a part of the wave coming from section 1. In other words, we can consider the circuit as comprising two waves moving in opposite direction:

(1) A main wave A_1B_1, giving a transmitted wave A_2B_2 and reflected wave C_1D_1.

(2) A main wave C_2D_2, giving a transmitted wave $C_1'D_1'$ and reflected wave $A_2'B_2'$.

The waves moving towards the transition point are single main waves, A_1B_1 and C_2D_2, and the waves moving away from the transition point are combinations of waves reflected in the section and waves transmitted from the other section.

89. Considering first the main wave moving towards rising λ: in this $C_2 = 0 = D_2$, hence, from (4)

$$\left.\begin{aligned} &a_1 C_1 + b_1 A_1 = 0 \\ \text{and} \quad &a_1 D_1 - b_1 B_1 = 0, \end{aligned}\right\} \qquad (6)$$

and herefrom

$$\left.\begin{aligned} &C_1 = - \frac{b_1}{a_1} A_1 = + \frac{c_2 - c_1}{c_2 + c} A_1 \\ \text{and} \quad &D_1 = + \frac{b_1}{a_1} B_1 = + \frac{c_1 - c_2}{c_1 + c_2} B_1; \end{aligned}\right\} \qquad (7)$$

which substituted in (349) gives

$$\left.\begin{aligned} A_2 &= a_1 A_1 - \frac{b_1^2}{a_1} A_1 = \frac{a_1^2 - b_1^2}{a_1} A_1 = \frac{2\,c_1}{c_1 + c_2} A_1 \\ &\text{and} \\ B_2 &= a_1 B_1 - \frac{b_1^2}{a_1} B_1 = \frac{a_1^2 - b_1^2}{a_1} B_1 = \frac{2\,c_1}{c_1 + c_2} B_1. \end{aligned}\right\} \quad (8)$$

Then for the *main wave* in section 1,

$$\left.\begin{aligned} i_1 &= \varepsilon^{-u_0 t}\, \varepsilon^{+s_1 \lambda} \left\{ A_1 \cos q\,(\lambda - t) + B_1 \sin q\,(\lambda - t) \right\} \\ &\text{and} \\ e_1 &= c_1\, \varepsilon^{-u_0 t}\, \varepsilon^{+s_1 \lambda} \left\{ A_1 \cos q\,(\lambda - t) + B_1 \sin q\,(\lambda - t) \right\}. \end{aligned}\right\} \quad (9)$$

When reaching a transition point $\lambda = 0$, the wave resolves into the *reflected wave,* turned back on section 1, thus:

$$\left.\begin{aligned} i_1' &= -\frac{c_2 - c_1}{c_2 + c_1}\, \varepsilon^{-u_0 t}\, \varepsilon^{-s_1 \lambda} \left\{ A_1 \cos q\,(\lambda + t) - B_1 \sin q\,(\lambda + t) \right\} \\ &\text{and} \\ e_1' &= +\, c_1 \frac{c_2 - c_1}{c_2 + c_1}\, \varepsilon^{-u_0 t}\, \varepsilon^{-s_1 \lambda} \left\{ A_1 \cos q\,(\lambda + t) - B_1 \sin q\,(\lambda + t) \right\}. \end{aligned}\right\} \quad (10)$$

The *transmitted wave,* which by passing over the transition point enters section 2, is given by

$$\left.\begin{aligned} i_2 &= \frac{2\,c_1}{c_1 + c_2}\, \varepsilon^{-u_0 t}\, \varepsilon^{+s_2 \lambda} \left\{ A_1 \cos q\,(\lambda - t) + B_1 \sin q\,(\lambda - t) \right\} \\ &\text{and} \\ e_2 &= c_2 \frac{2\,c_1}{c_1 + c_2}\, \varepsilon^{-u_0 t}\, \varepsilon^{+s_2 \lambda} \left\{ A_1 \cos q\,(\lambda - t) + B_1 \sin q\,(\lambda - t) \right\}. \end{aligned}\right\} \quad (11)$$

The reflection angle, $\tan\,(i_1') = -\frac{B_1}{A_1}$, is supplementary to the impact angle, $\tan\,(i_1) = +\frac{B_1}{A_1}$, and transmission angle, $\tan\,(i_2) = +\frac{B_1}{A_1}$.

Reversing the sign of λ in the equation (10) of the reflected wave, that is, counting the distance for the reflected wave also in the direction of its propagation, and so in opposite direction as in the main wave and the transmitted wave, equations (10) become

$$\left.\begin{aligned} i_1'' &= +\frac{c_2-c_1}{c_2+c_1}\varepsilon^{-u_0 t}\varepsilon^{+s_1\lambda'}\{A_1\cos q(\lambda'-t)+B_1\sin q(\lambda-t)\}, \\ e_1'' &= +c_1\frac{c_2-c_1}{c_2+c_1}\varepsilon^{-u_0 t}\varepsilon^{+s_1\lambda'}\{A_1\cos q(\lambda'-t)+B_1\sin q(\lambda'-t)\}, \end{aligned}\right\} \quad (12)$$

and then

$$\left.\begin{aligned} i_2 + i_1'' &= i_1, \\ \text{or} \qquad i_1 + i_1' &= i_2, \\ \frac{c_1}{c_2}e_2 + e_1'' &= e_1, \\ e_1 + e_1'' &= e_2. \end{aligned}\right\} \quad (13)$$

(1) In a single electric wave, current and e.m.f. are in phase with each other. Phase displacements between current and e.m.f. thus can occur only in resultant waves, that is, in the combination of the main and the reflected wave, and then are a function of the distance λ, as the two waves travel in opposite direction.

(2) When reaching a transition point, a wave splits up into a reflected wave and a transmitted wave, the former returning in opposite direction over the same section, the latter entering the adjoining section of the circuit.

(3) Reflection and transmission occur without change of the phase angle; that is, the phase of the current and of the voltage in the reflected wave and in the transmitted wave, at the transition point, is the same as the phase of the main wave or incoming wave. Reflection and transmission with a change of phase angle can occur only by the combination of two waves traveling in opposite direction over a circuit; that is, in a resultant wave, but not in a single wave.

(4) The sum of the transmitted and the reflected current equals the main current, when considering these currents in their respective direction of propagation.

The sum of the voltage of the main wave and the reflected wave equals the voltage of the transmitted wave.

The sum of the voltage of the reflected wave and the voltage of the transmitted wave reduced to the first section by the ratio of voltage transformation $\frac{c_2}{c_1}$, equals the voltage of the main wave.

(5) Therefore a voltage transformation by the factor $\frac{c_2}{c_1} = \sqrt{\frac{L_2}{C_2}\frac{C_1}{L_1}}$ occurs at the transition point; that is, the transmitted wave of voltage equals the difference between main wave and reflected wave multiplied by the transformation ratio $\frac{c_2}{c_1}$; $e_2 = \frac{c_1}{c_2}(e_1 - e_1'')$. As result thereof, in passing from one section of a circuit to another section, the voltage of the wave may decrease or may increase. If $\frac{c_2}{c_1} > 1$, that is, when passing from a section of low inductance and high capacity into a section of high inductance and low capacity, as from a transmission line into a transformer or a reactive coil, the voltage of the wave is increased; if $\frac{c_2}{c_1} < 1$, that is, when passing from a section of high inductance and low capacity into a section of low inductance and high capacity, as from a transformer to a transmission line, the voltage of the wave is decreased.

This explains the frequent increase to destructive voltages, when entering a station from the transmission line or cable, of an impulse or a wave which in the transmission line is of relatively harmless voltage.

The ratio of the transmitted to the reflected wave is given by

$$\left.\begin{aligned} \frac{i_2}{i_1''} &= \frac{2c_1}{c_2 - c_1} = \frac{2\sqrt{L_1C_2}}{\sqrt{L_2C_1} - \sqrt{L_1C_2}} = \frac{2}{\sqrt{\frac{L_2}{L_1}\frac{C_1}{C_2}} - 1} \\ &\text{and} \\ \frac{e_2}{e_1''} &= \frac{2c_2}{c_2 - c_1} = \frac{2\sqrt{L_2C_1}}{\sqrt{L_2C_1} - \sqrt{L_1C_2}} = \frac{2}{1 - \sqrt{\frac{L_1}{L_2}\frac{C_2}{C_1}}}. \end{aligned}\right\} \qquad (14)$$

90. Example:

Transmission line	Transformer
$L_1 = 1.95 \times 10^{-3}$	$L_2 = 1$
$C_1 = 0.0162 \times 10^{-6}$	$C_2 = 0.4 \times 10^{-6}$
$c_1 = 346$	$c = 1580$
$\frac{i_2}{i_1''} = 0.56$	$\frac{e_2}{e_1''} = 2.56$

And in the opposite direction

$$\frac{i_2}{i_1''} = -2.56 \qquad \frac{e_2}{e_1''} = -0.56.$$

The ratio $\frac{e_2}{e_1''}$ becomes a maximum, $= \infty$, for $\frac{L_1}{C_1} = \frac{L_2}{C_2}$, but in this case $e_1'' = 0$; that is, no reflection occurs, and the reflected wave equals zero, the transmitted wave equals the incoming wave.

$$\frac{i_2}{i_1} = \frac{2\,c_1}{c_2 + c_1};$$

hence, becomes a maximum for $c_2 = 0$, or $c_1 = \infty$ and then $= 2$; in which case $e_2 = 0$. (15)

$$\frac{e_2}{e_1} = \frac{2\,c_2}{c_2 + c_1};$$

hence, becomes a maximum for $c_1 = 0$, or $c_2 = \infty$ and then $= 2$; in which case $i_2 = 0$. From the above it is seen that the maximum value to which the voltage can build up at a single transition point is twice the voltage of the incoming wave, and this occurs at the open end of the circuit, or, approximately, at a point where the ratio of inductance to capacity very greatly increases.

$$\frac{i_1''}{i_1} = \frac{c_2 - c_1}{c_2 + c_1};$$

hence, becomes a maximum, and equal to 1, for $c_1 = 0$, or $c_2 = \infty$. (16)

$$\frac{e_1''}{e_1} = \frac{c_2 - c_1}{c_2 + c_1}$$

has the same value as the current-ratio.

91. Consider now a wave traversing the circuit in opposite direction; that is, C_2D_2 is the main wave, A_2B_2 the reflected wave, C_1D_1 the transmitted wave, and $A_1 = 0 = B_1$. In equation (4) this gives

$$A_2 = b_1 C_1;$$
$$B_2 = -b_1 D_1;$$
$$C_2 = a_1 C_1,$$

and

$$D_2 = a_1 D_1;$$

hence,

$$\left.\begin{aligned} C_1 &= \frac{1}{a_1} C_2 = \frac{2\,c_2}{c_1 + c_2} C_2; \\ D_1 &= \frac{1}{a_1} D_2 = \frac{2\,c_2}{c_1 + c_2} D_2; \\ A_2 &= \frac{b_1}{a_1} C_2 = \frac{c_1 - c_2}{c_1 + c_2} C_2, \end{aligned}\right\} \quad (17)$$

and

$$B_2 = -\frac{b_1}{a_1} D_2 = -\frac{c_1 - c_2}{c_1 + c_2} D_2 = \frac{c_2 - c_1}{c_2 + c_1} D_2;$$

that is, the same relations as expressed by equations (7) and (8) for the wave traveling in opposite direction.

The equations of the components of the wave then are:

Main wave:

$$\left.\begin{aligned} \bar{i}_2 &= -\,\varepsilon^{-u_0 t}\,\varepsilon^{-s_2 \lambda}\,\{C_2 \cos q\,(\lambda + t) + D_2 \sin q\,(\lambda + t)\} \\ \bar{e}_2 &= +c_2 \varepsilon^{-u_0 t}\,\varepsilon^{-s_2 \lambda}\,\{C_2 \cos q\,(\lambda + t) + D_2 \sin q\,(\lambda + t)\}; \end{aligned}\right\} \quad (18)$$

Transmitted wave:

$$\left.\begin{aligned} \bar{i}_1 &= -\frac{2\,c_2}{c_1 + c_2}\varepsilon^{-u_0 t}\,\varepsilon^{-s_1 \lambda}\{C_2 \cos q\,(\lambda+t) + D_2 \sin q\,(\lambda+t)\} \\ \bar{e}_1 &= +c_1 \frac{2\,c_2}{c_1+c_2}\varepsilon^{-u_0 t}\,\varepsilon^{-s_1 \lambda}\{C_2 \cos q\,(\lambda+t) + D_2 \sin q\,(\lambda+t)\}; \end{aligned}\right\} \quad (19)$$

Reflected wave:

$$\left.\begin{aligned} \bar{i}_2' &= \frac{c_1 - c_2}{c_1 + c_2}\varepsilon^{-u_0 t}\,\varepsilon^{+s_1 \lambda}\;\{C_2 \cos q\,(\lambda - t) - D_2 \sin q\,(\lambda - t)\} \\ \bar{e}_2' &= c_2 \frac{c_1 - c_2}{c_1 + c_2}\varepsilon^{-u_0 t}\,\varepsilon^{+s_1 \lambda}\{C_2 \cos q\,(\lambda - t) - D_2 \sin q\,(\lambda - t)\}; \end{aligned}\right\} \quad (20)$$

or, in the direction of propagation, that is, reversing the sign of λ:

$$\left.\begin{aligned}\bar{i}_2'' &= -\frac{c_1 - c_2}{c_1 + c_2}\varepsilon^{-u_0 t}\varepsilon^{-s_1\lambda'}\{C_2 \cos q(\lambda'+t) + D_2 \sin q(\lambda'+t)\}\\ \bar{e}_2'' &= c_2\frac{c_1 - c_2}{c_1 + c_2}\varepsilon^{-u_0 t}\varepsilon^{-s_1\lambda'}\{C_2 \cos q(\lambda'+t) + D_2 \sin q(\lambda'+t)\}.\end{aligned}\right\} \tag{21}$$

92. The compound wave, that is, the resultant of waves passing the transition point in both directions, then is

$$\left.\begin{aligned}&\left.\begin{aligned}i_1^0 &= \bar{i}_1 + i_1' + i_1\\ e_1^0 &= \bar{e}_1 + e_1' + e_1\end{aligned}\right\}\\ &\left.\begin{aligned}i_2^0 &= i_2 + \bar{i}_2 + \bar{i}_2'\\ e_2^0 &= e_2 + \bar{e}_2 + \bar{e}_2'\end{aligned}\right\}\end{aligned}\right\} \tag{22}$$

In the neighborhood of the transition point, that is, for values λ which are sufficiently small, so that $\varepsilon^{+s\lambda}$ and $\varepsilon^{-s\lambda}$ can be dropped as being approximately equal to 1, by substituting equations (9) to (11) and (18) to (21) into (22) we have

$$\left.\begin{aligned}i_1^0 = \varepsilon^{-u_0 t}\quad&[\{A_1 \cos q(\lambda - t) + B_1 \sin q(\lambda - t)\}\\ &-\frac{c_2 - c_1}{c_1 + c_2}\{A_1 \cos q(\lambda + t) - B_1 \sin q(\lambda + t)\}\\ &-\frac{2c_2}{c_1 + c_2}\{C_2 \cos q(\lambda + t) + D_2 \sin q(\lambda + t)\}];\\ e_1^0 = c_1\varepsilon^{-u_0 t}\quad&[\{A_1 \cos q(\lambda - t) + B_1 \sin q(\lambda - t)\}\\ &+\frac{c_2 - c_1}{c_2 + c_1}\{A_1 \cos q(\lambda + t) - B_1 \sin q(\lambda + t)\}\\ &+\frac{2c_2}{c_1 + c_2}\{C_2 \cos q(\lambda + t) + D_2 \sin q(\lambda + t)\}];\\ i_2^0 = -\varepsilon^{-u_0 t}\quad&[\{C_2 \cos q(\lambda + t) + D_2 \sin q(\lambda + t)\}\\ &-\frac{c_1 - c_2}{c_1 + c_2}\{C_2 \cos q(\lambda - t) - D_2 \sin q(\lambda - t)\}\\ &-\frac{2c_1}{c_1 + c_2}\{A_1 \cos q(\lambda - t) + B_1 \sin q(\lambda - t)\}];\end{aligned}\right\} \tag{23}$$

$$
\left.\begin{aligned}
e_2{}^0 = c_2\varepsilon^{-u_0 t} \; & [\{C_2 \cos q(\lambda + t) + D_2 \sin q(\lambda + t)\} \\
& + \frac{c_1 - c_2}{c_1 + c_2}\{C_2 \cos q(\lambda - t) - D_2 \sin q(\lambda - t)\} \\
& + \frac{2\,c_1}{c_1 + c_2}\{A_1 \cos q(\lambda - t) + B_1 \sin q(\lambda - t)\}].
\end{aligned}\right|
$$

In these equations the first term is the main wave, the second term its reflected wave, and the third term the wave transmitted from the adjoining section over the transition point.

Expanding and rearranging equations (23) gives

$$
\begin{aligned}
i_1{}^0 &= \frac{2\,\varepsilon^{-u_0 t}}{c_1 + c_2}[\{(c_1 A_1 - c_2 C_2)\cos q\lambda + c_2(B_1 - D_2)\sin q\lambda\}\cos qt \\
&\qquad - \{(c_1 B_1 + c_2 D_2)\cos q\lambda - c_2(A_1 + C_2)\sin q\lambda\}\sin qt]; \\
e_1{}^0 &= \frac{2\,c_1}{c_1 + c_2}\varepsilon^{-u_0 t}[\{c_2(A_1 + C_2)\cos q\lambda + (c_1 B_1 + c_2 D_2)\sin q\lambda\}\cos qt \\
&\qquad - \{c_2(B_1 - D_2)\cos q\lambda - (c_1 A_1 - c_2 C_2)\sin q\lambda\}\sin qt]; \qquad (24) \\
i_2{}^0 &= \frac{2\,\varepsilon^{-u_0 t}}{c_1 + c_2}[\{(c_1 A_1 - c_2 C_2)\cos q\lambda + c_1(B_1 - D_2)\sin q\lambda\}\cos qt \\
&\qquad - \{(c_1 B_1 + c_2 D_2)\cos q\lambda - c_1(A_1 + C_2)\sin q\lambda\}\sin qt]; \\
e_2{}^0 &= \frac{2\,c_2}{c_1 + c_2}\varepsilon^{-u_0 t}[\{c_1(A_1 + C_1)\cos q\lambda + (c_1 B_1 + c_2 D_2)\sin q\lambda\}\cos qt \\
&\qquad - \{c_1(B_1 - D_2)\cos q\lambda - (c_1 A_1 - c_2 C_2)\sin q\lambda\}\sin qt]
\end{aligned}
$$

93. This gives the distance phase angle of the waves:

$$
\left.\begin{aligned}
\tan i_1{}^0 &= \frac{c_2\{B_1 - D_2)\cos qt + (A_1 + C_2)\sin qt\}}{(c_1 A_1 - c_2 C_2)\cos qt - (c_1 B_1 + c_2 D_2)\sin qt}, \\
\tan i_2{}^0 &= \frac{c_1\{(B_1 - D_2)\cos qt + (A_1 + C_2)\sin qt\}}{(c_1 A_1 - c_2 C_2)\cos qt - (c_1 B_1 + c_2 D_2)\sin qt};
\end{aligned}\right\} \qquad (25)
$$

hence,

$$
\frac{\tan i_2{}^0}{\tan i_1{}^0} = \frac{c_1}{c_2} = \sqrt{\frac{L_1 C_2}{L_2 C_1}}; \qquad (26)
$$

$$
\left.\begin{aligned}
\tan e_1{}^0 &= \frac{(c_1 B_1 + c_2 D_2)\cos qt + (c_1 A_1 - c_2 C_2)\sin qt}{c_2\{(A_1 + C_2)\cos qt - (B_1 - D_2)\sin qt\}}, \\
\tan e_2{}^0 &= \frac{(c_1 B_1 + c_2 D_2)\cos qt + (c_1 A_1 - c_2 C_2)\sin qt}{c_1\{(A_1 + C_2)\cos qt - c_1(B_1 - D_2)\sin qt\}};
\end{aligned}\right\} \qquad (27)
$$

hence,

$$\frac{\tan e_2{}^0}{\tan e_1{}^0} = \frac{c_2}{c_1} = \sqrt{\frac{L_2C_1}{L_1C_2}}; \tag{28}$$

that is, at a transition point the distance phase angle of the wave changes so that the ratio of the tangent functions of the phase angle is constant, and the ratio of the tangent functions of the phase angle of the voltages is proportional, of the currents inversely proportional to the circuit constants $c = \sqrt{\frac{L}{C}}$.

In other words, the transition of an electric wave or impulse from one section of a circuit to another takes place at a constant ratio of the tangent functions of the phase angle, which ratio is a constant of the circuit sections between which the transition occurs.

This law is analogous to the law of refraction in optics, except that in the electric wave it is the ratio of the tangent functions, while in optics it is the ratio of the sine functions, which is constant and a characteristic of the media between which the transition occurs.

Therefore this law may be called the *law of refraction of a wave* at the boundary between two circuits, or at a transition point.

The law of refraction of an electric wave at the boundary between two media, that is, at a transition point between two circuit sections, is given by the constancy of the ratio of the tangent functions of the incoming and refracted wave.

CHAPTER XI

INDUCTIVE DISCHARGES

94. The discharge of an inductance into a transmission line may be considered as an illustration of the phenomena in a compound circuit comprising sections of very different constants; that is, a combination of a circuit section of high inductance and small resistance and negligible capacity and conductance, as a generating station, with a circuit of distributed capacity and inductance, as a transmission line. The extreme case of such a discharge would occur if a short circuit at the busbars of a generating station opens while the transmission line is connected to the generating station.

Let r = the total resistance and L = the total inductance of the inductive section of the circuit; also let $g = 0, C = 0$, and L_0 = inductance, C_0 = capacity, r_0 = resistance, g_0 = conductance of the total transmission line connected to the inductive circuit.

In either of the two circuit sections the total length of the section is chosen as unit distance, and, translated to the velocity measure, the length of the transmission line is

$$\lambda_0 = \sigma = \sqrt{L_0 C_0},$$

and the length of the inductive circuit is

$$\lambda_1 = \sigma_1 = \sqrt{L_1 C_1} = 0; \tag{1}$$

that is, the inductive section of zero capacity has zero length when denoted by the velocity measure λ, or is a "massed inductance."

It follows herefrom that throughout the entire inductive section $\lambda = 0$, and current i_1 therefore is constant throughout this section.

Choosing now the transition point between the inductance and the transmission line as zero of distance, $\lambda = 0$, the inductance is massed at point $\lambda = 0$, and the transmission line extends from $\lambda = 0$ to $\lambda = \lambda_0$.

Denoting the constants of the inductive section by index 1, those of the transmission line by index 2, the equations of the two circuit sections, from (33) of Chapter VIII, are

$$\left.\begin{aligned} i_1 &= \varepsilon^{-u_0 t}\{(A_1 - C_1)\cos qt - (B_1 + D_1)\sin qt\}, \\ e_1 &= c_1\varepsilon^{-u_0 t}\{(A_1 + C_1)\cos qt - (B_1 - D_1)\sin qt\}; \end{aligned}\right\} \quad (2)$$

$$\left.\begin{aligned} i_2 = \varepsilon^{-u_0 t}\{&\varepsilon^{+s\lambda}[A_2\cos q(\lambda - t) + B_2\sin q(\lambda - t)] \\ &- \varepsilon^{-s\lambda}[C_2\cos q(\lambda + t) + D_2\sin q(\lambda + t)]\}, \\ e_2 = c_2\varepsilon^{-u_0 t}\{[&\varepsilon^{+s\lambda}[A_2\cos q(\lambda - t) + B_2\sin q(\lambda - t)] \\ &+ \varepsilon^{-s\lambda}[C_2\cos q(\lambda + t) + D_2\sin q(\lambda + t)]\}, \end{aligned}\right\} \quad (3)$$

and the constants of the second section are related on those of the first section by the equations (28) to (30) of Chapter VIII:

$$\left.\begin{aligned} A_2 &= a_1 A_1 + b_1 C_1, & C_2 &= a_1 C_1 + b_1 A_1, \\ B_2 &= a_1 B_1 - b_1 D_1, & D_2 &= a_1 D_1 - b_1 B_1, \end{aligned}\right\} \quad (4)$$

where

$$\left.\begin{aligned} a_1 &= \frac{c_1 + c_2}{2\,c_2}, \\ b_1 &= \frac{c_1 - c_2}{2\,c_2}, \end{aligned}\right\} \quad (5)$$

and

$$c = \sqrt{\frac{L}{C}}. \quad (6)$$

95. In the inductive section having the constants L and r, that is, at the point $\lambda = 0$ of the circuit, current i_1 and voltage e_1 must be related by the equation of inductance,

$$e_1 = ri_1 - L\frac{di_1}{dt}. \quad (7)$$

Substituting (2) in (7), and expanding, gives

$$\begin{aligned} c_1\{(A_1 + C_1)\cos qt &- (B_1 - D_1)\sin qt\} \\ = (r + u_0 L)&\{(A_1 - C_1)\cos qt - (B_1 + D_1)\sin qt\} \\ &+ qL\{(A_1 - C_1)\sin qt + (B_1 + D_1)\cos qt\}, \end{aligned}$$

and herefrom the identities

$$\left.\begin{aligned} c_1 (A_1 + C_1) &= (r + u_0 L)(A_1 - C_1) + qL(B_1 + D_1), \\ c_1 (B_1 - D_1) &= (r + u_0 L)(B_1 + D_1) - qL(A_1 - C_1). \end{aligned}\right\} \quad (8)$$

Writing

$$A_1 - C_1 = M$$

and

$$B_1 + D_1 = N \quad (9)$$

gives

$$c_1 (A_1 + C_1) = (r + u_0 L) M + qLN$$

and

$$c_1 (B_1 - D_1) = (r + u_0 L) N - qLM, \quad (10)$$

which substituted in (4) of Chapter X gives

$$\left.\begin{aligned} A_2 &= \frac{1}{2c}\{(c + r + u_0 L) M + qLN\} \cong \frac{1}{2}(M + pN), \\ B_2 &= \frac{1}{2c}\{(c + r + u_0 L) N - qLM\} \cong \frac{1}{2}(N - pM), \\ C_2 &= \frac{1}{2c}\{qLN - (c - r - m_0 L) M\} \cong \frac{1}{2}(pN - M), \\ D_2 &= \frac{1}{2c}\{qLM + (c - r - m_0 L) N\} \cong \frac{1}{2}(pM + N), \end{aligned}\right\} \quad (11)$$

where in the second expression terms of secondary order have been dropped.

$$c = \sqrt{\frac{L_0}{C_0}}, \qquad p = \frac{qL}{c}.$$

Then substituting in (2) gives the equations of massed inductance:

$$\left.\begin{aligned} i_1 &= \varepsilon^{-u_0 t}\{M \cos qt - N \sin qt\} \\ e_1 &= \varepsilon^{-u_0 t}\{[(r + u_0 L) M + qLN] \cos qt - [(r + u_0 L) N - qLM] \sin qt\}. \end{aligned}\right\} \quad (12)$$

If at $t = 0$, $e_1 = 0$, that is, if at the beginning of the transient discharge the voltage at the inductance is zero, as for instance the inductance had been short-circuited, then, substituting in

(12), and denoting by i_0 the current at the moment $t = 0$, or at the moment of start, we have

$$t = 0,\ i_1 = i_0,\ e_1 = 0; \text{ hence,}$$

$$\left.\begin{aligned} M &= i_0, \\ N &= -\frac{r + u_0 L}{qL}\, i_0, \end{aligned}\right\} \tag{13}$$

and

$$\left.\begin{aligned} i_1 &= i_0 \varepsilon^{-u_0 t} \left\{ \cos qt + \frac{r + u_0 L}{qL} \sin qt \right\}, \\ e_1 &= \frac{(qL)^2 + (r + u_0 L)^2}{qL}\, i_0 \varepsilon^{-u_0 t} \sin qt. \end{aligned}\right\} \tag{14}$$

In this case

$$\left.\begin{aligned} A_2 &= \frac{i_0}{2}; \\ B_2 &= -\frac{i_0}{2\,cqL}\left\{ (qL)^2 + (r + u_0 L)^2 + c\,(r + u_0 L) \right\}; \\ C_2 &= -\frac{i_0}{2}; \\ D_2 &= \frac{i_0}{2\,cqL}\left\{ (qL)^2 + (r + u_0 L)^2 - c\,(r + u_0 L) \right\}. \end{aligned}\right\} \tag{15}$$

96. In the case that the transmission line is open at its end, at point $\lambda = \lambda_0$,

$$\lambda = \lambda_0,$$
$$i_2 = 0;$$

hence, this substituted in (3), expanded and rearranged as function of cos qt and sin qt, gives the two identities

$$\left.\begin{aligned} &\varepsilon^{+s\lambda_0}(A_2 \cos q\lambda_0 + B_2 \sin q\lambda_0) = \varepsilon^{-s\lambda_0}(C_2 \cos q\lambda_0 + D_2 \sin q\lambda_0) \\ &\text{and} \\ &\varepsilon^{+s\lambda_0}(A_2 \sin q\lambda_0 - B_2 \cos q\lambda_0) = -\varepsilon^{-s\lambda_0}(C_2 \sin q\lambda_0 - D_2 \sin q\lambda_0). \end{aligned}\right\} \tag{16}$$

Squared and added these two equations (16) give

$$\varepsilon^{+2s\lambda_0}(A_2^2 + B_2^2) = \varepsilon^{-2s\lambda_0}(C_2^2 + D_2^2). \tag{17}$$

Divided by each other and expanded equations (16) give

$$(A_2C_1 - B_2D_2) \sin 2\, q\lambda_0 = (A_2D_2 + B_2B_2) \cos 2\, q\lambda_0. \quad (18)$$

Substituting (381) into equations (17) and (18) gives

$$\epsilon^{+2\, s\lambda_0}\left\{(qL)^2 + (c + r + u_0L)^2\right\} = \epsilon^{-2\, s\lambda_0}\left\{(qL)^2 + (c - r - u_0L)^2\right. \quad (19)$$

$$\left\{(qL)^2 + (r + u_0L)^2 - c^2\right\} \sin 2\, q\lambda_0 = 2\, cqL \cos 2\, q\lambda_0. \quad (20)$$

Since $2\, s\lambda_0$ is a small quantity, in equation (19) we can substitute

$$\epsilon^{\pm 2\, s\lambda_0} = 1 \pm 2\, s\lambda_0;$$

hence, rearranging (19) and substituting

$$s = u_0 - u$$

gives

$$c\,(r + u_0L) - (u - u_0)\,\lambda_0 \left\{(qL)^2 + (r + u_0L)^2 + c^2\right\} = 0. \quad (21)$$

Since $(r + u_0L)$ is a small quantity compared with $c^2\,(qL)^2$, it can be neglected, and equation (20) and (21) assume the form

$$\left\{(qL)^2 - c^2\right\} \sin 2\, q\lambda_0 = 2\, cqL \cos 2\, q\lambda_0 \quad (22)$$

$$c\,(r + u_0L) - (u - u_0)\,\lambda_0 \left\{(qL)^2 + c^2\right\} = 0, \quad (23)$$

and, transformed, equation (22) assumes the form

$$\left.\begin{aligned} &\tan 2\, q\lambda_0 = \frac{2\, cqL}{(qL)^2 - c^2}; \\ &\text{or} \\ &q = -\frac{c}{L} \tan q\lambda_0, \\ &\text{or} \\ &q = +\frac{c}{L} \cot q\lambda_0, \end{aligned}\right\} \quad (24)$$

hence $\tan 2\, q\lambda_0$ is *positive* if $qL > c$, as is usually the case.

Expanded for u_0, equation (23) assumes the form

$$\left.\begin{aligned} &u_0 = \frac{u\lambda_0 \left\{(qL)^2 + c^2\right\} - cr}{\lambda_0 \left\{(qL)^2 + c^2\right\} + cL}, \\ &\text{or} \\ &1 - \frac{u_0}{u} = \frac{c\left(L + \frac{r}{w}\right)}{cL + \lambda_0 \left\{(qL)^2 + c^2\right\}}. \end{aligned}\right\} \quad (25)$$

$$s = -(u - u_0).$$

From equations (24) q is calculated by approximation, and then from (25) u_0 and s.

As seen, in all these expressions of q, u_0, s, etc., the integration constants M and N eliminate; that is, the frequency, time attenuation constant, power transfer, etc., depend on the circuit constants only, but not on the distribution of current and voltage in the circuit.

97. At any point λ of the circuit, the voltage is given by equation (3), which, transposed, gives

$$\begin{aligned} e = c\varepsilon^{-u_0t}\{\varepsilon^{+s\lambda}&[(A_2\cos q\lambda + B_2\sin q\lambda)\cos qt \\ &+ (A_2\sin q\lambda - B_2\cos q\lambda)\sin qt] \\ + \varepsilon^{-s\lambda}&[(C_2\cos q\lambda + D_2\sin q\lambda)\cos qt \\ &- (C_2\sin q\lambda - D_2\cos q\lambda)\sin qt]\}, \end{aligned}$$

or approximately,

$$\begin{aligned} e = c\varepsilon^{-u_0t}\{&[(A_2 + C_2)\cos q\lambda + (B_2 + D_2)\sin q\lambda]\cos qt \\ &+ [(A_2 - C_2)\sin q\lambda - (B_2 - D_2)\cos q\lambda]\sin qt\}. \end{aligned}$$

Similarly to equation (381),

$$\left.\begin{aligned} A_2 + C_2 &= pN; \\ A_2 - C_2 &= M; \\ B_2 + D_2 &= N; \\ B_2 - D_2 &= -pM, \end{aligned}\right\} \tag{26}$$

where

$$p = \frac{qL}{c},$$

then

$$\left.\begin{aligned} e_2 &= \varepsilon^{-u_0t}(qL\cos q\lambda + c\sin q\lambda)(N\cos qt + M\sin qt), \\ i_2 &= \varepsilon^{-u_0t}\left(\cos q\lambda - \frac{qL}{c}\sin q\lambda\right)(M\cos qt - N\sin qt); \end{aligned}\right\} \tag{27}$$

$$\left.\begin{aligned} e_1 &= qL\varepsilon^{-u_0t}(N\cos qt + M\sin qt), \\ i_1 &= \varepsilon^{-u_0t}(M\cos qt - N\sin qt). \end{aligned}\right\} \tag{28}$$

If
$$e_1 = 0 \text{ for } t = 0,$$
$$N = 0;$$

hence,

$$\left.\begin{aligned} i_1 &= i_0 \varepsilon^{-u_0 t} \cos qt, \\ e_1 &= qLi_0 \varepsilon^{-u_0 t} \sin qt; \\ i_2 &= i_0 \varepsilon^{-u_0 t} \left(\cos q\lambda - \frac{qL}{c} \sin q\lambda\right) \cos qt, \\ e_2 &= i_0 \varepsilon^{-u_0 t} (qL \cos q\lambda + c \sin q\lambda) \sin qt. \end{aligned}\right\} \tag{29}$$

Writing

$$\sqrt{\tfrac{1}{2}(M^2 + N^2)} = I_0, \tag{30}$$

the effective values of the quantities are

$$\left.\begin{aligned} I_1 &= I_0 \varepsilon^{-u_0 t}, \\ E_1 &= qLI_0 \varepsilon^{-u_0 t}; \\ I_2 &= I_0 \varepsilon^{-u_0 t} \left(\cos q\lambda - \frac{qL}{c} \sin q\lambda\right), \\ E_2 &= I_0 \varepsilon^{-u_0 t} (qL \cos q\lambda + c \sin q\lambda). \end{aligned}\right\} \tag{31}$$

Herefrom it follows that

$$I_2 = 0 \text{ for } \lambda = \lambda_0 \text{ by the equation}$$

$$\cos q\lambda_0 - q\frac{L}{c} \sin q\lambda_0 = 0,$$

or

$$q = \frac{c}{L} \cot q\lambda_0, \tag{32}$$

while
$$q = -\frac{c}{L} \tan q\lambda_0 \tag{33}$$

gives

$$qL \cos q\lambda_0 + c \sin q\lambda_0 = 0;$$

that is,

$$E_2 = 0 \text{ at } \lambda = \lambda_0.$$

At the open end of the line $\lambda = \lambda_0$ the voltage E_2 by substituting (32) into (31) is

$$E_2^0 = \frac{qL}{\cos q\lambda_0} I_0 \varepsilon^{-u_0 t}. \tag{34}$$

At the grounded end of the line $\lambda = \lambda_0$ the current I_2, by substituting (403) into (401), is

$$I_2^0 = \frac{I_0 \varepsilon^{-u_0 t}}{\cos q\lambda_0}. \tag{35}$$

An inductance discharging into the transmission line thus gives an oscillatory distribution of voltage and current along the line.

98. As example may be considered the three-phase high-potential circuit, comprising a generating system of $r = 2$ ohms and $L = 0.5$ henry per phase and connected to a long-distance transmission line of $r_0 = 0.4$ ohm, $L_0 = 0.002$ henry, $g_0 = 0.2 \times 10^{-6}$ mho, $C_0 = 0.016 \times 10^{-6}$ farad per mile of conductor or phase, and of $l_0 = 80$ miles length.

$$c = \sqrt{\frac{L_0}{C_0}} = 354, \qquad c^2 = 125{,}300;$$

$$\sigma_0 = \sqrt{L_0 C_0} = 5.66 \times 10^{-6};$$

$$\lambda_0 = l_0 \sigma_0 = 0.453 \times 10^{-3};$$

$$u = \frac{1}{2}\left(\frac{r_0}{L_0} + \frac{g_0}{C_0}\right) = 106;$$

$$\frac{c}{L} = 708,$$

and herefrom, substituting in equations (34) and (35),

$$q = -708 \tan (0.0259\, q)^\circ \quad \text{(zero voltage)}$$

$$= +708 \cot (0.0259\, q)^\circ \quad \text{(zero current)},$$

$$1 - \frac{u_0}{u} = \frac{1}{0.618\, q^2 10^{-6} + 1.28}$$

$q\lambda_0 =$	100.35°	185.64°	273.83°	362.89°	452.32°	541.94°
$q =$	3875	7168	10,572	14,010	17,463	20,920
$1 - \frac{u_0}{u} =$	0.0946	0.0302	0.0142	0.00816	0.0047	0.0037
$u_0 =$	95.8	102.8	104.5	105.1	105.5	105.6
$-s =$	10.2	3.2	1.5	0.87	0.5	0.4

By equation (31) the effective values of the first six harmonics are given as

(1) Quarter-wave: 100.35°.

$$q_1 = 3875;$$
$$u_0 = 95.8;$$
$$I = i_0 \varepsilon^{-u_0 t} (\cos q\lambda - 5.48 \sin q\lambda),$$
$$E = 1939\, i_0 \varepsilon^{-u_0 t} (\cos q\lambda + 0.182 \sin q\lambda).$$

(2) Half-wave: 185.64°.

$$q_2 = 7168;$$
$$u_0 = 102.8;$$
$$I = i_0 \varepsilon^{-u_0 t} (\cos q\lambda - 10.14 \sin q\lambda);$$
$$E = 3585\, i_0 \varepsilon^{-u_0 t} (\cos q\lambda + 0.098 \sin q\lambda).$$

(3) Three-quarter wave: 273.83°.

$$q_3 = 10{,}572;$$
$$u_0 = 104.5;$$
$$I = i_0 \varepsilon^{-u_0 t} (\cos q\lambda - 14.90 \sin q\lambda);$$
$$E = 5287\, i_0 \varepsilon^{-u_0 t} (\cos q\lambda + 0.067 \sin q\lambda).$$

(4) Full wave: 362.89°.

$$q_4 = 14{,}010;$$
$$u_0 = 105.1;$$
$$I = i_0 \varepsilon^{-u_0 t} (\cos q\lambda - 19.8 \sin q\lambda);$$
$$E = 7005\, i_0 \varepsilon^{-u_0 t} (\cos q\lambda + 0.050 \sin q\lambda).$$

(5) Five-quarter wave: 452.32°.

$$q_5 = 17{,}463;$$
$$u_0 = 105.5;$$
$$I = i_0 \varepsilon^{-u_0 t} (\cos q\lambda - 24.65 \sin q\lambda);$$
$$E = 8732\, i_0 \varepsilon^{-u_0 t} (\cos q\lambda + 0.040 \sin q\lambda).$$

(6) Three-half wave: 541.94°.

$$q_6 = 20{,}920;$$
$$u_0 = 105.6;$$
$$I = i_0\varepsilon^{-u_0t}\,(\cos q\lambda - 29.6 \sin q\lambda);$$
$$E = 10{,}460\, i_0\varepsilon^{-u_0t}\,(\cos q\lambda + 0.033 \sin q\lambda).$$

PART IV
VARIATION OF CIRCUIT CONSTANTS

CHAPTER I

VARIATION OF CIRCUIT CONSTANTS

1. In the preceding investigations on transients, the usual assumption is made, that the circuit constants: resistance r, inductance L, capacity C and shunted conductance g, are constant. While this is true, with sufficient approximation, for the usual machine frequencies and for moderately high frequencies, experience shows that it is not even approximately true for very high frequencies and for very sudden circuit changes, as steep wave front impulses, etc.

If r, L, C and g are assumed as constant, it follows that the attenuation is independent of the frequency, that is, waves of all frequencies decay at the same rate, and as the result, a complex wave or an impulse traversing a circuit dies out without changing its wave shape or the steepness of its wave front.

Experience, however, shows that steep wave fronts are dangerous only near their origin, and rapidly lose their destructiveness by the flattening of the wave front when running along the circuit. Experimentally, small inductances shunted by a spark gap, inserted in transmission lines for testing for high frequencies or steep wave fronts, show appreciable spark lengths, that is, high voltage gradients, only near the origin of the disturbance.

The rectangular wave of starting a transmission line by connecting it to a source of voltage, which is given by the theory under the assumption of constant r, L, C and g, is not shown by oscillograms of transmission lines.

If r and L are constant, the power factor of the line conductor,

$\frac{r}{\sqrt{r^2 + (2\pi f L)^2}}$, should with increasing frequency continuously

decrease, and reach extremely low values, at very high frequencies, so that at these, an oscillatory disturbance should be sustained over very many cycles, and show with increasing frequency an increasing liability to become a sustained or cumulative oscillation. Experience, however, shows that high frequency oscillations die out much more rapidly than accounted for by the standard theory, and show at very high frequency practically no tendency to become cumulative.

Therefore, when dealing with transients containing very high frequencies or steep wave fronts, the previous theory, which is based on the assumption of constant r, L, C and g, correctly represents the transient only in its initial stage and near its origin, but less so its course after its initial stage and at some distance from the origin, especially with high frequency transients or steep wave fronts.

It therefore, is of importance to investigate the factors which cause a change of the line constants r, L, C and g, with increasing frequency or steepness of wave front, and the effect produced on the course of the transient as regard to duration and wave shape, by the variation of the line constants.

The two most important factors in the variation of the circuit constants r, L, C and g seem to be the *unequal current distribution in the conductor* and the *finite velocity of the electric field.*

UNEQUAL CURRENT DISTRIBUTION IN THE CONDUCTOR.

2. The magnetic field of the current surrounds this current and fills all the space outside thereof, up to the return current. Some of the magnetic field due to the current in the interior and in the center of a conductor carrying current thus is inside of the conductor, while all the magnetic field of the current in the outer layer of the conductor is outside of it. Therefore, more magnetic field surrounds the current in the interior of the conductor than the current in its outer layer, and the inductance therefore increases from the outer layer of the conductor toward its interior, by the "internal magnetic field." In the interior of the conductor, the reactance voltage thus is higher than on the outside.

At low frequency, with moderate size of conductor, this difference is inappreciable in its effect. At higher frequencies, however, the higher reactance in the interior of the conductor, due to this internal magnetic field, causes the current density to decrease toward the interior of the conductor, and the current to lag, until finally the current flows practically only through a thin layer of the conductor surface.

As the result thereof, the effective resistance of the conductor is increased, due to the uneconomical use of the conductor material caused by the lower current density in the interior, and due to the phase displacement, which results in the sum of the currents in the successive layers of the conductor being larger than the resultant current. Due to this unequal current distribution, the internal reactance of the conductor is decreased, as less current penetrates to the interior of the conductor, and thus produces less magnetic field inside of the conductor.

The derivation of the equations of the effective resistance of unequal current distribution in the conductor, r_1, and of the internal reactance x_1 under these conditions is given in Part II, Chapter VII. It is interesting to note that effective resistance and internal reactance, with increasing frequency, approach the same limit, and become proportional to the square root of the frequency, the square root of the permeability, and the square root of the resistivity of the conductor material, while at low frequencies the resistance is independent of the frequency and directly proportional to the resistivity, and the internal reactance is independent of the resistivity and directly proportional to the frequency.

FINITE VELOCITY OF THE ELECTRIC FIELD.

3. The derivation of the equations of the effective resistance of magnetic radiation, and in general of the effects of the finite velocity of the electric field on the line constants, are given in Part II, Chapter VIII.

The magnetic radiation resistance is proportional to the square of the frequency (except at extremely high frequencies). It therefore is negligible at low and medium frequencies, but becomes the dominating factor at high frequencies. It is proportional to the distance of the return conductor, but entirely independent of size, shape, or material of the conductor, as is to be expected, since it represents the energy dissipated into space. Only at extremely high frequencies the rise of radiation resistance becomes less than proportional to the square of the frequency. It becomes practically independent of the distance of the return conductor, when the latter becomes of the magnitude of the quarter wave length.

The same applies to the capacity. Due to the finite velocity of propagation, the dielectric or electrostatic field lags behind the voltage which produces it, by the same angle by which the magnetic field lags behind the current, and the capacity current or charging current thus is not in quadrature with the voltage, or reactive, but displaced in phase by more than 90°, thus contains a—negative—energy component, which gives rise to a shunted conductance of dielectric radiation g. This gives rise to an energy dissipation by the conductor, at high frequencies, by dielectric radiation into space, of the same magnitude as the energy dissipation by magnetic radiation, above considered.

The term "shunted conductance" g has been introduced into the general equations of the electric circuit largely from theoretical reasons, as representing the power consumption proportional to the voltage. Most theoretical investigations of transmission circuits consider only r, L and C as the circuit constants, and omit g, since under average transmission line conditions, at low and moderate frequencies, g usually is negligible. In communication circuits, as telegraph and telephone, there is a "leakage," which would be represented by a shunted conductance, and in underground cables there is a considerable energy consumption by dielectric losses in the insulation, as the investigations of the last years have shown, which gives a shunted conductance. In overhead power lines, however, energy losses depending on the voltage—and leading to a term g—have been known only at such high voltages where corona appears.

It is interesting, therefore, to note that at high frequencies "shunted conductance" g may reach very formidable values even in transmission lines, due to electrostatic radiation.

In investigating the effect of the finite velocity of the electric field on the inductance L and the capacity C, it is seen that the equations of external inductance and of capacity are not affected, but remain the same as the usual values derived by neglecting the velocity of the electric field, except at extremely high frequencies, when the distance of the return conductor approaches quarter wave length.

EQUATIONS OF ELECTRICAL CONSTANTS, AND NUMERICAL VALUES.

4. In the following are given, compiled from Part II, Chapters VII to IX, the equations of the components of the electrical constants, as functions of the frequency, for conductors with return conductor, and also for conductors without return conductor (as approximated by lightning strokes or wireless antennæ):

Resistance: True ohmic resistance or effective resistance of unequal current distribution, and magnetic radiation resistance.

Reactance: Low frequency internal reactance or internal reactance of unequal current distribution, and external reactance.

Inductance: Low frequency internal inductance or internal inductance of unequal current distribution, and external inductance.

Shunted conductance and capacity are not so satisfactorily represented, and therefore, instead of representing energy storage and power dissipation depending on the voltage by a conductance g and a capacity C or susceptance b, in shunt with each other, it is more convenient to represent them by an effective resistance, the dielectric radiation resistance r_c, and a capacity reactance x_c, in series with each other.

Equations of Electrical Constants.

Let l = length of conductor, cm.
l_r = radius of conductor, cm.
l_1 = circumference of conductor, cm.
l_2 = shortest circumference of conductor, cm.
l' = distance of return conductor
f = frequency, cycles per second
λ = electrical conductivity, mhos per cm.^{3}
μ = magnetic permeability
$S = 3 \times 10^{10}$ = velocity of radiation in empty space;
it is, then, log denoting the natural logarithm.

Resistances:

True ohmic resistance (thermal):

$$r_0 = \frac{l}{\lambda \pi l_r^2} \text{ ohms.}$$

Effective resistance of unequal current distribution (thermal):

$$r_1 = \frac{\pi l}{l_1}\sqrt{\frac{0.4\,\mu f}{\lambda}}\,10^{-4} \text{ ohms.}$$

Effective magnetic radiation resistance:

(*a*) Return conductor at distance l':

$$r_3 = \frac{8\,\pi^2 f^2 l' l}{S}\,10^{-9} \text{ ohms;}$$

at extremely high frequencies:

$$r_4 = 4\,\pi f l \left\{\frac{\pi}{2} - \text{col}\,\frac{2\,\pi f l'}{S}\right\} 10^{-9} \text{ ohms.}$$

(*b*) Conductor without return conductor:

$$r_2 = 2\,\pi^2 f l\, 10^{-9} \text{ ohms.}$$

Effective dielectric radiation (shunted) resistance:*

(*a*) Return conductor at distance l':

$$r_c = \frac{2\,l' S}{l}\,10^{-9} \text{ ohms,}$$

at extremely high frequencies:

$$r_c = \frac{S^2}{\pi f l}\left\{\frac{\pi}{2} - \text{col}\,\frac{2\,\pi f l'}{S}\right\} 10^{-9} \text{ ohms.}$$

(*b*) Conductor without return conductor:

$$r_c = \frac{S}{2\,fl}\,10^{-9} \text{ ohms.}$$

Reactances:

Low frequency internal reactance:

$$x_{10} = \pi f l \mu\, 10^{-9} \text{ ohms.}$$

Internal reactance of unequal current distribution:

$$x_1 = \frac{\pi l}{l_1}\sqrt{\frac{0.4\,\mu f}{\lambda}}\,10^{-4} \text{ ohms} = r_1.$$

*As shunted resistance and reactance, r_c and x_c are inverse proportional to the length of the conductor l, that is, the longer the conductor, the more current is shunted across, and the lower therefore are r_c and x_c. For this reason, the shunt constants usually are given as conductance g and susceptance b. In the present case, however, r and x give simpler expressions.

External reactance:

(*a*) Return conductor at distance l':

$$x_0 = 4\,\pi f l \log \frac{l'}{l_r}\, 10^{-9} = 4\,\pi f l \log \frac{2\,\pi l'}{l_2}\, 10^{-9} \text{ ohms};$$

at extremely high frequencies:

$$x_4 = 4\,\pi f l \left\{ \log \frac{S}{2\,\pi f l_r} - 0.5772 - \text{sil} \frac{2\,\pi f l'}{S} \right\} 10^{-9} \text{ ohms.}$$

(*b*) Conductor without return conductor:

$$x_2 = 4\,\pi f l \left\{ \log \frac{S}{2\,\pi f l_r} - 0.5772 \right\} 10^{-9} \text{ ohms.}$$

Shunted capacity reactance:

(*a*) Return conductor at distance l':

$$x_c = \frac{S^2 \log \frac{l'}{l_r}}{\pi f l}\, 10^{-9} = \frac{S^2 \log \frac{2\,\pi l'}{l_2}}{\pi f l}\, 10^{-9} \text{ ohms};$$

at extremely high frequencies:

$$x_c = \frac{S^2}{\pi f l} \left\{ \log \frac{S}{2\,\pi f l_r} - 0.5772 - \text{sil} \frac{2\,\pi f l'}{S} \right\} 10^{-9} \text{ ohms.}$$

(*b*) Conductor without return conductor:

$$x_c = \frac{S^2}{\pi f l} \left\{ \log \frac{S}{2\,\pi f l_r} - 0.5772 \right\} 10^{-9} \text{ ohms.}$$

Inductances:

Low frequency internal inductance:

$$L_{10} = \frac{l\mu}{2}\, 10^{-9} \text{ henrys.}$$

Internal inductance of unequal current distribution:

$$L_1 = \frac{l}{l_1} \sqrt{\frac{0.1\,\mu}{\lambda f}}\, 10^{-4} \text{ henrys.}$$

External inductance:

(*a*) Return conductor at distance l':

$$L_0 = 2\,l \log \frac{l'}{l_r}\, 10^{-9} = 2\,l \log \frac{2\,\pi l'}{l_2}\, 10^{-9} \text{ henrys};$$

at extremely high frequencies:

$$L_4 = 2\,l \left\{ \log \frac{S}{2\,\pi f l_r} - 0.5772 - \text{sil} \frac{2\,\pi f l'}{S} \right\} 10^{-9} \text{ henrys.}$$

(*b*) Conductor without return conductor:

$$L_2 = 2\,l\left\{\log\frac{S}{2\,\pi f l_r} - 0.5772\right\}10^{-9}\text{ henrys.}$$

Capacity:

(*a*) Return conductor at distance l':

$$C_0 = \frac{10^9}{2\,S^2\log\frac{l'}{l_r}}\text{ farads;}$$

at extremely high frequencies:

$$C_4 = \frac{10^9}{2\,S^2\left\{\log\frac{S}{2\,\pi f l_r} - 0.5772 - \text{sil}\,\frac{2\,\pi f l'}{S}\right\}}\text{ farads.}$$

(*b*) Conductor without return conductor:

$$C_2 = \frac{10^9}{2\,S^2\left\{\log\frac{S}{2\,\pi f l_r} - 0.5772\right\}}\text{ farads.}$$

5. Herefrom then follow the Circuit Constants:

At Low Frequencies (Machine Frequencies up to 10^3 cycles, approx.):

$$r = r_0$$
$$x = x_1 + x_0$$
$$L = L_{10} + L_0$$
$$C = C_0$$
$$g = 0$$

At Medium Frequencies (10^3 to 10^5 cycles, approx.):

$$r = r_1$$
$$x = x_{10} + x_0$$
$$L = L_1 + L_0$$
$$C = C_0$$
$$g = 0$$

At High Frequencies (10^5 to 10^7 cycles, approx.):

$r = r_1 + r_3$ (with return conductor)
$ = r_1 + r_2$ (without return conductor)
$x = x_1 + x_0$ (with return conductor)
$ = x_1 + x_2$ (without return conductor)
$L = L_1 + L_0$ (with return conductor)
$ = L_1 + L_2$ (without return conductor)
$C = C_0$ (with return conductor)
$ = C_2$ (without return conductor)

(approximately, or represented by x_c and r_c).

g represented by r_c and x_c:

$$g = \frac{r_c}{r_c^2 + x_c^2}; \quad b = \frac{x_c}{r_c^2 + x_c^2} = 2\pi fC;$$

at extremely high frequencies (above 10^7 cycles, approx.):

$$\begin{aligned} r &= r_1 + r_4 \text{ (with return conductor)} \\ &= r_1 + r_2 \text{ (without return conductor)} \\ x &= x_1 + x_4 \text{ (with return conductor)} \\ &= x_1 + x_2 \text{ (without return conductor)} \end{aligned}$$

C and g represented by r_c and x_c, thus:

$$g = \frac{r_c}{r_c^2 + x_c^2}.$$

$$C = \frac{x_c}{2\pi f(r_c^2 + x_c^2)}.$$

From these follow the derived circuit constants:
Magnetic attenuation:

$$u_1 = \frac{r}{2L}$$

Dielectric attenuation:

Usually zero at low and medium frequencies,

$$u_2 = \frac{g}{2C} = \frac{\pi f r_c}{x_c}$$

at high and very high frequencies.

Attenuation constant:

$$u = u_1 + u_2 = \tfrac{1}{2}\left(\frac{r}{L} + \frac{g}{C}\right)$$

Series power factor:

$$\cos \omega = \frac{r}{\sqrt{r^2 + x^2}}.$$

Shunt power factor:

Zero at low and medium frequencies,

$$\cos \omega_c = \frac{r_c}{\sqrt{r_c^2 + x_c^2}}$$

at high and very high frequencies.

Duration of oscillation:*

$$t_0 = \frac{1}{u} \text{ seconds}$$

$$N_0 = \frac{f}{u} \text{ cycles.}$$

6. As seen, four successive stages may be distinguished in the expressions of the circuit constants as functions of the frequency.

1. Low frequencies, such as the machine frequencies of 25 and 60 cycles. The resistance is the true ohmic resistance, the internal reactance and inductance that corresponding to uniform current density throughout the conductor, with conductors of moderate size, and of non-magnetic material.

2. Medium frequencies, of the magnitude of a thousand to ten thousand cycles. Resistance and internal reactance or inductance are those due to unequal current distribution in the conductor, that is, the resistance is rapidly increasing, and the internal inductance decreasing. The conductance g is still negligible, radiation effects still absent, and all the energy loss that of thermal resistance.

3. High frequencies, of the magnitude of one hundred thousand to one million cycles. The radiation resistance is appreciable and becomes the dominating factor in the energy dissipation. The internal inductance has practically disappeared, due to the current penetrating only a thin surface layer. A considerable shunted conductance exists due to the dielectric radiation.

4. Extremely high frequencies, of the magnitude of many millions of cycles, when the quarter wave length has become of the same magnitude or less than the distance of the return conductor. Radiation effects entirely dominate, and the usual expressions of inductance and of capacity have ceased to apply.

This last case is of little industrial importance, as such extremely high frequencies propagate only over short distances.

* Under "Duration" of a transient is understood the time (or the number of cycles), which the transient would last, that is, the time in which it would expend its energy, if continuing at its initial intensity. With a simple exponential transient, this is the time during which it decreases to $\frac{1}{\epsilon}$ or 36.8 per cent. of its initial value. It decreases to one-tenth of its initial value in 2.3 times this time.

It would come into consideration only in calculating the flattening of the wave front of a rectangular impulse in the immediate neighborhood of its origin, and similar problems.

Thus far, a general investigation does not seem feasible. Substituting the equations of the circuit constants, as functions of the frequency, into the general equations of the electric circuit, leads to expressions too complex for general utility, and the investigation thus must largely be made by numerical calculations.

Only when the frequencies which are of importance in the problem lie fairly well in one of the four ranges above discussed —as is the case in the investigation of the flattening of a steep wave front in moderate distances from its origin—a more general theoretical investigation becomes possible at present.

CHAPTER II.

WAVE DECAY IN TRANSMISSION LINES.

7. From the equations given in Chapter I, numerical values of the line constants are calculated and given in Table IV, for average transmission line conditions, that is, a copper wire No. 00, with 6 ft. = 182 cm. between the conductors, and an average height of 30 ft. = 910 cm. above ground, for the two conditions:

(*a*) A high frequency oscillation between two line conductors.

(*b*) A high frequency oscillation between one line conductor and the ground.

The table gives:

The thermal resistance r_1, the radiation resistance r_3, and the total resistance $r = r_1 + r_3$.

The internal reactance x_1, the external reactance x_3, and the total reactance $x = x_1 + x_3$.

The magnetic attenuation $u_1 = \frac{r}{2L}$, the dielectric attenuation $u_2 = \frac{g}{2C}$, and the total attenuation $u = u_1 + u_2$.

The table also gives the duration of a transient in micro-seconds t and in cycles N, that is, the time which a high frequency oscillation of the frequency f would last, if continuing with its initial intensity and the number of cycles which it would perform. It also gives the power factor, in per cent, of the series circuit, as determined by resistance and inductance, and of the shunt circuit, as determined by shunted conductance and capacity.

As seen, the attenuation constant u is constant up to nearly one thousand cycles. Thus in this range, all the frequencies die out at the same rate. From about one thousand cycles up to about 100,000 cycles, the attenuation constant gradually increases, and thus oscillations die out the more rapidly, the higher the frequency, as seen by the gradual decrease of the duration t. However, as the increase of the attenuation constant and thus the increase of the rapidity of the decay of the disturbance, in this range, is smaller than the increase of frequency, the number of cycles performed by the oscillation increases. Thus, at 25 or 60 cycles, the stored energy, which supplies the oscillating power,

TABLE IV.—ATTENUATION CONSTANTS OF COPPER WIRE NO. 00 B. & S. G.

(Per kilometer of conductor.)

$f =$	25	60	500	750	1000	2000	5000	7500	10,000	15,000
$\sqrt{f} =$	1.398	1.778	2.699	2.875	3	3.301	3.699	3.875	4	4.176
$r_1 =$	0.24	0.24	0.24	0.24	0.273	0.386	0.612	0.750	0.865	1.060
x_1	0.008	0.019	0.157	0.236	0.273	0.386	0.612	0.750	0.865	1.060
(a) Return conductor at 6 ft = 182 cm. distance.										
r_3	....	....	....	....	....	....	0.0012	0.0027	0.0048	0.0108
$r =$	0.24	0.24	0.24	0.24	0.273	0.386	0.613	0.753	0.870	1.071
$x_3 =$	0.187	0.45	3.75	5.625	7.5	15.0	37.5	56.25	75.00	112.5
$x =$	0.195	0.469	3.907	5.861	7.773	15.39	38.11	57.0	75.86	113.6
$L =$	1.245	1.245	1.245	1.245	1.24	1.23	1.22	1.215	1.21	1.205
$\mu_1 = \frac{r}{2L} =$	96	96	96	96	110	157	251	310	360	445
$\mu_2 = \frac{g}{2C} =$	....	....	0.005	0.011	0.020	0.080	0.5	1.125	2.0	4.5
$\mu = \mu_1 + \mu_2 =$	96	96	96	96	110	157	252	311	362	450
$\sqrt{\mu} =$	1.982	1.982	1.982	1.982	2.041	2.196	2.401	2.493	2.559	2.653
$t =$	10,400	10,400	10,400	10,400	9,090	6,370	3,970	3,210	2,760	2,220
$N =$	0.26	0.62	5.2	7.8	9.09	12.74	19.85	24.1	27.6	33.3
$\cos \omega =$	....	....	6.16	4.10	3.51	2.52	1.61	1.32	1.15	0.94
$\cos \omega_c =$	....	....	....	....	....	....	....	....	0.01	0.01
(b) Conductor 30 ft. above ground as return; $l' = 1820$ cm.										
$r_3 =$	....	....	....	....	....	....	0.012	.027	0.48	0.108
$r =$	0.24	0.24	0.24	0.24	0.273	0.386	0.624	0.777	0.913	1.168
$x_3 =$	0.26	0.624	5.2	7.8	10.4	20.8	52.0	78.0	104.0	156.0
$x =$	0.268	0.643	5.357	8.036	10.67	21.16	52.61	78.75	104.9	157.1
$L =$	1.705	1.705	1.705	1.705	1.70	1.69	1.68	1.675	1.67	1.665
$\mu_1 = \frac{r}{2L} =$	70	70	70	70	80	114	186	232	273	350
$\mu_2 = \frac{g}{2C} =$	....	....	0.04	0.08	0.14	0.58	3.6	8.1	14.4	32.5
$\mu = \mu_1 + \mu_2 =$	70	70	70	70	80	115	190	240	287	383
$\sqrt{\mu} =$	1.845	1.845	1.845	1.845	1.903	2.061	2.279	2.380	2.458	2.583
$t =$	14,300	14,300	14,300	14,300	12,500	8,700	5,260	4,170	3,480	2,610
$N =$	0.36	0.86	7.15	10.7	12.5	17.4	26.3	31.3	34.8	39.1
$\cos \omega =$	....	....	4.48	2.98	2.56	1.89	1.19	0.98	0.87	0.74
$\cos \omega_c =$	....	....	....	....	....	0.01	0.02	0.03	0.05	0.07

TABLE IV.—*Continued.*

f =	20,000	50,000	75,000	10^5	2×10^5	5×10^5	7.5×10^5	10^6	2×10^6	5×10^6	Cycles
$/f$ =	4.301	4.699	4.875	5	5.301	5.699	5.875	6	6.301	6.699	
r_1 =	1.225	1.93	2.37	2.73	3.86	6.12	7.50	8.65	12.25	19.3	Ohms
r_t =	1.225	1.93	2.37	2.73	3.86	6.12	7.50	8.65	12.25	19.3	Ohms
(*a*) Return conductor at 6 ft. = 182 cm. distance.											
r_3 =	0.0192	0.12	0.27	0.48	1.92	12	27	48	192	1,200	Ohms
r =	1.243	2.05	2.64	3.21	5.78	18.1	34.5	56.6	204	1,219	Ohms
x_3 =	150	375	562.5	750	1,500	3,750	5,625	7,500	15,000	37,500	Ohms.
x =	151.2	376.9	564.9	752.7	1,504	3,756	5,632	7,509	15,012	37,520	Ohms
L =	1.205	1.20	1.20	1.20	1.20	1.195	1.195	1.195	1.195	1.195	Henrys
$\mu_1 = \frac{r}{2L}$ =	516	855	1,100	1,340	2,410	7,570	14,400	23,700	85,600	570,000	
$\mu_2 = \frac{g}{2C}$ =	8.0	50	112.5	200	800	5,000	11,200	20,000	80,000	500,000	
$\mu = \mu_1 + \mu_2$ =	524	905	1,212	1,540	3,210	12,570	25,650	43,700	165,600	1,010,000	
$/\mu$ =	2.719	2.957	3.084	3.187	3.506	4.099	4.409	4.640	5.219	6.004	
t =	1,910	1,100	825	649	311	79.6	39.0	22.9	6.04	1.00	Micro-seconds
N =	38.2	55.0	61.9	64.9	62.2	39.8	29.2	22.9	12.1	5.0	Cycles
$\cos \omega$ =	0.82	0.54	0.47	0.43	0.38	0.48	0.61	0.75	1.36	3.25	Per cent.
$\cos \omega_c$ =	0.01	0.03	0.05	0.06	0.13	0.32	0.48	0.64	1.27	3.18	Per cent.
(*b*) Conductor 30 ft. above ground as return; l' = 1820 cm.											
r_3 =	0.192	1.2	2.7	4.8	19.2	120	270	480	1,920	12,000	Ohms
r =	1.417	3.13	5.07	7.53	23.06	126.1	277.5	488.6	1,932	12,020	Ohms
x_3 =	208	520	780	1,040	2,080	5,200	7,800	10,400	20,800	52,000	Ohms
x =	209.2	252	782	1,043	2,084	5,206	7,807	10,409	20,812	52,020	Ohms
L =	1.665	1.66	1.66	1.66	1.66	1.655	1.655	1.655	1.655	1.655	Henrys
$\mu_1 = \frac{r}{2L}$ =	424	942	1,520	2,270	6,930	38,200	83,800	147,500	583,000	3,630,000	
$\mu_2 = \frac{g}{2C}$ =	57.6	360	810	1,440	5,760	36,000	81,000	144,000	576,000	3,600,000	
$\mu = \mu_1 + \mu_2$ =	482	1,302	2,330	3,710	12,690	74,200	164,800	291,500	1,159,000	7,230,000	
$/\mu$ =	2.683	3.115	3.367	3.569	4.103	4.870	5.217	5.465	6.064	6.859	
t =	2,070	768	429	269	78.8	13.5	6.07	3.43	0.86	0.14	Micro-seconds
N =	41.4	38.4	31.2	26.9	15.8	6.75	4.55	3.43	1.72	0.69	Cycles
$\cos \omega$ =	0.68	0.60	0.65	0.71	1.10	2.42	3.55	4.70	9.24	22.5	Per cent.
$\cos \omega_c$ =	.09	0.23	0.35	0.46	0.92	2.31	3.46	4.62	9.22	22.5	Per cent.

would be expended in less than one cycle, that is, a real oscillation would hardly materialize (except by other sources of energy, as the stored magnetic energy of a transformer connected to the line). At 1000 cycles, the oscillation would already last 9 to 12 cycles, and at still higher frequencies reach a maximum of 41.4

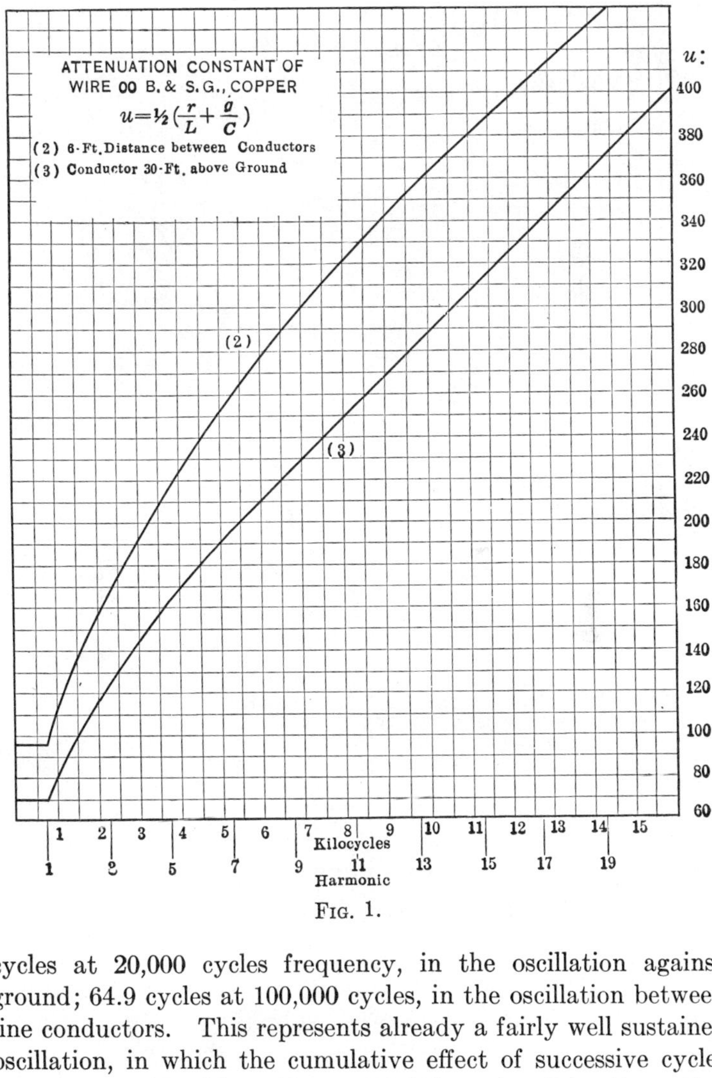

FIG. 1.

cycles at 20,000 cycles frequency, in the oscillation against ground; 64.9 cycles at 100,000 cycles, in the oscillation between line conductors. This represents already a fairly well sustained oscillation, in which the cumulative effect of successive cycles may be considerable. Above 100,000 cycles the attenuation constant begins to rise rapidly, and reaches enormous values, due to the rapidly increasing energy dissipation by radiation. As

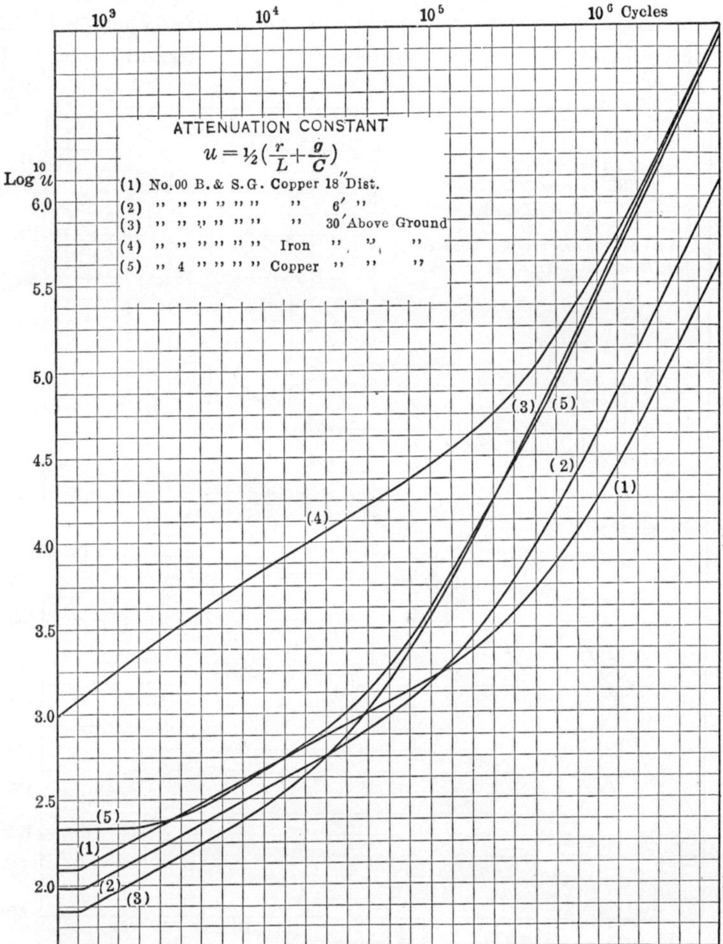

FIG. 2.

the result, the duration of the oscillation very rapidly decreases, and the number of cycles performed by the oscillation decreases, until, beyond a million cycles, the energy dissipation is so rapid that practically no oscillation can occur; the oscillation dying out in a cycle or less, thus being practically harmless.

The attenuation constant is plotted, up to 15,000 cycles, in Fig. 1, with the frequency as abscissæ, and is plotted in Fig. 2 in logarithmic scale, as (2) and (3).

Noteworthy is the great difference between the oscillation against ground, and the oscillation between line conductors; the oscillation against ground is more persistent at low frequencies, due to the greater amount of stored energy in the electric field of the conductor, which reaches all the distance to the ground. When reaching into very high frequencies, however, the energy dissipation by radiation becomes appreciable at lower frequencies in the oscillation against ground than in the oscillation between line conductors, and reaches much higher values, with the result that the decay of an oscillation between line and ground is much more rapid at high frequencies than the decay of an oscillation between line conductors. For instance, at 100,000 cycles, the latter performs 65 cycles before dying out, while the former has dissipated its energy in 27 cycles, that is, less than half the time.

8. To further investigate this, in Tables V and VI the numerical values of effective resistance, power factor, attenuation constant and duration of a transient oscillation, in cycles, are given for six typical conductors and circuits, for frequencies from 10 cycles to five million cycles, and plotted in Figs. 2, 3 and 4 in logarithmic scale.

1, 2 and 3 are lines of high power; copper conductor No. 00 B. & S. G., in 1 with 18 in. = 45.5 cm. between conductors, corresponding about to average distribution conductors; in 2 with 6 ft. = 182 cm., between conductors, corresponding to about average transmission line conductors with the oscillation between two lines, and in 3 with 60 ft. = 1830 cm. between conductor and return conductor, corresponding to an oscillation between line and ground, under average transmission line conditions, with the conductor 30 ft. above ground. 4, 5 and 6 give the same condition of an oscillation between line and ground, but in 4 an iron wire of the size of No. 00 B. & S. G., such as has been proposed for the station end of transmission lines, to oppose the approach of high frequency disturbances. In 5 a copper wire No. 4 B. & S. G., that is, a low power transmission line, is represented, and in 6 a stranded aluminum conductor of the same conductivity as copper wire No. 00 B. & S. G.

The equations of the constants for these six circuits are given in

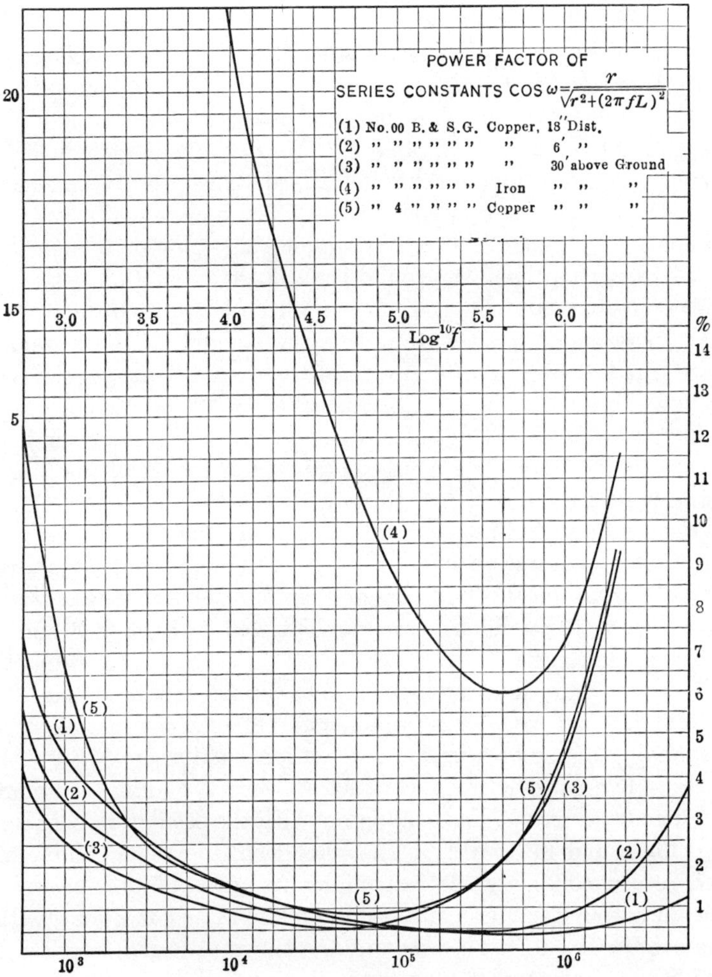

FIG. 3.

Table V. This table also gives the limiting frequencies, between which the various formulas apply with sufficient accuracy for practical purposes, and the lower limits, where the effects become appreciable, in the various conductors.

In Fig. 2 the attenuation constants are plotted, in Fig. 3 the power factors and in Fig. 4 the duration, in cycles.

As such a transient oscillation dies out exponentially, theoret-

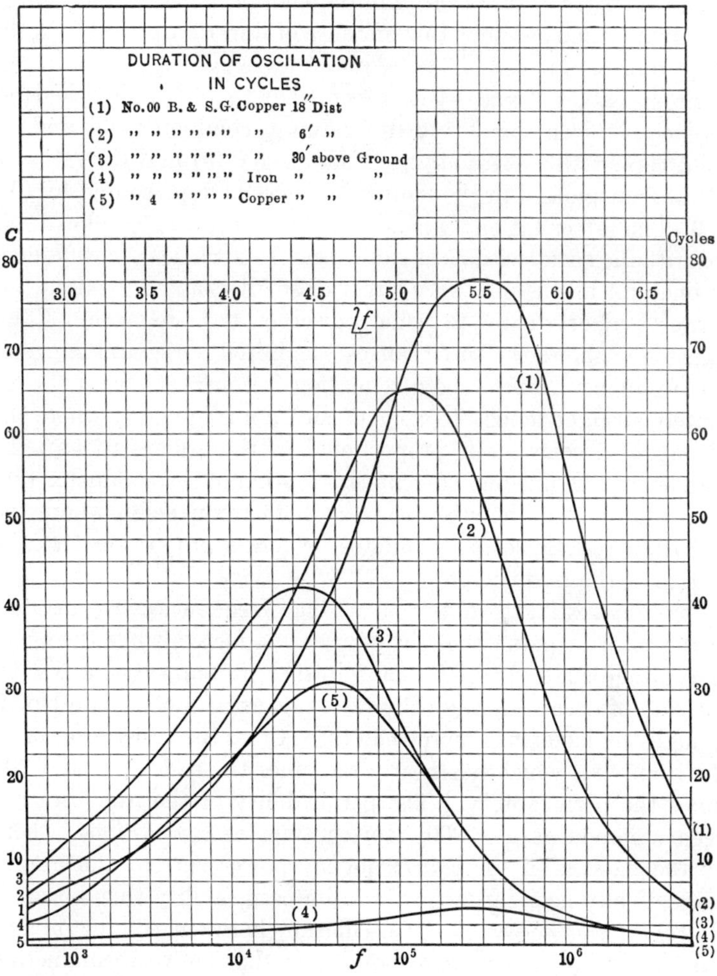

FIG. 4.

ically it has no definite duration, but lasts forever, though practically it may have ceased in a few micro-seconds. Thus as duration is defined the time, or the number of cycles, which the oscillation would last if maintaining its initial intensity. In reality, in this time the duration has decreased to $\frac{1}{\epsilon}$, or 37 per cent. of its initial value. Physically, at 37 per cent. of its initial value, or $0.37^2 = 0.135$ of its initial energy, it has become practically harmless, so that this measure of duration probably is the most representative.

From Tables V and VI it is seen, that there is no marked difference between the stranded aluminum conductor (6), and the solid copper wire of the same conductivity (3) and the values of (6) are not plotted in the figures, but may be represented by (3).

9. The attenuation constant μ, in Fig. 2, is plotted in logarithmic scale, with $\underline{/f}$ as abscissæ. In such scale, a difference of one unit means ten times larger or smaller, and a straight line means proportionally to some power of the frequency. This figure well shows the three ranges: the initial horizontal range at low frequency, where the attenuation is constant; the approximately straight moderate slope of medium frequency, where the attenuation constant is proportional to the square root of the frequency, the unequal current distribution in the conductor predominating, and the steep slope at high frequencies, where the radiation resistance predominates, which is proportional to the square of the frequency.

It is interesting to note that at high frequencies the distance of the return conductor is the dominating factor, while the effect of conductor size and material vanishes; in copper wire No. 00 the rate of decay is practically the same as in copper wire No. 4, though the latter has more than three times the resistance; or in the iron wire, which has nearly six times the resistance and 200 times the permeability. The permeability of the iron wire has been assumed as $\mu = 200$, representing load conditions, where by the passage of the low frequency power current the iron is magnetically near saturation, and its permeability thus lowered. However, the decay of the oscillation between conductor and ground is 6 to 7 times more rapid than that between conductors 6 ft. apart, and that between conductors 18 in. apart about 3 times less.

This shows, that to produce quicker damping of high frequency waves, such as are instrumental in steep wave fronts, the most effective way is to separate the conductors as far as possible, perhaps even lead them to the station by separate single conductor

TABLE V.—LINE CONSTANTS.

f = frequency in cycles per second.
$S = 3 \times 10^{10}$ cm. sec. = propagation velocity in empty space.
$\kappa = 1$ = specific capacity.
l_0 = length of conductor, in kilometers $= 10^5$ cm.

				Per kilometer of conductor:					
				(1)	(2)	(3)	(4)	(5)	(6)
	Size of round conductor, B. & S. G.			00	00	00	00	4	(00)
	Radius of conductor, cm.		$l_r =$	0.463	0.463	0.463	0.463	0.258	0.65
	Distance of return conductor, cm.		$l' =$	45.5	182	1820	1820	1820	1820
	Material of conductor			Copp	Copper	Copper	Iron	Copper	Aluminum
	Conductivity		$\lambda = 10^5 \times$	6.2	6.2	6.2	1.1	6.2	3.8
	Permeability		$\mu =$	1	1	1	200	1	1
Low frequency	Ohmic resistance	$r_0 = \frac{10^5 l_0}{\pi l_r^2 \lambda}$ ohms =		0.240	0.240	0.240	1.35	0.770	0.240
	Internal reactance	$x_0' = 0.1\, \pi\mu f l_0\, 10^{-3}$ ohms =	$f \times 10^{-3} \times$	0.314	0.314	0.314	62.83	0.314	0.314
	Limit, cycles		$f =$	10^3	10^3	10^3	10	2×10^3	10^3
High frequency	Thermal resistance } Internal reactance }	$r_1 = x_1 = \frac{5\, l_0}{l_r}\sqrt{\frac{0.4\, \mu f}{\lambda}}$ ohms =	$10^{-3} \times \sqrt{f} \times$	8.65	8.65	8.65	290	15.5	7.86
	Lower limit, cycles		$f =$	2×10^5	10^5	2×10^4	2×10^5	2×10^4	2×10^4
	Radiation resistance	$r_2 = \frac{0.8\, \pi^2 f^2 l' l_0}{S}\, 10^{-3}$ ohms =	$f^2 \times 10^{-9} \times$	0.012	0.048	0.48	0.48	0.48	0.48
	Upper limit, cycles		$f =$	10^8	3×10^7	3×10^6	3×10^6	3×10^6	3×10^6
	External reactance	$X_0 = 0.4\, \pi f l_0 \log \frac{l'}{l_r}\, 10^{-3}$ ohms =	$f \times 10^{-3} \times$	5.76	7.5	10.4	10.4	11.1	9.95
	Total resistance Total reactance	$r = r_1$ (or r_0) $+ r_2$ $x = x_1$ (or x_0') $+ x_0$							

	Inductance	$L = \frac{x}{2\pi f}$ henrys							
	Magnetic attenuation	$\frac{r}{2L}$							
	Dielectric radiation resistance	$r_c = \frac{2Sl'}{\kappa l_0} 10^{-14}$ ohms =	....	0.0273	0.1092	1.092	1.092	1.092	1.092
	Capacity reactance	$x_c = \frac{S^2}{\pi f \kappa l_0} \log \frac{l'}{l_r} 10^{-14}$ ohms =	$\frac{10^6}{f} \times$	13.2	17.2	23.7	23.7	25.4	22.8
	Dielectric attenuation	$\frac{g}{2C} = \pi f \frac{r_c}{x_c} = \frac{2\pi^2 f^2 l'}{S \log \frac{l'}{l_r}} =$	$f^2 \times 10^{-6} \times$	0.0065	0.020	0.144	0.144	0.135	0.151
	Attenuation constant	$u = \frac{1}{2}\left(\frac{r}{L} + \frac{g}{C}\right)$							

lines; but the use of high resistance conductors, or of magnetic material, as iron, offers little or practically no advantage in damping very high frequencies or flattening steep wave fronts.

At medium and low frequencies, however, the relation reverses, and the decay of the wave is the smaller, the greater the distance of the return conductor. The reason is, that in this range the effective resistance is still independent of the conductor distance, while the inductance increases with increasing distance. At medium and low frequencies, the iron conductor offers an enormously increased attenuation: from 10 to 20 times that of non-magnetic conductors.

10. The power factor of the conductor is plotted in Fig. 3. As seen, it decreases, from unity at very low frequencies, to a minimum at medium high frequencies, and then increases again to very high values at very high frequencies. The minimum value is a fraction of one per cent. except with the iron conductor, where the minimum is very much higher. The power factor is of importance as it indicates the percentage of the oscillating energy, which is dissipated per wave of oscillation. This is represented still better by Fig. 4, the duration of the oscillation in cycles, that is, the number of cycles which an oscillation lasts before dissipating the stored energy which causes it.

At medium high frequencies, the oscillation is the more persistent the lower the ohmic resistance of the conductor and the further away the return conductor, while at very high frequencies the reverse is the case, and the oscillation is the more persistent, the shorter the distance of the return conductor, while the size and material of the conductor ceases to have any effect.

The maximum number of cycles is reached at medium high frequencies, in the range between 20,000 and 100,000 cycles—depending on conductor size and distance of return conductor. It thus is in this range of frequencies, where an oscillation caused by some disturbance lasts the greatest number of cycles, that the possibility, by some energy supply by means of an arc, etc., to form a stationary oscillation or even a cumulative oscillation, thus to become continuous or "undamped," is greatest.

It would thus appear, that this range of frequencies, of 20,000 to 100,000, represents what may be called the "danger frequencies" of transmission systems. It is interesting to note, that experimental investigations have shown, that the natural fre-

TABLE VI.—ATTENUATION CONSTANTS.

		$f =$	10	60	500	1000	2000	5000 *c*	10 *kc*	20	50	100	200	500	1000	2000	5000 *kc*
1	No. 00 B. & S. G. l_r = 0.463 cm. l' = 45.5 cm. Copper.	$r =$	0.24	0.24	0.24	0.273	0.386	0.612	0.866	1.230	1.96	2.85	4.34	9.12	20.65	60.25	319.3
		$\cos\omega =$	96.9	54.9	7.86	4.53	3.26	2.07	1.48	1.055	0.674	0.494	0.376	0.316	0.358	0.524	1.105 %
		$u =$	124	124	124	142	204	327	466	666	1,081	1,615	2,620	6,595	17,740	58,800	336,500
		$N =$	0.08	0.48	4.03	7.04	9.80	15.3	21.5	30.2	46.2	64.5	76.4	76.0	56.4	34.0	14.9 *c*
2	l' = 182 cm. Copper.	$r =$	0.24	0.24	0.24	0.273	0.386	0.613	0.870	1.249	2.05	3.21	5.78	18.1	56.6	204.2	1,219
		$\cos\omega =$	95.1	45.4	6.16	3.51	2.52	1.61	1.15	0.822	0.544	0.427	0.384	0.483	0.753	1.36	3.25 %
		$u =$	96	96	96	110	157	252	362	524	905	1,540	3,210	12,570	43,700	165,600	1,010,000
		$N =$	0.10	0.62	5.20	9.09	12.74	19.85	27.6	38.2	55.0	64.9	62.2	39.8	22.9	12.1	5.0 *c*
3	l' = 1,820 cm. Copper.	$r =$	0.24	0.24	0.24	0.273	0.386	0.624	0.913	1.417	3.13	7.53	23.06	126.1	488.6	1,932	12,020
		$\cos\omega =$	91.3	34.9	4.48	2.56	1.83	1.19	0.870	0.677	0.600	0.706	1.105	2.42	4.70	9.24	22.5 %
		$u =$	70	70	70	80	115	190	287	482	1,302	3,710	12,690	74,200	291,500	1,159,000	7,230,000
		$N =$	0.14	0.86	7.15	12.5	17.4	26.3	34.8	41.4	38.4	26.9	15.8	6.75	3.43	1.72	0.69 *c*
4	Iron.	$r =$	1.35	2.25	6.48	9.16	12.95	20.51	29.05	41.29	66.0	96.4	148.9	325	770	2,331	12,650
		$\cos\omega =$	(80)	61.7	48.5	42.4	35.8	27.1	21.5	16.0	11.3	8.5	6.7	6.0	7.2	11.0	23.3 %
		$u =$	(25)	148	873	1,470	2,410	4,444	6,914	10,210	18,060	28,200	48,200	130,400	371,000	1,266,000	7,370,000
		$N =$	(0.40)	0.40	0.57	0.68	0.83	1.13	1.45	1.96	2.77	3.55	4.17	3.83	2.70	1.58	0.68 *c*
5	No. 4 B. & S. G. l_r = 0.258 cm. l' = 1,820 cm. Copper.	$r =$	0.77	0.77	0.77	0.77	0.772	1.112	1.598	2.382	4.66	9.70	26.1	131	495	1,940	12,030
		$\cos\omega =$	98.9	74.5	13.4	6.75	3.38	1.96	1.42	1.06	0.84	0.87	1.17	2.34	4.44	8.7	21.1 %
		$u =$	212	212	212	212	213	311	458	721	1,647	4,080	12,750	70,550	274,500	1,087,000	6,775,000
		$N =$	0.047	0.28	2.36	4.72	9.38	16.05	21.8	27.8	30.4	24.5	15.7	7.1	3.65	1.84	0.75
6	7 strand Al. equivalent to No. 00 Copper. l_r = 0.65 cm. l' = 1,820 cm.	$r =$	0.24	0.24	0.24	0.248	0.351	0.568	0.834	1.302	2.950	7.28	22.71	125.6	487.9	1,931	12,018
		$\cos\omega =$	91.9	36.0	4.64	2.42	1.72	1.12	0.83	0.65	0.59	0.73	1.14	2.5	4.9	9.8	23.7
		$u =$	73	73	73	76	109	181	275	467	1,298	3,780	13,130	77,000	303,000	1,207,000	7,580,000
		$N =$	0.14	0.82	6.85	13.2	18.3	27.6	36.4	42.8	38.5	26.5	15.2	6.5	3.3	1.66	0.65

quency of oscillation of the high voltage windings of large power transformers usually is within this range of danger frequencies. The possibility of the formation of destructive cumulative oscillations or stationary waves in the high voltage windings of large power transformers thus is greater than probably with any other class of circuits, so that such high potential transformer windings require specially high disruptive strength and protection. This accounts for the not infrequent disastrous experience with such transformers, before this matter was realized.

11. Figure 4 also shows, that the duration of an oscillation in iron wire, in cycles, is very low at all frequencies. Thus the formation of a stationary oscillation in an iron conductor is practically excluded, but such conductors would act as a dead resistance, damping any oscillation by rapid energy dissipation.

The duration of high frequency oscillation, in cycles, increases with increasing frequency, to a maximum at medium frequencies. This obviously does not mean that the time during which the oscillation lasts increases; the time, in micro-seconds, naturally decreases with increasing frequency, due to the increasing attenuation constant. Thus in conductor (2) for instance, the oscillation lasts 65 cycles at a frequency of 100,000 cycles, but only 9 cycles at 1000-cycle frequency. However, the 65 cycles are traversed in 650 micro-seconds, while the 9 cycles last 9000 microseconds, or 14 times as long. It is not the total time of oscillation, but the cumulative effect due to the numerous and only slowly decreasing successive waves, which increases as represented in Fig. 4.

An oscillation between copper wires No. 4 B. & S. G., 6 ft. apart, would give a duration curve, which at moderate frequencies follows (5) of Fig. 4, but at high frequencies follows (2). Thus the average duration and average rate of decay would be about the same as (3), an oscillation between copper wire No. 00 and ground. However, the oscillation would be more persistent in such a conducto at high, and less persistent at low frequencies. A complex wave, containing all the harmonics from low to very high ones, such as a steep wave front impulse or an approximately rectangular wave, as may be produced by a spark discharge, etc.; such a wave would have about the same average rate of decay in a copper wire No. 4 with return at 6 ft., as in a copper wire No. 00 with ground return. The wave front would flatten, and the wave round off, approach more and more sine shape, due to the more rapid disappearance of the higher frequencies, while at the same time decreasing in amplitude. The wave thus would pass through many intermediate shapes. But these intermediate shapes would be materially different with wire No. 4 and return at 6 ft., as with No. 00 and ground return; in the latter, the flattening of the steep wave front, and rounding of the wave, would be much more rapid at the beginning, due to the shorter duration of the transient, and while such wave would last about the same time, that is, pass over the lines to about the same distance, it would carry steep wave fronts to a much lesser distances, that is, its danger zone would be materially less than that of the wave in copper wire No. 4 with return at 6 ft.

It therefore, is of great interest to further investigate the effect of the changing attenuation constant on complex waves, and more particularly those with steep wave fronts, as the rectangular waves of starting or disconnecting lines, etc.

CHAPTER III.

ATTENUATION OF RECTANGULAR WAVES.

12. The destructiveness of high frequencies or step wave fronts in industrial circuits is rarely due to over-voltage between the circuit conductors or between conductor and ground, but is due to the piling up of the voltage locally, in inductive parts of the circuit, such as end turns of transformers or generators, current transformers, potential regulators, etc., or inside of inductive windings as the high potential coils of power transformers, by the formation of nodes and wave crests. Such effects may be produced by high frequency oscillations sustained over a number of cycles, as discussed in Chapter II, by oscillations lasting only a very few cycles or a fraction of a cycle, or due to non-oscillating transients, as single impulses, etc. As the high rate of change of voltage with the time, and the correspondingly high voltage gradients along the conductor are the source of danger, to calculate and compare oscillatory and non-oscillatory effects in this respect, it has become customary in the last years, to speak of an "equivalent frequency" of impulses, wave fronts or other non-oscillatory transients.

As "effective" or "equivalent" frequency of an impulse, wave front, etc., is understood the frequency of an oscillation, which has the same maximum amplitude, e or i, and the same maximum gradient $\frac{de}{dt}$ or $\frac{di}{dt}$. Thus assuming an impulse which reaches a maximum voltage $e = 60{,}000$, and has a maximum rate of increase of voltage of $\frac{de}{dt} = 10^{11}$, that is, a maximum voltage rise at the rate of 10^{11} volts per second, or 10,000 volts per micro-second. As the average voltage rise of a sine wave is $\frac{2}{\pi}$ times the maximum, the average rise of an oscillation of the same maximum gradient as the impulse, would be $\frac{2}{\pi}\frac{de}{dt} = \frac{20{,}000}{\pi}$ volts per micro-second and the total voltage rise of $e = 60{,}000$ thus would occur in $\frac{e}{\frac{2}{\pi}\frac{de}{dt}} = \frac{60{,}000\,\pi}{20{,}000} = 9.4$ micro-seconds. A complete cycle of this oscillation thus would last $4 \times 9.4 = 37.6$ micro-seconds, and the equivalent frequency of the impulse would be $f = \frac{10^6}{37.6} =$ 26,600 cycles or 26.6 kilo cycles. The equivalent frequency of a perfectly rectangular wave front, if such could exist, obviously would be infinity.

QUARTER WAVE CHARGING OR DISCHARGING OSCILLATION OF A LINE.

13. Considering first the theoretically rectangular wave of connecting a transmission line to a circuit, or disconnecting it from the circuit.

Suppose a transmission line, open at the distant end, is connected to a voltage E. At this moment, the voltage of the line is zero. It should be, in permanent conditions, E. Thus the circuit voltage consists of a permanent voltage E (which is the instantaneous value of the alternating supply voltage at this moment, $E_0 \sin \varphi$) and the transient voltage $-E$. We thus have a transient voltage, which is uniformly $= -E$ all along the line except at the switching point $l = 0$, where the transient voltage is zero.

Or, suppose a transmission line, open at the far end, is connected to a source of voltage, and at the moment where this voltage is E, the line short circuits at some point, by a spark discharge, flash-over, etc. Thus at this moment, the voltage $= 0$ at the point of short circuit, and is $= E$ everywhere between this point and the end of the line. Thus we get a line discharge leading to the same transient, a theoretically rectangular wave. In the part of the line between generator and short circuit, we have a different transient, a circuit of voltage $e = 0$ at one end, $e = E$ throughout the entire length at time $t = 0$, and $e = E$ continuously at the other end, where the generator maintains the voltage.

However, this again leads to the same transient, of a theoretically rectangular wave.

Assuming thus, as an instance, a transmission line of 100 km. length, of copper wire No. 00 B. & S. G., 30 ft. above ground, open circuited at the other end $l = 100$ km., and connected to a source of voltage E at the beginning, $l = 0$.

Then the beginning of the line, $l = 0$, is grounded, at the time $t = 0$, thus giving a quarter wave oscillation, with the terminal conditions:

Voltage along the line constant $= E$, at time $t = 0$, except at the beginning of the line, $l = 0$, where the voltage is 0.

Current along the line $= 0$ at $t = 0$, except at $l = 0$, where the current is indefinite.

14. The equation of the quarter wave oscillation of the line conductor against ground, then is (Part III, Chapter VII):

$$e = \frac{4E}{\pi} \epsilon^{-ut} \sum_0^{\infty}{}_n \frac{\sin (2n + 1)\tau \cos (2n + 1)\vartheta}{n + 1} \tag{1}$$

where:

ϑ is the time angle of the fundamental wave of oscillation, of frequency:

$$f_0 = \frac{S}{4l_0} = \frac{3 \times 10^{10}}{4 \times 100 \times 10^5} = 750 \text{ cycles.} \tag{2}$$

τ is the distance angle, for $l_0 = 100$ km. $= 90° = \frac{\pi}{2}$, that is,

$$\left.\begin{aligned} \vartheta &= 2\pi f_0 t \\ \tau &= \frac{\pi l}{2l_0} \end{aligned}\right\} \tag{3}$$

Equation (1), however, assumes that u, and thus r, L, C and g are constant for all frequencies. As this is not the case, but u is a function of the frequency, and thus of n : u_n, ϵ^{-ut} can not be taken out of the summation sign. Equation (1) thus must be written:

$$e = \frac{4E}{\pi} \sum_0^{\infty}{}_n \epsilon^{-u_n t} \frac{\sin (2n + 1) \tau \cos (2\pi + 1) \vartheta}{2n + 1} \tag{4}$$

where u_n is the value of u for the frequency: $f = (2\,n + 1)\,f_0$.

From (4) follows, as the voltage gradient along the line:

$$\frac{de}{dt} = \frac{4E}{\pi} \sum_0^\infty {}^n \epsilon^{-u_n t} \cos (2n + 1) \tau \cos (2n + 1) \vartheta \qquad (5)$$

$$= \frac{2E}{\pi} \left\{ \sum_0^\infty {}^n \epsilon^{-\mu_n t} \cos (2n + 1)(\tau + \vartheta) \right.$$

$$\left. + \sum_0^\infty {}^n \epsilon^{-\mu_n t} \cos (2n - 1) (\tau - \vartheta) \right\} \qquad (6)$$

The maximum voltage gradient occurs at the wave front, that is, for $\vartheta = \tau$. Substituting this, and substituting further, from (3):

$$d\tau = \frac{\pi}{2l_0} dl \qquad (7)$$

gives, as the maximum voltage gradient,

$$G = \frac{de}{dl} = \frac{E}{l_0} \left\{ \sum_0^\infty {}^n \epsilon^{-u_n t} \cos (2n + 1) 2\tau + \sum_0^\infty {}^u \epsilon^{-\mu_n t} \right\} \qquad (8)$$

It is, however,

$$\sum_0^\infty {}^n \cos (2n + 1) 2\tau = 0 \qquad (9)$$

for all values of τ except $\tau = 0$ and $\tau = \pi$, that is, the beginning of the line, $l = 0$.

Thus, approximately, as μ_n varies gradually:

$$\sum_0^\infty {}^n \epsilon^{-u_n t} \cos (2\pi + 1) 2\tau = 0 \qquad (10)$$

except for values of $\tau = 0$ or very near thereto.

Substituting (10) into (8) gives

$$G = \frac{E}{l_0} \sum_0^\infty {}^n \epsilon^{-u_n t} \qquad (11)$$

as the approximate expression of the maximum voltage gradient, that is, the steepness of the wave front, at time t, that is, at distance from the origin of the wave.

$$l = St = 3 \times 10^{10} t \qquad (12)$$

If ϑ differs materially from τ, the term with $(\tau - \vartheta)$ in equation (6) also vanishes, and $\frac{de}{dt} = 0$, that is, there is no voltage gradient except at and near the wave front.

From (11) are now calculated numerical values of the steepness of the wave front G, for various times t after its origin, and thus (by 12), various distances l from the origin. These numerical values are given in table VII, for $E = 60{,}000$ volts.

At a fundamental frequency of $f_0 = 750$ cycles, successive harmonics differ from each other by 1500 cycles, and for every value of t, values of $\epsilon^{-u_n t}$ thus have to be calculated for the frequencies:

$n =$	0	1	2	3	4	5	6	etc.
$f =$	750	2250	3750	5250	6750	8250	9750	cycles, etc.,

until the further terms add no further appreciable amount to $\Sigma\epsilon^{-u_n t}$. In calculating, it is found that for instance for $t = 5$ micro-seconds, or $l = 1.5$ km., this occurs at $f = 5 \times 10^6$ cycles, thus beyond the $\frac{5 \times 10^6}{750} = 6670$th harmonic. Thus 6670 terms of the series would have to be calculated to get this one point of the wave gradient: more terms for shorter, less terms for longer distances l of wave travel. This obviously is impossible, and some simpler approximation, of sufficient occuracy, thus is required.

This may be done as follows:

In the range from 5×10^5 to 10^6 cycles for instance, there are $\frac{10^6 - 5 \times 10^5}{1500} = 333$ harmonics. Instead of calculating u and ϵ^{-ut} for each of these harmonics, calculate ϵ^{-ut} for the average value of these 333 harmonics, and multiply by 333.

Thus, dividing the entire frequency range (beyond the lowest harmonics, which are calculated separately), into groups, and calculating one average value for each group, the calculation of $\Sigma\epsilon^{-ut}$ becomes feasible.

Since ϵ^{-ut} is calculated through $lg\epsilon^{-ut} = -utlg\epsilon$, as values of t, multiples of $\frac{1}{lg\epsilon}$ have been chosen, to still further simplify the calculation, in deriving a curve of gradients G.

15. Table VII gives the values of the maximum voltage gradient of the wave front, in volts per meter at 60,000 volts maximum initial line voltage, the equivalent frequency of the wave front, in kilocycles, and the length of the wave front, in meters, for various times of wave travel, from .03 micro-seconds up, and corresponding distances of wave travel, from 10 meters from the origin as rectangular wave, up to thousands of km., for copper wire No. 00 B. & S. G., 30 ft. = 910 cm. above ground, with the ground as return.

TABLE VII.—ATTENUATION OF WAVE FRONT OF QUARTER-WAVE OSCILLATION, OF 100 KM. LINE, 60,000 VOLTS.

Time t, Micro-seconds.	Distance l		f, Where Σ^{-u_i} Vanishes.	Voltage Gradient.		Wave Front.	
	Km.	Miles.		$\frac{1}{E}\frac{de}{dt}$	G, Volts per Meter.	Equivalent Kilo-cycles.	Length, Meters.
From origin of wave.							
(3)	Copper wire No. 00 B. & S. G., 30 ft. = 910 cm. above ground.						
0	0	0	....	∞	60,000	∞	0
0.03	0.01	0.0062	....	5,800	5,500	8,800	6
0.575	0.1725	0.107	10×10^6	1,370	1,290	2,060	73
1.15	0.345	0.215	8×10^6	1,025	967	1,540	98
2.3	0.69	0.43	6×10^6	670	630	1,000	150
11.5	3.45	2.15	4×10^6	300	280	450	330
23	6.9	4.3	2×10^6	218	205	330	460
92	27.6	17.2	10^6	104	98	156	960
230	69	43	0.7×10^6	62	58	92	1,630
2,300	690	430	0.2×10^6	13	12	19	7,900
23,000	6,900	4,300	50,000	1.3	1.2	1.9	79,000
(1)	Copper wire No. 00 B. & S. G., return conductor at 18 in. = 45.5 cm. distance.						
2.3	0.69	0.43	20×10^6	3,450	3,240	5,160	29
23	6.9	4.3	10×10^6	1,020	960	1,530	98
230	69	43	4×10^6	250	234	370	400
(5)	Copper wire No. 4 B. & S. G., 30 ft. = 910 cm. above ground.						
23	6.9	4.3	2×10^6	227	213	340	440
230	69	43	0.7×10^6	64	60	94	1,570
(4)	Iron wire No. 00 B. & S. G., 30 ft. = 910 cm. above ground.						
23	6.9	4.3	2×10^6	143	135	215	700
230	69	43	0.3×10^6	11	10	16	9,400

For comparison are given some data of the same conductor, with the return conductor at 18 in. = 45.5 cm., and also for a copper wire No. 4, and an iron wire No. 00, with the ground as return.

These data are plotted in Fig. 5, showing the wave front, as it gradually flattens out in its travel over the line, from the very steep wave at 170 meters from the origin, to the wave with a front of 1630 meters, 69 km. away.

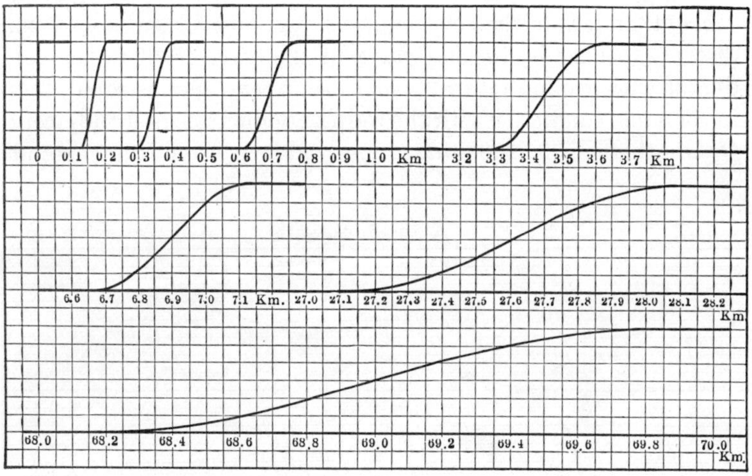

FIG. 5.

A comparison of the data of the four circuit conditions is given in Table VIII, and plotted in Fig. 6.

TABLE VIII.—ATTENUATION OF WAVE FRONT OF QUARTER-WAVE OSCILLATION, OF 100 KM. LINE, 60,000 VOLTS.

Conductor	(3)	(1)	(5)	(4)
Size No	00	00	4	00
Material	Copper	Copper	Copper	Iron
Distance of return conductor, cm	1820	45.5	1820	1820
After 23 microseconds, 6.9 km.:				
Gradient, volts per meter	205	960	213	135
Length of wave front, meters	460	98	440	700
Equivalent kilocycles	330	1530	340	215
After 230 microseconds, 69 km.:				
Gradient, volts per meter	58	234	60	10
Length of wave front, meters	1630	400	1570	9400
Equivalent kilocycles	92	370	94	16

It is interesting to note, that there is practically no difference in the flattening of the wave front on a low resistance conductor, No. 00, and a high resistance conductor, No. 4. There is, however, an enormous difference due to the effect of the closeness of the return conductor: with the return conductor at 18 in. distance, the wave front is still materially steeper at 6.9 km. distance, than it is in the conductor with ground return at 0.69 km. distance. That is, the flattening of the wave front in the conductor with ground return, is more than ten times as rapid, than in the same conductor with the return conductor closely adjacent. Or in other words, the danger zone of steep wave front, extends

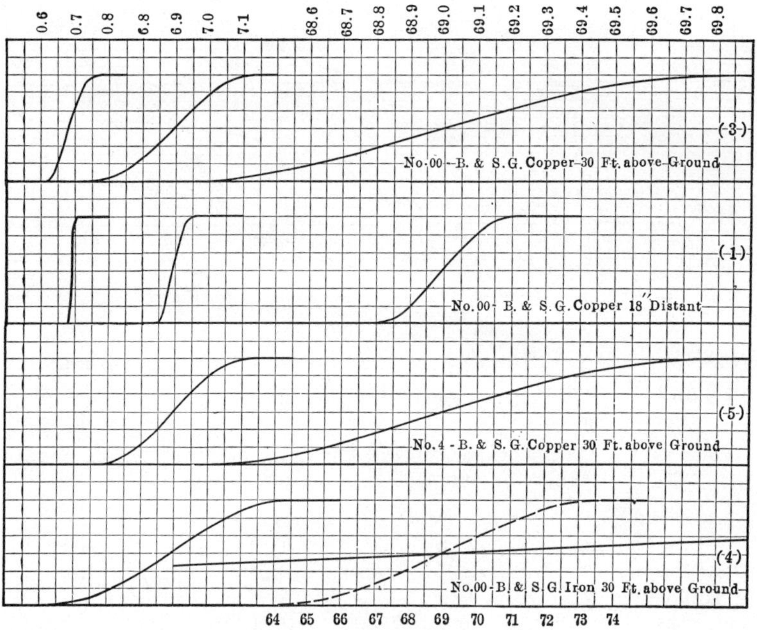

Fig. 6.

in a conductor with the return conductor closely adjacent, to more than ten times the distance than in the conductor with ground return.

This means, where it is desired to transmit a high-frequency impulse or steep wave front to the greatest possible distance, it is essential to arrange conductor and return conductor as closely adjacent as possible. But where it is essential to limit the harmful effect of very high frequency or steep wave front as much as possible to the immediate neighborhood of its origin, the return conductor should be separated as far as possible.

The data on iron wire are very disappointing: there is an enormous increase in the flattening of the wave front at great dis-

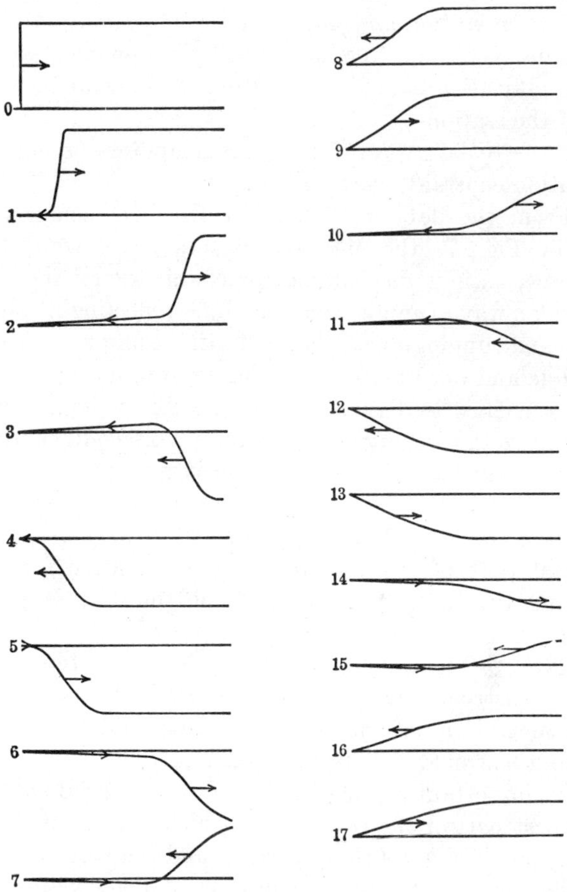

FIG. 7.

tances, by the use of iron as conductor material, so much so that the wave front of the iron conductor at 69 km. distance had to be shown (dotted) at one-tenth the scale as for the other conductors, in Fig. 6. But at moderate distances, 6.9 km. from the origin, the flattening of the wave front in the iron conductor is only little greater than in a copper conductor of the same size: 215 kilocycles against 330 kilocycles. At short distances, the difference almost entirely ceases, and within 1 km. from the origin, the wave front in an iron conductor is nearly as steep as in a copper conductor of the same size. Thus a short length of iron wire between station and line would exert practically no protection against very high-frequency oscillations or steep wave fronts, such as may be produced by lightning strokes in the neighborhood of the station.

This was to be expected from the shape of the curve of the attenuation constant, shown in Fig. 2.

16. From the data in Table VII then are constructed and shown in Fig. 7, the successive curves of voltage distribution in the line, as it is discharging (or charging), by the originally rectangular wave running over the line, reflecting at the end of the line and running back, then reflecting again at the beginning of the line and once more traversing it, etc., until gradually the transient energy is dissipated and the line voltage reaches its average, zero in discharge, or the supply voltage in charge. The direction of the wave travel in the successive positions is shown by the arrows in the center of the wave front; the existence and direction of current flow in the line by the arrows in the (nearly) horizontal part of the diagram. As seen, after four to five reflections, the voltage distribution in the line is practically sinoidal.

However, in these diagrams, Figs. 5 to 7, the wave front has been constructed from the maximum voltage gradient derived by the calculation, assuming as approximation the shape of the wave front as sinoid. This usually is a sufficient approximation, since the important feature is the maximum gradient, that is, the steepest part of the wave front, which was given by the calculation; but it is not strictly correct, and the wave front differs from sine shape. It thus is of interest to investigate the exact shape of the wave front, in its successful stages of flattening.

RECTANGULAR TRAVELING WAVE.

17. For this purpose may be investigated as further instance the gradual destruction of wave shape and decay of a 60,000-cycle rectangular traveling wave, during its passage over a transmission line, changing from the original rectangular wave shape produced at its origin by lightning stroke, spark discharge, etc., into practically a sine wave.

For this purpose, for the elementary symmetrical traveling wave, the equation is

$$e = \frac{4\,E}{\pi} \sum_{0}^{\infty} {}_{n}\, \epsilon^{-u_n t} \sin\,(2\,n + 1)\varphi \tag{13}$$

where $\varphi = \vartheta = \tau$ is the running time coordinate, in angular expression.

Values of e are calculated, for various times t, from 1 microsecond to 360 microseconds, for all the angles φ, where e has not yet become constant and equal to E.

Table IX.—Attenuation of 60,000-Cycle Rectangular Wave, in Line of No. 00 B. & S. G., Copper, 30 Ft. above Ground.

Time, t = Wave travel, l =	0 0	1 0.3	2 0.6	5 1.5	10 3	20 6	40 12	100 30	360 108	ms. km.
φ = 0 degrees	0.785	0	0	0	0	0	0	0	0	
φ = 1 degree	.....	0.250	0.170	0.106	0.074	0.052	0.347	0.022	0.010	
φ = 5 degrees	.....	0.785	0.645	0.475						
φ = 10 degrees	.....	.....	0.785	.....	0.606	0.468	0.582	0.217	0.100	
φ = 20 degrees	.....	.....	.....	0.783	0.770	0.707	0.709	0.405	0.193	
φ = 30 degrees	.....	.....	.....	.....	0.780	0.762	0.740	0.547	0.282	
φ = 40 degrees	.....	.....	.....	.....	.....	0.775	0.758	0.639	0.360	
φ = 50 degrees	.....	.....	.....	.....	.....	.....	0.765	0.691	0.426	
φ = 60 degrees	.....	.....	.....	.....	.....	.....	.....	0.722	0.476	
φ = 70 degrees	.....	.....	.....	.....	.....	.....	.....	0.730	0.513	
φ = 80 degrees	.....	.....	.....	.....	.....	.....	.....	.....	0.535	
φ = 90 degrees	0.785	0.785	0.785	0.783	0.780	0.775	0.765	0.730	0.540	
Wave Front — Degrees........	0	10	16	26	44	70	90	140	180	
Wave Front — Meters.........	0	140	220	340	620	980	1250	1940	2500	
Wave Front — Kilo-cycles. Max.	∞	1110	750	470	325	230	165	103	63	
Wave Front — Kilo-cycles. Avg.	∞	1080	670	420	268	155	120	77	60	
Highest apprec. harmonic.	∞	61	45	33	15	9	7	3	1	
Decrease of wave max....	0	.....	.....	.....	0.6	1.3	2.6	7.0	31.2	%
Maximum gradient, volts per meter............	∞	690	470	295	205	140	100	65	40	

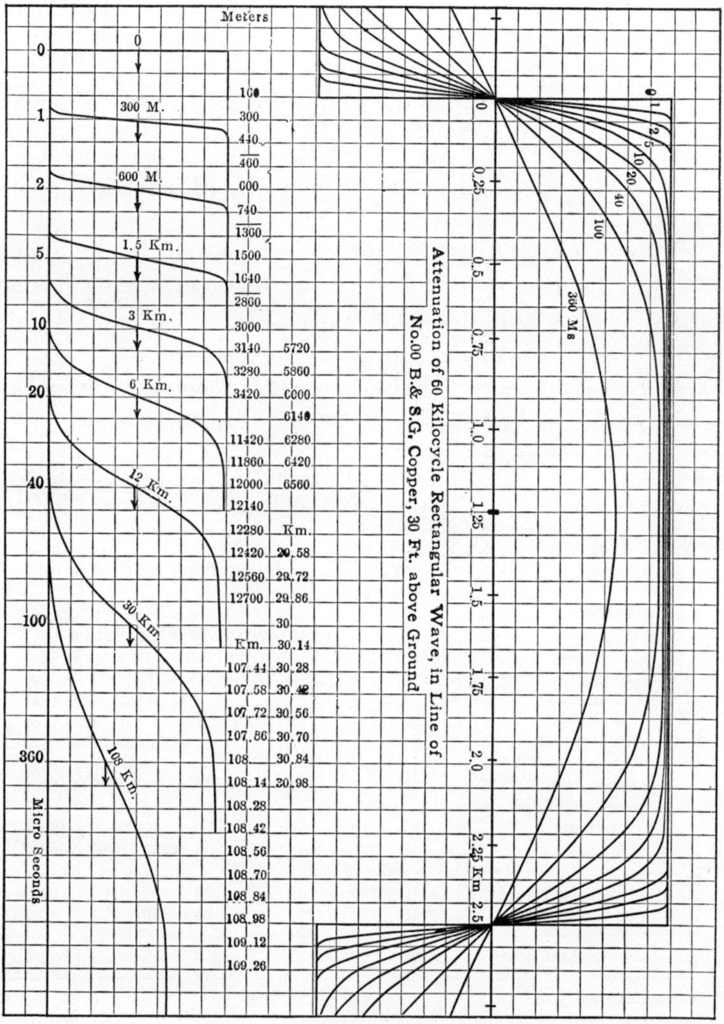
Attenuation of 60 Kilocycle Rectangular Wave, in Line of No.00 B.& S.G. Copper, 30 Ft. above Ground
Meters
Micro Seconds
300 M.
600 M.
1.5 Km.
3 Km.
6 Km.
12 Km.
30 Km.
108 Km.
360 Ms
2.25 Km

FIG. 8.

In Table IX are given numerical values, and plotted in Fig. 8, of

$$e \frac{\pi}{4 E} = \sum_0^\infty n \epsilon^{-u_n t} \sin (2 n + 1)\varphi,$$

from 0 to 90°, and from 1 to 360 microseconds, and derived therefrom the values of the length of wave front, in degrees and in meters, and the equivalent frequency, in kilocycles. Two values are given, the maximum kilocycles, derived from the steepest part of the wave front, and the average kilocycles, derived from the total wave front. The difference indicates the deviation of the shape of the wave front from a sinoid.

18. As seen, due to the high fundamental frequency, 60,000 cycles, the number of significant harmonics is very greatly reduced, until frequencies are reached, where the attenuation is so enormous, that the destruction of the wave by its energy dissipation occurs in a few meters.

Thus already after 1 microsecond or 300-meter wave travel, the calculation needs to extend only to the 61st harmonic; after 2 microseconds, to the 45th harmonic, etc., while after 100 microseconds, or 30-km. travel, only the third harmonic is still appreciable, and after 360 microseconds, or 108-km. travel, even this has disappeared, and the wave is essentially a sine wave.

In 30-km. travel, the wave maximum has decreased 7 per cent; in 108 km., it has decreased 31.2 per cent.

It is important to note, however, that waves of relatively high frequency, within the range of the danger frequencies between 20 and 100 kilocycles, can travel considerable distances, if no other causes of rapid attenuation are at work, but those considered here, and still retain a large part of their energy. Thus the 60,000-cycle wave, after traversing 100 km. of line, still retains about 70 per cent of its amplitude, that is, about one-half of its energy. Thus the danger from resonance of power transformer windings with such frequencies is not local but rather extends over a large part of the system.

The lower part of Fig. 8 shows the shape of the wave front at various distances from the origin; the upper part shows the gradual change of the wave from rectangular to flat top to sine wave.

Plotting for the 750-cycle quarter-wave oscillation and the 60,000-cycle traveling wave, the logarithm of the length of the wave front, and of the equivalent frequency of the wave front, against the logarithm of the distance of wave travel, gives practically straight lines (except for very great distances of wave travel, where the lower harmonics predominate), and from the slope of these lines follows that:

The length of wave front is approximately proportional, and the equivalent frequency of the wave front approximately inverse proportional to the square root of the distance of wave travel:

$$l_f = c_1 \sqrt{l} \tag{14}$$

$$f_f = \frac{c_2}{\sqrt{l}} \tag{15}$$

This would give a wave front constant c_1 and an equivalent frequency constant c_2 of the circuit.

CHAPTER IV

FLATTENING OF STEEP WAVE FRONTS

19. A rectangular wave is represented by the equation

$$e = \frac{4E}{\pi}\sum_0^{\infty}{}_n\, \epsilon^{-ut}\,\frac{\sin(2n+1)\tau\cos(2n+1)\vartheta}{2n+1} \tag{1}$$

where

$$\vartheta = 2\,\pi f_0 t = \text{time angle}, \tag{2}$$

$$\tau = 2\pi\frac{l}{l_0} = \text{distance angle}, \tag{3}$$

$$f_0 = \text{fundamental frequency},$$

$$l_0 = \frac{S}{f_0} = \text{wave length}, \tag{4}$$

$$S = 3 \times 10^{10} = \text{velocity of propagation},$$

$$E = \text{maximum voltage of wave}.$$

The voltage gradient is

$$\begin{aligned}\frac{de}{dl} &= \frac{de}{d\tau}\frac{d\tau}{dl}\\ &= \frac{4E}{\pi}\frac{d\tau}{dl}\sum_0^{\infty}{}_n\,\epsilon^{-ut}\cos(2n+1)\tau\cos(2n+1)\vartheta.\end{aligned} \tag{5}$$

Substituting for $\frac{d\tau}{dl}$ from (3), and resolving the cos-product, gives

$$\frac{de}{dl} = \frac{4\,E}{l_0}\left\{\sum_0^{\infty}{}_n\,\epsilon^{-ut}\cos(2n+1)(\tau-\vartheta) + \sum_0^{\infty}{}_n\,\epsilon^{-ut}\cos(2n+1)(\tau+\vartheta)\right\} \tag{6}$$

$$\sum_0^{\infty}{}_n\,\epsilon^{-ut}\cos(2n+1)(\tau+\vartheta) \text{ approaches zero for } \tau+\vartheta \neq 0, \tag{7}$$

thus,

$$\frac{de}{dl} = \frac{4\,E}{l_0}\sum_0^{\infty}{}_n\,\epsilon^{-ut}\cos(2n+1)(\tau-\vartheta). \tag{8}$$

The maximum voltage gradient occurs for $\vartheta = \tau$, and thus is

$$G = \frac{de}{dl} = \frac{4\,E}{l_0} \sum_0^\infty{}^n \epsilon^{-ut} \tag{9}$$

This can be written in the form

$$G = \frac{2\,E}{f_0\,l_0} \sum_0^\infty{}^n 2\,f_0\,\epsilon^{-ut} \tag{10}$$

It is, however, by (4),

$$f_0 l_0 = S \tag{11}$$

thus,

$$G = \frac{2\,E}{S} \sum_0^\infty{}^n 2\,f_0\,\epsilon^{-ut} \tag{12}$$

The values of ϵ^{-ut} are taken for all values of frequency, differing from each other by $2\,f_0$, and, at and near the wave front, the value of $\sum_0^\infty{}^n 2\,f_0\,\epsilon^{-ut}$ thus approaches the value of $\int_0^\infty \epsilon^{-ut} df$.

Substituting this into (10), gives as the equation of the maximum steepness of wave front

$$G = \frac{de}{dl} = \frac{2\,E}{S} \int_0^\infty \epsilon^{-ut}\,df. \tag{12}$$

As seen, in this expression (12), wave length and frequency have disappeared. Equation (12) thus applies to any wave, whether of finite length or not. That is, it represents broadly (though only approximately) the maximum gradient of any steep wave front or impulse, at the time t after its origin as rectangular wave front or impulse.

As u is not a simple function of f, the integration of $\int_0^\infty \epsilon^{-ut}\,df$ in its general form meets with difficulties, and the integral may be evaluated thusly:

Equation (12) can be written in the form

$$G = \frac{2\,E}{S} \int_{-\infty}^{+\infty} f\epsilon^{-ut}\,d\log f. \tag{13}$$

Plotting then $f\epsilon^{-ut}$ as ordinates, with $\log f$ as abscissæ, gives a curve, and the area of this curve gives the integral.

20. Instead of using $\log f$ as abscissæ, it is more convenient to use the common logarithm $\rceil\underline{f}$, and divide the measured area by $\rceil\underline{\epsilon} = 0.4343$.

Numerically, the integration is done by calculating $f\epsilon^{-ut}$ for (approximately) constant intervals of $\rceil\underline{f}$, adding all the values,

TABLE X.—ATTENUATION OF WAVE FRONT OF RECTANGULAR IMPULSE.
Copper Wire No. 00 B. & S. G., 30 Ft. above Ground, Ground Return.

f	u	$f\epsilon^{-ut}$ $t=10^{-3}$	10^{-4}	10^{-5}	10^{-6}	10^{-7}
1	70	1	1			
2	70	2	2			
5	70	5	5			
10	70	9	10			
20	70	19	20	88		
50	70	47	50			
10^2	70	93	99	100		
2×10^2	70	186	198	200		
5×10^2	70	468	496	500		
10^3	80	923	992	1,000	8,888	
2×10^3	115	1,760	1,970	2,000		
5×10^3	190	4,150	4,920	5,000		
10^4	287	7,500	2,720	9,999	10,000	
2×10^4	482	10,200	19,000	19,900	20,000	88,888
5×10^4	1,300	10,800	44,000	49,400	50,000	
10^5	3,710	2,450	69,000	96,400	99,500	100,000
2×10^5	12,690	1	56,100	176,000	197,000	200,000
5×10^5	74,200	0	240	190,000	466,000	498,000
10^6	291,500	....	0	54,300	748,000	970,000
2×10^6	1,160,000	....	....	18	625,000	1,780,000
5×10^6	7,230,000	....	....	0	3,660	1,860,000
10^7	28,800,000	....	....	....	1	562,000
2×10^7	115,000,000	....	....	....	0	200
	$\Sigma =$	38,660	206,800	605,000	2,230,000	6,060,000
	$\div 3 \times 0.4343 =$	29,700	158,700	464,000	1,710,000	4,650,000
	$G = \frac{2E}{S} \times i =$	0.119	0.635	1.86	6.84	18.6
	[$l =$	300	30	3	0.3	0.3 km.]

and multiplying the sum by the (average) difference between successive $\rceil\underline{f}$.

As an instance is given, in Table X, the calculation for circuit (3) of the preceding, that is, copper wire No. 00 B. & S. G., 30 ft. above ground, with the ground as return conductor, for the times $t = 0.1$, 1, 10, 100 and 1000 microseconds, corresponding to the distance of travel of the wave front of 0.03, 0.3, 3, 30 and 300 km. As frequency intervals are selected: 1, 2, 5, 10, 20, 50, 100, 200, etc., giving the logarithms: 0, 0.3, 0.6, 1, 1.3, etc.

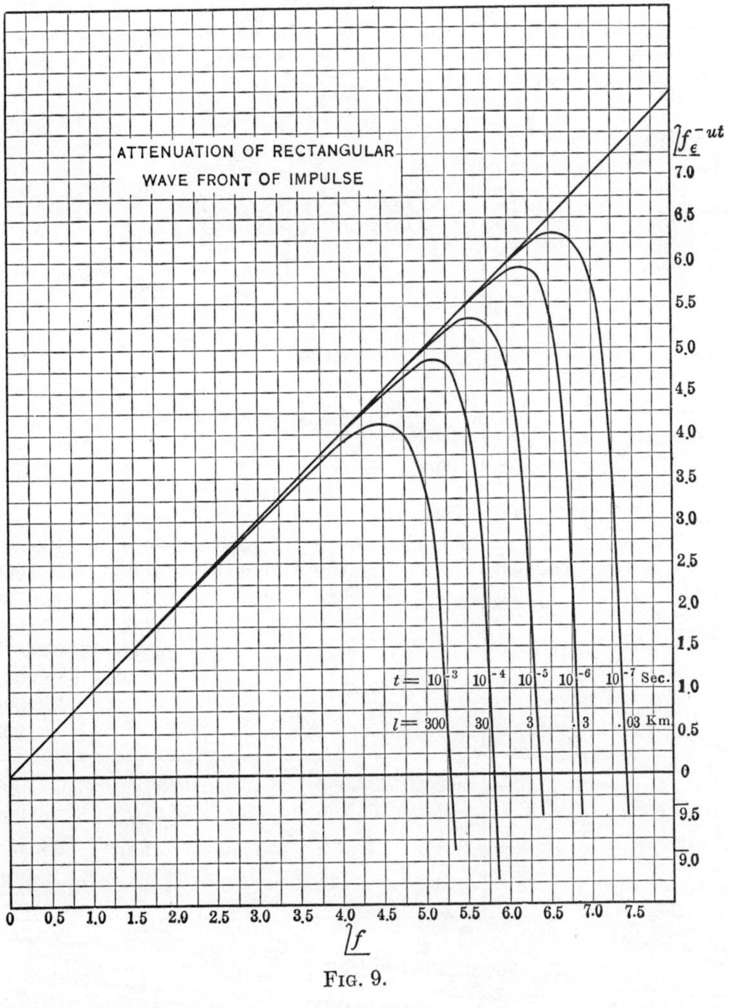

FIG. 9.

These curves are plotted, in logarithmic scale, in Fig. 9.

As seen, all these curves rise as straight lines under 45 degrees, and then very abruptly drop to negligible values.

In Table X, the values of $f\epsilon^{-ut}$ are added, then divided by 3, since there are three intervals for each unit of $\rceil\underline{f}$, and multiplied by $\rceil\underline{\epsilon}$, to reduce to natural logarithms Multiplying by $\frac{E}{S}$ then gives the gradient G.

For medium and high frequencies, the attenuation constant u is given by the preceding equations as

$$u = \frac{r_1 + r_3}{2\,(L_1 + L_0)} + \frac{g}{2\,C}. \tag{14}$$

Neglecting the internal inductance L_1, as small compared with the external inductance, this gives

$$\left.\begin{aligned} u &= \frac{\frac{\pi}{l_1}\sqrt{2.5\,\frac{\mu}{\lambda}}\,10^4}{\log\frac{l'}{l_r}}\sqrt{f} + \frac{4\,\pi^2 l'}{S\log\frac{l'}{l_r}}\,f^2 \\ &= m_1\sqrt{f} + m_2 f^2 \end{aligned}\right\} \tag{15}$$

where

$$m_1 = \frac{\pi\sqrt{2.5\,\frac{\mu}{\lambda}}}{l_1\log\frac{l'}{l_r}}\,10^4 \tag{16}$$

$$m_2 = \frac{4\,\pi^2 l'}{S\log\frac{l'}{l_r}} \tag{17}$$

For the conductor (31) in Table X, it is

$$\left.\begin{aligned} m_1 &= 2.6 \\ m_2 &= 0.29 \times 10^{-6} \end{aligned}\right\} \tag{18}$$

This expression (15) of u holds for all frequencies except very low frequencies—below 1000 cycles—and extremely high frequencies—many millions of cycles. The latter are of little importance, as they are wiped out in the immediate neighborhood of the origin of the rectangular impulse. At the low frequencies, the attenuation is so small, within the distances which come into consideration in the wave travel, and these low frequencies give such a small part of the wave front gradient, that the error made by the use of (15) is negligible. For instance, in the case of Table X, even at t = 100 microseconds, or 10 km. wave travel, the error made in the voltage gradient by altogether neglecting the attenuation of the frequencies up to 1000 cycles, would be only 0.01 per cent.

Thus the equation (15) of the attenuation constant can safely be used for all practical purposes.

As the first term in equation (15) is proportional to $\sqrt{f}$, the second term to f^2, the second term is negligible at low and medium frequencies, while the first term is negligible at high frequencies.

Both terms are equal at

$$f = \left(\frac{m_1}{m_2}\right)^{2/3} \tag{19}$$

That is, in the above instance, at

$$f = 43{,}000 \text{ cycles.}$$

21. Thus, for high frequencies, that is, within moderate distances from the origin of the rectangular impulse—up to some kilometers—the first term can be neglected and the attenuation constant expressed by

$$u = m_2 f^2 \tag{20}$$

The integral in equation (12) then becomes

$$F = \int_0^{\infty} \epsilon^{-m_2 t\, f^2}\, df. \tag{21}$$

Substituting,

$$x^2 = m_2 t f^2$$

gives

$$df = \frac{dx}{\sqrt{m_2 t}}$$

and

$$F = \frac{1}{\sqrt{m_2 t}} \int_0^{\infty} \epsilon^{-x^2}\, dx.$$

It is, however,

$$\int_0^{\infty} \epsilon^{-x^2}\, dx = \frac{1}{2}\sqrt{\pi} \tag{22}$$

thus,

$$F = \frac{1}{2}\sqrt{\frac{\pi}{m_2 t}} \tag{23}$$

and

$$G = \frac{E\sqrt{\pi}}{S\sqrt{m_2 t}} \tag{24}$$

or, since

$$l = St: \tag{25}$$

$$G = \frac{E\sqrt{\pi}}{\sqrt{m_2 Sl}} \tag{26}$$

Substituting (17) into (26), gives, in cm. and volts per cm.

$$\begin{aligned} G &= \frac{E}{2}\sqrt{\frac{\log \frac{l'}{l_r}}{\pi l' l}} \\ &= 0.282\, E \sqrt{\frac{1}{l'l} \log \frac{l'}{l_r}} \end{aligned} \tag{27}$$

Thus, approximately:

The maximum gradient, or steepness of the wave front of a rectangular impulse, in the neighborhood and at moderate distances from its origin, decreases inverse proportional with the square root of the distance or time of wave travel.

It decreases with increasing distance l' of the return conductor, nearly inverse proportional to the square root of l'.

22. For the six types of circuits considered in the previous instances, it is:

	$G =$	$\rceil\underline{G\sqrt{l'}} =$
(1) Copper wire 00 B. & S. G., 18 in. = 45.5 cm. from return conductor:	$\frac{1}{\sqrt{l}} \times 1700$	3.230
(2) Copper wire 00 B. & S. G., 6 ft. = 182 cm. from return conductor:	968	2.986
(3) Copper wire 00 B. & S. G., 60 ft. = 1820 cm. from return conductor:	360	2.556
(4) Iron wire 00 B. & S. G., 60 ft. = 1820 cm. from return conductor:	360	2.556
(5) Copper wire 4 B. & S. G., 60 ft. = 1820 cm. from return conductor:	373	2.572
(6) Aluminum, stranded, same conductivity and arrangement as (3):	353	2.548

where G is given in volts per meter, at $E = 60{,}000$ volts, and l in kilometers:

$$\left.\begin{aligned} G &= 89 \times 10^{-3}\, E \sqrt{\frac{1}{l'l} \log \frac{l'}{l_r}} \\ &= 135 \times 10^{-3}\, E \sqrt{\frac{1}{l'l} \rceil\underline{\frac{l'}{l_r}}} \end{aligned}\right\} \tag{28}$$

$$= 8100 \sqrt{\frac{1}{l'l} \rceil\underline{\frac{l'}{l_r}}}$$

From Tables X, IX, and VII are collected the values of wave gradients G, and given in Table XI, together with their logarithms and the $\rceil\underline{G_0}$, calculated from equation (28), for comparison of the different methods of calculation. Table XI then gives the difference Δ, and its value in per cent.

The values of G are plotted in Fig. 10. The drawn line gives the values calculated by equation (28); the three-cornered stars the values from Table X, the crosses the values from Table IX, and the circles the values from Table VII.

TABLE XI.—CALCULATION OF WAVE FRONT.

Copper Wire No. 00 B. & S. G., 30 Ft. above Ground.

Dist. l km.	Gradient, Volts per meter at E = 60,000 V.				
	G	$\log G$	$\log G_o$*	Δ	= %
	Impulse, fo = 0 (Table X).				
0.03	1860	3.269	3.317	+0.048	+11.7
0.3	684	2.835	2.817	−0.018	− 4.3
3	186	2.269	2.317	+0.048	+11.7
30	63.5	1.803	1.817	+0.014	+ 3.3
300	11.9	1.075	1.317	(+0.242)	
	Traveling Wave, fo = 60,000 (Table IX).				
0.3	690	2.839	2.817	−0.022	−5.2
0.6	470	2.672	2.667	−0.005	−1.1
1.5	295	2.470	2.468	−0.002	−0.5
3	205	2.312	2.317	+0.005	+1.1
6	140	2.146	2.167	+0.021	+5.0
12	100	2.000	2.016	+0.016	+3.7
30	65	1.813	1.817	+0.004	+0.9
108	40	1.602	1.539	(−0.063)	
	Quarter Wave, fo = 750 (Table VII).				
0.01	5500	3.740	3.556	−0.184	
0.1725	1290	3.111	2.938	−0.173	
0.345	967	2.985	2.787	−0.198	
0.69	630	2.799	2.637	+0.162	
3.45	280	2.447	2.287	−0.160	
6.9	205**	2.312	2.137	−0.175	
27.6	98	1.991	1.836	−0.155	
69	58	1.763	1.637	(−0.125)	
690	12	1.079	1.137	(+0.058)	
6900	1.2	0.079	0.637	(+0.558)	
	Same, 18 in. = 45.5 cm. between conductors.				
0.69	3240	3.510	3.311	−0.199	
6.9	960	2.982	2.811	−0.171	
69.0	234	2.369	2.311	(−0.058)	

* Calculated by equation (28).

** 131 to 293

23. As seen from Table XI and Fig. 10, the agreement of the equations (27) and (28) is satisfactory with the values of the wave front gradient taken from Tables X and IX.

The values of G from Table X differ erratically. This table was calculated by graphical integration, and it is probable that the intervals have been chosen too large, in that range where the curve drops very abruptly, as seen in Fig. 9.

The agreement with the values from Table IX is very close. In this table, representing the course of 60.000-cycle rectangular

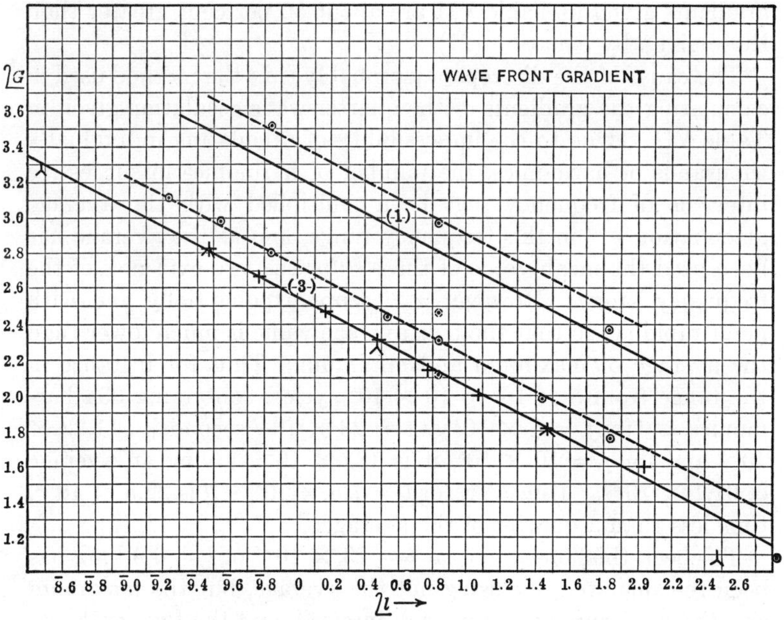

FIG. 10.

traveling wave, the individual harmonics have been calculated, as they were relatively few, due to the high frequency of the fundamental. This agreement is important as it justifies equation (27). In deriving (27), we have substituted integration for summation, that is, have replaced the discontinuous values of the individual harmonics by a continuous curve. Any error resulting from this should be greatest where the number of discontinuous harmonics is the least. Table XI and Fig. 10, however, show that the agreement of the gradient of the 60,000-cycle traveling wave with equation (27), is good even at $l = 30$ km., where only two significant harmonics are left, the fundamental and the third harmonic.

Thus the method of deriving an equation for G by integration is justified.

Unsatisfactory, however, is the agreement of equation (27) with the values of the gradient of the quarter-wave oscillation, taken from Table VII. These values lay on a straight line in Fig. 10, given as dotted line, showing proportionality with the square root of the distance, but there is a constant error, and the gradients given in Table VII are about 50 per cent greater than those given by equation (27).

The cause of the discrepancy probably is in the method used in calculating the gradients of the quarter-wave oscillation given in Table VII. Due to the impossibility of calculating individually thousands of harmonics, the number of harmonics in successive intervals of frequency, has been multiplied with the average attenuation ϵ^{-ut} in this frequency range, and as the average has been used the mean value of the ϵ^{-ut} for the two extremes of the frequency range. However, ϵ^{-ut} drops exponentially and with great rapidity, so that the true average value is much lower than the average of maximum and minimum. Thus for instance in the range from 5×10^5 to 10×10^5 cycles, containing 334 harmonics, ϵ^{-ut} for $f = 5 \times 10^5$ [at $t = 23 \times 10^{-6}$ seconds], is 1.22×10^{-3} and for $f = 10 \times 10^5$, it is 181×10^{-3}, giving an average of 91.1×10^{-3}. However, for $f = 7.5 \times 10^5$, ϵ^{-ut} is 22.5×10^{-3}, thus less than a quarter of the average.

To check this, one value, at $t = 23 \times 10^{-6}$ or $l = 6.9$ km., has been re-calculated, by using not the average, but the maximum and the minimum of ϵ^{-ut} in each interval, and the two gradients derived therefrom: $G = 293$ and 131, are marked with dotted circles in Fig. 10. As seen, the use of the minimum value of ϵ^{-ut} agrees nearer with equation (27), as was to be expected.

It appears probable that the equation (27) gives more reliable values of wave front gradient, within the range of its applicability, than the method of calculation used in Table VII, and as it is much simpler, it is preferable.

As seen from Table XI and Fig. 10, the parabolic law of wave front flattening, given by equation (27) and (28), holds good up to about 30 km. distance of wave travel, and with fair approximation even to 100 km. In the range beyond this—which is of lesser importance, as the flattening of the wave front has greatly reduced its danger at these distances—the values of the gradient G decrease with increasing rapidity, with distance and time, due to the medium high harmonics showing in the attenuation.

24. At great distances from the origin of the rectangular wave, when the very high harmonics have practically died out and the wave attenuation is determined by the medium frequency harmonics, the second term of u in equation (15) becomes negligible, and u can be approximated by

$$u = m_1\sqrt{f} \tag{37}$$

The integral in equation (12) then becomes

$$F = \int_0^\infty \epsilon^{-m_1t\sqrt{f}}\,df \tag{38}$$

Substituting,

$$m_1t\sqrt{f} = x$$

gives

$$df = \frac{2\ x\,dx}{m_1{}^2t^2}$$

and

$$F = \frac{2}{m_1{}^2t^2}\int_0^\infty \epsilon\, x^{-x}\,dx$$

It is, however,

$$\int_0^\infty x\,\epsilon^{-x}\,dx = 1 \tag{39}$$

thus,

$$F = \frac{2}{m_1{}^2t^2} \tag{40}$$

and

$$G = \frac{4\ E}{Sm_1{}^2t^2} \tag{41}$$

or,

$$G = \frac{4\ SE}{m_1{}^2l^2} \tag{42}$$

Substituting (16) gives

$$G = \frac{1.6\, S\lambda l_1^2 \left(\log \frac{l'}{l_r}\right)^2}{\pi^2 \mu l^2} E \tag{43}$$

That is, at great distances (or considerable time) from the origin, the flattening of the wave front approaches inverse proportionality with the square of the distance (or time) of the wave travel.

However, this range is of little importance.

APPENDIX

VELOCITY FUNCTIONS OF THE ELECTRIC FIELD

1. In the study of the propagation of the electric field through space (wireless telegraphy and telephony, lightning discharges and other very high-frequency phenomena), a number of new functions appear (Part II, Chapter VIII).

By the following equations these functions are defined, and related to the "Sine-Integral" Si x, the "Cosine-Integral" Ci x, and the "Exponential Integral," Ei x, of which tables were calculated by J. W. L. Glaisher (Philosophical Transactions of the Royal Society of London, 1870, Vol. 160):

[$\underline{|n}$ here denotes $1 \times 2 \times 3 \times \ldots \times n$]

$$\text{col } x = \int_x^\infty \frac{\sin u}{u}\, du$$

$$= \int_x^\infty \left\{ u - \frac{u^3}{\underline{|3}} + \frac{u^5}{\underline{|5}} - \frac{u^7}{\underline{|7}} + - \ldots \right\} \frac{du}{u}$$

$$= \Big/ u - \frac{1}{3}\frac{u^3}{\underline{|3}} + \frac{1}{5}\frac{u^5}{\underline{|5}} - \frac{1}{7}\frac{u^7}{\underline{|7}} + - \ldots \Big/_x^\infty$$

$$= \frac{\pi}{2} - \left\{ x - \frac{1}{3}\frac{x^3}{\underline{|3}} + \frac{1}{5}\frac{x^5}{\underline{|5}} - \frac{1}{7}\frac{x^7}{\underline{|7}} + - \ldots \right\}$$

$$= \cos x \left\{ \frac{1}{x} - \frac{\underline{|2}}{x^3} + \frac{\underline{|4}}{x^5} - \frac{\underline{|6}}{x^7} + - \ldots \right\}$$

$$+ \sin x \left\{ \frac{1}{x^2} - \frac{\underline{|3}}{x^4} + \frac{\underline{|5}}{x^6} - \frac{\underline{|7}}{x^8} + - \ldots \right\}$$

$$= \frac{\pi}{2} - \text{Si } x$$

$$= \frac{\pi}{2} - \int_0^x \frac{\sin u}{u}\, du.$$

Herefrom follows:

$$\int_x^\infty \frac{\sin au}{u}\, du = \text{col } ax$$

$$\begin{aligned}
\text{sil}\ x &= \int_x^\infty \frac{\cos u}{u} du \\
&= \int_x^\infty \left\{ 1 - \frac{u^2}{\underline{|2}} + \frac{u^4}{\underline{|4}} - \frac{u^6}{\underline{|6}} + - \dots \right\} \frac{du}{u} \\
&= \Big/ \log u - \frac{1}{2}\frac{u^2}{\underline{|2}} + \frac{1}{4}\frac{u^4}{\underline{|4}} - \frac{1}{6}\frac{u^6}{\underline{|6}} + - \dots \Big/_x^\infty \\
&= -\left\{ \gamma + \frac{1}{4}\log x^4 \right\} + \left\{ \frac{1}{2}\frac{x^2}{\underline{|2}} - \frac{1}{4}\frac{x^4}{\underline{|4}} + \frac{1}{6}\frac{x^6}{\underline{|6}} - \frac{1}{8}\frac{x^8}{\underline{|8}} + - \dots \right\} \\
&= \cos x \left\{ \frac{1}{x^2} - \frac{\underline{|3}}{x^4} + \frac{\underline{|5}}{x^6} - \frac{\underline{|7}}{x^8} + - \dots \right\} \\
&\qquad - \sin x \left\{ \frac{1}{x} - \frac{\underline{|2}}{x^3} + \frac{\underline{|4}}{x^5} - \frac{\underline{|6}}{x^7} + - \dots \right\} \\
&= -\text{Ci}\ (x) \\
&= -\int_\infty^x \frac{\cos u}{u} du;
\end{aligned}$$

where

$$\gamma = 0.5772156 \dots$$

herefrom follows:

$$\int_x^\infty \frac{\cos au}{u} du = \text{sil}\ ax$$

$$\begin{aligned}
\text{Expl}\ x &= \int_x^q \frac{\varepsilon^u}{u} du = \int_{-x}^\infty \frac{\varepsilon^{-u}}{u} du \\
&= \int_x^q \left\{ 1 + u + \frac{u^2}{\underline{|2}} + \frac{u^3}{\underline{|3}} + \frac{u^4}{\underline{|4}} + \dots \right\} \frac{du}{u} \\
&= \Big/ \log u + u + \frac{1}{2}\frac{u^2}{\underline{|2}} + \frac{1}{3}\frac{u^3}{\underline{|3}} + \frac{1}{4}\frac{u^4}{\underline{|4}} + \dots \Big/_x^q \\
&= -\left\{ \gamma + \frac{1}{4}\log x^4 + x + \frac{1}{2}\frac{x^2}{\underline{|2}} + \frac{1}{3}\frac{x^3}{\underline{|3}} + \frac{1}{4}\frac{x^4}{\underline{|4}} + \dots \right\} \\
&= \varepsilon^x \left\{ \frac{1}{x} + \frac{1}{x^2} + \frac{\underline{|2}}{x^3} + \frac{\underline{|3}}{x^4} + \frac{\underline{|4}}{x^5} + \dots \right\} \\
&= -\text{Ei}\ x \\
&= -\int_\infty^{-x} \frac{\varepsilon^{-u}}{u} du;
\end{aligned}$$

where

q is given by: expl $q = 0$ as: $q = 0.37249680\ldots$

herefrom follows:

$$\int_x^q \frac{\epsilon^{au}}{u}\,du = expl\; ax$$

$$\text{Expl}\,(-x) = \int_x^\infty \frac{\varepsilon^{-u}}{u}du$$

$$= \int_x^\infty \left\{1 - u + \frac{u^2}{\underline{|2}} - \frac{u^3}{\underline{|3}} + \frac{u^4}{\underline{|4}} - + \ldots\right\}\frac{du}{u}$$

$$= \Big/ \log u - u + \frac{1}{2}\frac{u^2}{\underline{|2}} - \frac{1}{3}\frac{u^3}{\underline{|3}} + \frac{1}{4}\frac{u^4}{\underline{|4}} - + \ldots \Big/_x^\infty$$

$$= -\left\{\gamma + \frac{1}{4}\log x^4\right\} + \left\{x - \frac{1}{2}\frac{x^2}{\underline{|2}} + \frac{1}{3}\frac{x^3}{\underline{|3}} - \frac{1}{4}\frac{x^4}{\underline{|4}} + - \ldots\right\}$$

$$= \varepsilon^{-x}\left\{\frac{1}{x} - \frac{1}{x^2} + \frac{\underline{|2}}{x^3} - \frac{\underline{|3}}{x^4} + \frac{\underline{|4}}{x^5} - + \ldots\right\}$$

$$= -\text{Ei}\,(-x)$$

$$= -\int_\infty^x \frac{\varepsilon^{-u}}{u}\,du.$$

herefrom follows:

$$\int_x^\infty \frac{\epsilon^{-au}}{u}\,du = expl\,(-\,ax)$$

Tables of these four functions, redured from the Glaisher tables, are given in the following for 6 decimals, and their first part plotted in Fig. 11.

2. As seen from the preceding equations, these functions have the following properties:

$\text{col}\,0 = \frac{\pi}{2}$	$\text{col}\,\infty = 0$
$\text{sil}\,0 = +\infty$	$\text{sil}\,\infty = 0$
$\text{expl}\,0 = +\infty$	$\text{expl}\,(+\infty) = -\infty$
$\text{expl}\,q = 0$	$\text{expl}\,(-\infty) = 0$

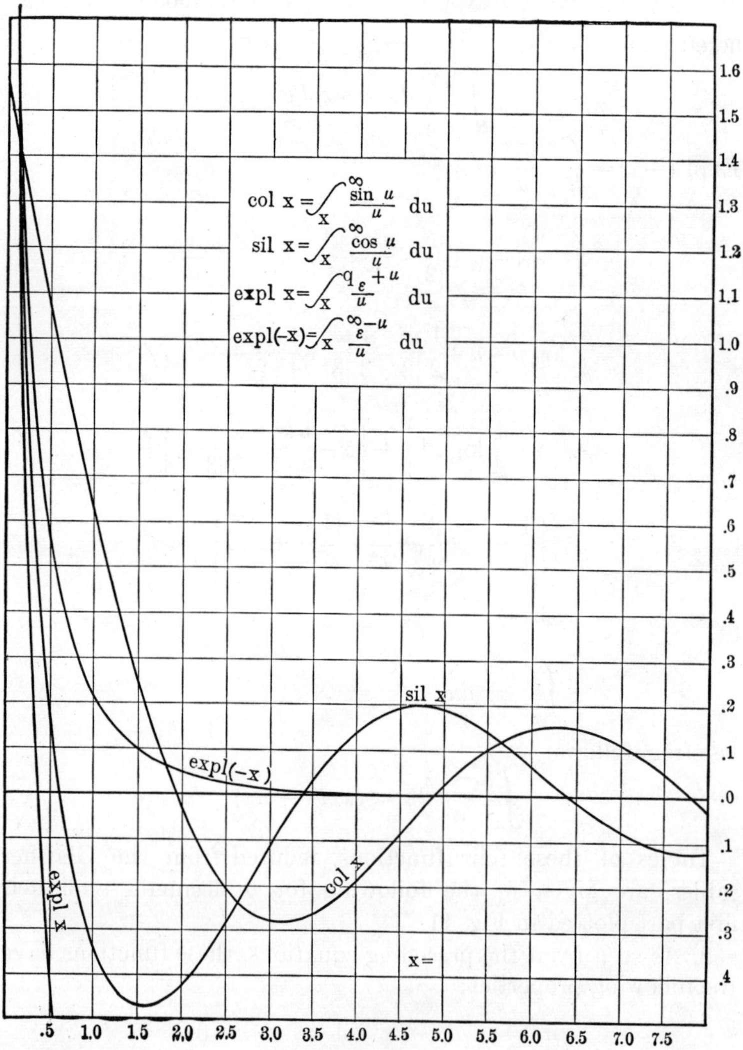

col x = ∫_x^∞ (sin u / u) du
sil x = ∫_x^∞ (cos u / u) du
expl x = ∫_x^q (ε^{+u} / u) du
expl(−x) = ∫_x^∞ (ε^{−u} / u) du
sil x
expl(−x)
col x
expl x
x=
1.6
1.5
1.4
1.3
1.2
1.1
1.0
.9
.8
.7
.6
.5
.4
.3
.2
.1
.0
.1
.2
.3
.4
.5
1.0
1.5
2.0
2.5
3.0
3.5
4.0
4.5
5.0
5.5
6.0
6.5
7.0
7.5

FIG. 11.

col x has maxima at the even, sil x at the odd quadrants, and these maxima are alternately positive and negative; that is,

$$\left.\begin{aligned} \operatorname{col}\frac{\pi}{2}2s &= \max. \\ \operatorname{sil}\frac{\pi}{2}(2s-1) &= \max. \end{aligned}\right\} s = \text{integer number.}$$

For large values of s, the numerical values of these maxima approach the values:

$$\operatorname{col}\frac{\pi}{2}\,2s = (-1)^s\frac{1}{\pi s};$$

$$\operatorname{sil}\frac{\pi}{2}(2s-1) = (-1)^s\frac{2}{\pi(2s-1)},$$

for: $s > 40$, the approximation is correct to the 6th decimal.

From the series expressions of these functions follows:

$$\operatorname{col}(-x) = \pi - \operatorname{col} x$$

$$\operatorname{sil}(-x) = \operatorname{sil} x$$

$$\operatorname{expl}(jx) = \operatorname{sil} x + j\operatorname{col} x - j\frac{\pi}{2}$$

$$\operatorname{expl}(-jx) = \operatorname{sil} x - j\operatorname{col} x + j\frac{\pi}{2}$$

$$\operatorname{col}(\pm jx) = \pm\frac{j}{2}\{\operatorname{expl} x - \operatorname{expl}(-x)\} + j\pi$$

$$\operatorname{sil}(\pm jx) = \frac{1}{2}\{\operatorname{expl} x + \operatorname{expl}(-x)\}$$

For small values of x, the approximations hold:

$$\begin{aligned} \operatorname{sil} x &= \log\frac{1}{x} - 0.5772156 + \frac{x^2}{4} \\ &= \log\frac{0.5615}{x} + \frac{x^2}{4} \end{aligned}$$

$$\operatorname{col} x = \frac{\pi}{2} - x$$

These approximations are accurate within one per cent for values of x up to 0.67.

For very small values of x, the approximations hold:

$$\text{sil } x = \log \frac{1}{x} - 0.5772156$$

$$= \log \frac{0.5615}{x}$$

$$\text{col } x = \frac{\pi}{2}$$

These are accurate within one per cent for values of x up to 0.016.

For moderately large values of x the approximations hold:

$$\text{sil } x = -\frac{\sin x}{x} + \frac{\cos x}{x^2}$$

$$\text{col } x = \frac{\cos x}{x} + \frac{\sin x}{x^2}$$

These are accurate within one per cent for values of x down to 8.

For large values of x, the approximations hold:

$$\text{sil } x = -\frac{\sin x}{x}$$

$$\text{col } x = \frac{\cos x}{x}$$

or:

$$\text{sil } x = -\frac{\sin v}{x}$$

$$\text{col } x = \frac{\cos v}{x}$$

where:

$$x = 2\,\pi u + v$$

and $v < 2\pi$, that is, u is the largest multiple of 2π, contained in x.

These approximations are accurate within one per cent for values of x down to 12.

In some problems on the velocity of propagation of the electric field through space, such as the mutual inductance of two finite conductors at considerable distance from each other, or the capacity of a sphere in space, two further functions appear, coll x and sill x, which by partial integration can be reduced to the functions col x and sil x.

It is:

$$\text{coll } x = \int_x^\infty \frac{\cos u}{u^2}\, du$$

$$= -\int_x^\infty \cos u \; d\left(\frac{1}{u}\right)$$

$$= \frac{\cos x}{x} - \text{col } x$$

$$\text{sill } x = \int_x^\infty \frac{\sin u}{u^2}\, du$$

$$= -\int_x^\infty \sin u \; d\left(\frac{1}{u}\right)$$

$$= \frac{\sin x}{x} + \text{sil } x$$

Herefrom follows:

$$\int_x^\infty \frac{\cos au}{u^2}\, du = a \text{ coll } ax$$

$$= \frac{\cos ax}{x} - a \text{ sill } ax$$

$$\int_x^\infty \frac{\sin au}{u^2}\, du = a \text{ sill } ax$$

$$= \frac{\sin ax}{x} + a \text{ sil } ax$$

It is:

$$\text{coll } o = \infty \qquad \text{coll } \infty = o$$
$$\text{sill } o = \infty \qquad \text{sill } \infty = o$$

With increasing values of x, coll x and sill x decrease far more rapidly than col x and sil x.

For moderately large values of x, the approximations hold:

$$\text{coll } x = -\frac{\sin x}{x^2} - \frac{2 \cos x}{x^3}$$

$$\text{sill } x = \frac{\cos x}{x^2} + \frac{2 \sin x}{x^3}$$

These approximations are accurate within one per cent for values of x up to respectively

For large values of x, the approximations hold:

$$\text{coll } x = -\frac{\sin x}{x^2}$$

$$\text{sill } x = \frac{\cos x}{x^2}$$

These are accurate within one per cent for values of x up to respectively 12.

For low and medium values of x, coll x and sill x are preferably reduced to col x and sil x by above given equations.

TABLE I

Col x and sil x from 0.00 to 1.00

x	col x	Δ	sil x	Δ
.00	+1.570 796	....	∞	....
		10 000		∞
.01	1.560 796		+4.027 980	
		9 999		693 073
.02	1.550 797		3.334 907	
		9 999		405 340
.03	1.540 798		2.929 567	
		9 998		287 507
.04	1.530 800		2.642 060	
		9 997		222 918
.05	1.520 803		2.419 142	
		9 995		182 047
.06	1.510 808		2.237 095	
		9 993		153 826
.07	1.500 815		2.083 269	
		9 990		133 156
.08	1.490 825		1.950 113	
		9 988		117 359
.09	1.480 837		1.832 754	
		9 985		104 886
.10	1.470 852		1.727 868	
		9 982		94 785
.11	1.460 870		1.633 083	
		9 978		86 437
.12	1.450 892		1.546 646	
		9 974		79 419
.13	1.440 918		1.467 227	
		9 969		73 434
.14	1.430 949		1.393 793	
		9 965		68 269
.15	1.420 984		1.325 524	
		9 960		63 765
.16	1.411 024		1.261 759	
		9 955		59 802
.17	1.401 069		1.201 957	
		9 949		56 285
.18	1.391 120		1.145 672	
		9 943		53 145
.19	1.381 177		1.092 527	
		9 937		50 321
.20	1.371 240		1.042 206	
		9 930		47 769
.21	1.361 310		.994 437	
		9 923		45 449
.22	1.351 387		.948 988	
		9 916		43 332
.23	1.341 471		.905 656	
		9 908		41 390
.24	1.331 563		.864 266	
		9 900		39 603
.25	1.321 663		.824 663	
		9 892		37 953
.26	1.311 771		.786 710	
		9 883		36 423
.27	1.301 888		.750 287	
		9 875		35 001
.28	1.292 013		.715 286	
		9 865		33 676
.29	1.282 148		.681 610	
		9 856		32 437
.30	1.272 292		.649 173	
		9 845		31 277
.31	1.262 447		.617 896	
		9 836		30 186
.32	1.252 611		.587 710	
		9 825		29 161
.33	1.242 786		.558 549	
		9 814		28 194
.34	1.232 972		.530 355	
		9 803		27 279
.35	1.223 169		.503 076	
		9 791		26 414
.36	1.213 378		.476 661	
		9 779		25 594
.37	1.203 599		.451 067	
		9 767		24 815
.38	1.193 832		.426 252	
		9 755		24 074
.39	1.184 077		.402 178	
		9 742		23 369
.40	1.174 335		.378 809	
		9 729		22 695
.41	1.164 606		.356 114	
		9 715		22 052
.42	1.154 891		.334 062	
		9 702		21 437
.43	1.145 189		.312 625	
		9 688		20 849
.44	1.135 501		.291 776	
		9 673		20 284
.45	1.125 828		.271 492	
		9 658		19 743
.46	1.116 170		.251 749	
		9 644		19 223
.47	1.106 526		.232 526	
		9 628		18 723
.48	1.096 898		.213 803	
		9 613		18 241
.49	1.087 285		.195 562	
		9 596		17 778
.50	+1.077 689		+0.177 784	

TABLE I—*Continued*

x	col x	Δ	sil x	Δ
.50	+1.077 689	9 596	+0.177 784	17 778
.51	1.068 108	9 581	.160 453	17 331
.52	1.058 545	9 563	.143 554	16 899
.53	1.048 998	9 547	.127 071	16 483
.54	1.039 448	9 530	.110 990	16 081
.55	1.029 956	9 512	.095 300	15 690
.56	1.020 461	9 495	.079 986	15 314
.57	1.010 985	9 476	.065 037	14 949
.58	1.001 527	9 458	.050 441	14 596
.59	.992 088	9 439	.036 190	14 251
.60	.982 667	9 421	.022 271	13 919
.61	.973 266	9 401	+ .008 675	13 516
.62	.963 885	9 381	− .004 606	13 281
.63	.954 523	9 362	− .017 582	12 976
.64	.945 182	9 341	− .030 260	12 678
.65	.935 861	9 321	− .042 650	12 390
.66	.926 561	9 300	− .054 758	12 108
.67	.917 282	9 279	− .066 591	11 833
.68	.908 024	9 258	− .078 158	11 567
.69	.898 788	9 236	− .089 463	11 305
.70	.889 574	9 214	− .100 515	11 052
.71	.880 382	9 192	− .111 318	10 803
.72	.871 213	9 169	− .121 879	10 561
.73	.862 066	9 147	− .132 203	10 324
.74	.852 942	9 124	− .142 296	10 093
.75	.843 842	9 100	− .152 164	9 868
.76	.834 765	9 077	− .161 810	9 646
.77	.825 713	9 052	− .171 240	9 430
.78	.816 684	9 029	− .180 458	9 218
.79	.807 680	9 004	− .189 470	9 012
.80	.798 700	8 980	− .198 279	8 809
.81	.789 746	8 954	− .206 889	8 601
.82	.780 817	8 929	− .215 305	8 416
.83	.771 913	8 904	− .223 530	8 225
.84	.763 035	8 878	− .231 568	8 038
.85	.754 184	8 851	− .239 423	7 855
.86	.745 358	8 825	− .247 098	7 675
.87	.736 560	8 798	− .254 597	7 499
.88	.727 788	8 772	− .261 923	7 326
.89	.719 043	8 745	− .269 079	7 156
.90	.710 325	8 718	− .276 068	6 989
.91	.701 636	8 689	− .282 893	6 825
.92	.692 974	8 662	− .289 558	6 665
.93	.684 340	8 634	− .296 064	6 506
.94	.675 735	8 605	− .302 415	6 351
.95	.667 158	8 577	− .308 614	6 199
.96	.658 610	8 548	− .314 662	6 048
.97	.650 092	8 518	− .320 563	5 901
.98	.641 602	8 490	− .326 319	5 756
.99	.633 143	8 459	− .331 931	5 612
1.00	+0.624 713	8 430	−0.337 404	5 473

TABLE II

Expl x and expl $(-x)$ from 0.00 to 1.00

x	expl x	Δ	expl $(-x)$	Δ
.00	$+\infty$	...	∞	...
.01	+4.017 929	∞	+4.037 930	∞
.02	3.314 707	703 223	3.354 708	684 222
.03	2.899 116	415 591	2.959 119	395 589
.04	2.601 257	297 859	2.681 264	277 855
.05	2.367 885	233 372	2.467 898	213 366
.06	2.175 283	192 602	2.295 307	172 591
.07	2.010 800	164 483	2.150 838	144 469
.08	1.886 884	143 916	2.026 941	123 897
.09	1.738 664	128 220	1.918 745	108 196
.10	1.622 813	115 851	1.822 924	95 821
.11	1.516 959	105 854	1.737 107	85 817
.12	1.419 350	97 609	1.659 542	77 565
.13	1.328 655	90 695	1.588 899	70 643
.14	1.243 841	84 814	1.524 146	64 753
.15	1.164 086	79 754	1.464 462	59 684
.16	1.088 731	75 355	1.409 187	55 275
.17	1.017 234	71 497	1.357 781	51 406
.18	.949 148	68 087	1.309 796	47 985
.19	.884 096	65 052	1.264 858	44 938
.20	.821 761	62 335	1.222 651	42 207
.21	.761 872	59 889	1.182 902	39 749
.22	.704 195	57 676	1.145 380	37 522
.23	.648 529	55 666	1.109 883	35 497
.24	.594 697	53 832	1.076 235	33 648
.25	.542 543	52 153	1.044 283	31 952
.26	.491 932	50 611	1.013 889	30 394
.27	.442 741	49 191	.984 933	28 956
.28	.394 863	47 878	.957 308	27 625
.29	.348 202	46 662	.930 918	26 390
.30	.302 669	45 533	.905 677	25 241
.31	.258 186	44 482	.881 506	24 171
.32	.214 683	43 503	.858 335	23 171
.33	.172 095	42 588	.836 101	22 234
.34	.130 363	41 732	.814 746	21 355
.35	.089 434	40 929	.794 215	20 531
.36	.049 258	40 176	.774 462	19 753
.37	+ .009 790	39 468	.755 441	19 021
.38	− .029 011	38 801	.737 112	18 329
.39	− .067 185	38 173	.719 437	17 675
.40	− .104 765	37 581	.702 380	17 057
.41	− .141 786	37 021	.685 910	16 470
.42	− .178 278	36 492	.669 997	15 913
.43	− .214 270	35 991	.654 613	15 384
.44	− .249 787	35 517	.639 733	14 880
.45	− .284 855	35 068	.625 331	14 402
.46	− .319 497	34 642	.611 387	13 944
.47	− .353 735	34 238	.597 877	13 510
.48	− .387 589	33 854	.584 784	13 093
.49	− .421 078	33 489	.572 089	12 695
.50	−0.454 220	33 142	+0.559 774	12 315

TABLE II—*Continued*

x	expl x	Δ	expl $(-x)$	Δ
.50	−0.454 220	33 142	+0.559 774	12 315
.51	− .487 032	32 812	.547 822	11 952
.52	− .519 531	32 498	.536 220	11 602
.53	− .551 730	32 200	.524 952	11 268
.54	− .583 646	31 915	.514 004	10 948
.55	− .615 291	31 645	.503 364	10 640
.56	− .646 677	31 387	.493 020	10 344
.57	− .677 819	31 141	.482 960	10 060
.58	− .708 726	30 907	.473 173	9 787
.59	− .739 410	30 684	.463 650	9 523
.60	− .769 881	30 472	.454 380	9 270
.61	− .800 150	30 269	.445 353	9 027
.62	− .830 226	30 076	.436 562	8 791
.63	− .860 119	29 892	.427 997	8 565
.64	− .889 836	29 717	.419 652	8 345
.65	− .919 386	29 550	.411 517	8 135
.66	− .948 778	29 392	.403 586	7 931
.67	− .978 019	29 241	.395 853	7 733
.68	−1.007 116	29 079	.388 309	7 544
.69	−1.036 077	28 960	.380 950	7 359
.70	−1.064 907	28 831	.373 769	7 181
.71	−1.093 615	28 707	.366 760	7 009
.72	−1.122 205	28 590	.359 918	6 842
.73	−1.150 684	28 479	.353 237	6 681
.74	−1.179 058	28 374	.346 713	6 524
.75	−1.207 333	28 275	.340 341	6 372
.76	−1.235 513	28 181	.334 115	6 226
.77	−1.263 605	28 092	.328 032	6 083
.78	−1.291 613	28 008	.322 088	5 944
.79	−1.319 542	27 929	.316 277	5 811
.80	−1.347 397	27 855	.310 597	5 680
.81	−1.375 182	27 785	.305 043	5 554
.82	−1.402 902	27 720	.299 611	5 432
.83	−1.430 561	27 659	.294 299	5 312
.84	−1.458 164	27 603	.289 103	5 196
.85	−1.485 714	27 550	.284 019	5 084
.86	−1.513 216	27 502	.279 045	4 974
.87	−1.540 673	27 457	.274 177	4 868
.88	−1.568 089	27 416	.269 413	4 764
.89	−1.595 467	27 379	.264 749	4 664
.90	−1.622 812	27 345	.260 184	4 565
.91	−1.650 126	27 314	.255 714	4 470
.92	−1.677 413	27 287	.251 336	4 378
.93	−1.704 677	27 264	.247 050	4 286
.94	−1.731 920	27 243	.242 851	4 199
.95	−1.759 146	27 226	.238 738	4 113
.96	−1.786 357	27 211	.234 708	4 030
.97	−1.813 557	27 200	.230 760	3 948
.98	−1.840 749	27 192	.226 891	3 869
.99	−1.867 935	27 186	.223 100	3 791
1.00	−1.895 118	27 183	+0.219 384	3 716

TABLE III

Col x and sil x from 0.0 to 5.0

x	col x	Δ	sil x	Δ
.0	+1.570 796	…	∞	…
.1	1.470 852		+1.727 868	
.2	1.371 240		1.042 206	
.3	1.272 292		.649 173	
.4	1.174 335		.378 809	
.5	1.077 689	…	.177 784	…
.6	.982 667		+ .022 271	
.7	.889 574		− .100 515	
.8	.798 700		− .198 279	
.9	.710 325		− .276 068	
1.0	.624 713	+82 602	− .337 404	+47 469
1.1	.542 111	79 362	− .384 873	35 586
1.2	.462 749	75 911	− .420 459	25 279
1.3	.387 838	72 269	− .445 739	16 268
1.4	.314 570	68 457	− .462 007	8 349
1.5	.246 113	64 497	− .470 356	+ 1 377
1.6	.181 616	60 412	− .471 733	− 4 765
1.7	.121 204	56 225	− .466 968	−10 157
1.8	.064 979	51 958	− .456 811	−14 871
1.9	+ .013 021	47 638	− .441 940	−18 959
2.0	− .034 617	43 285	− .422 981	−22 469
2.1	− .077 902	38 927	− .400 512	−25 437
2.2	− .116 839	34 582	− .375 075	−27 899
2.3	− .151 411	30 278	− .347 176	−29 884
2.4	− .181 689	26 035	− .317 292	−31 421
2.5	− .207 724	21 874	− .285 871	−32 534
2.6	− .229 598	17 818	− .253 337	−33 252
2.7	− .247 416	13 884	− .220 085	−33 597
2.8	− .261 300	10 094	− .186 488	−33 593
2.9	− .271 394	6 462	− .152 895	−33 265
3.0	− .277 856	+ 3 007	− .119 630	−32 638
3.1	− .280 863	− 258	− .086 992	−31 735
3.2	− .280 605	− 3 319	− .055 257	−30 579
3.3	− .277 284	− 6 166	− .024 678	−29 196
3.4	− .271 118	− 8 789	+ .004 518	−27 610
3.5	− .262 329	−11 177	.032 129	−25 846
3.6	− .251 152	−13 323	.057 974	−23 927
3.7	− .237 825	−15 231	.081 901	−21 877
3.8	− .222 594	−16 889	.103 778	−19 721
3.9	− .205 705	−18 298	.123 499	−17 483
4.0	− .187 407	−19 460	.140 982	−15 183
4.1	− .167 947	−20 375	.156 165	−12 848
4.2	− .147 572	−21 048	.169 013	−10 497
4.3	− .126 524	−21 486	.179 510	− 8 150
4.4	− .105 038	−21 694	.187 660	− 5 831
4.5	− .083 344	−21 680	.193 491	− 3 556
4.6	− .061 664	−21 455	.197 047	− 1 344
4.7	− .040 209	−21 030	.198 391	+ 787
4.8	− .019 179	−20 416	.197 604	2 824
4.9	+ .001 237	−19 628	.194 780	+ 4 750
5.0	+0.020 865		+0.190 030	

TABLE IV

expl x and expl $(-x)$ from 0.0 to 5.0

x	expl x	Δ	expl $(-x)$	Δ
.0	+ ∞	....	∞	....
.1	+ 1.622 813		+1.822 924	
.2	+ .821 761		1.222 651	
.3	+ .302 669		.905 677	
.4	− .104 765		.702 380	
.5	− .454 220	....	.559 774	....
.6	− .769 881		.454 380	
.7	− 1.064 907		.373 769	
.8	− 1.347 397		.310 597	
.9	− 1.622 812		.260 184	
1.0	− 1.895 118	....	.219 384	....
1.1	− 2.167 378	272 260	.185 991	33 393
1.2	− 2.442 092	274 714	.158 408	27 583
1.3	− 2.721 399	279 306	.135 451	22 957
1.4	− 3.007 207	285 809	.116 219	19 232
1.5	− 3.301 285	294 078	.100 020	16 199
1.6	− 3.605 320	304 034	.086 308	13 712
1.7	− 3.920 963	315 643	.074 655	11 653
1.8	− 4.249 868	328 904	.064 713	9 942
1.9	− 4.593 714	343 846	.056 204	8 509
2.0	− 4.954 234	360 521	.048 901	7 303
2.1	− 5.333 235	379 001	.042 614	6 287
2.2	− 5.732 615	399 379	.037 191	5 423
2.3	− 6.154 381	421 766	.032 502	4 689
2.4	− 6.600 670	446 289	.028 440	4 062
2.5	− 7.073 766	473 096	.024 915	3 525
2.6	− 7.576 115	502 349	.021 850	3 065
2.7	− 8.110 347	534 233	.019 182	2 668
2.8	− 8.679 298	568 950	.016 855	2 327
2.9	− 9.286 024	606 726	.014 824	2 031
3.0	− 9.933 833	647 808	.013 048	1 776
3.1	−10.626 300	692 468	.011 494	1 554
3.2	−11.367 303	741 002	.010 133	1 361
3.3	−12 161 041	793 739	.008 939	1 194
3.4	−13.012 075	851 034	.007 891	1 048
3.5	−13 925 354	913 279	.006 970	921
3.6	−14.906 254	980 900	.006 160	810
3.7	−15.960 619	1.054 365	.005 448	712
3.8	−17.094 802	1.134 183	.004 820	628
3.9	−18.315 714	1.220 912	.004 267	553
4.0	−19.630 874	1.315 160	.003 779	488
4.1	−21.048 467	1.417 592	.003 349	430
4.2	−22.577 401	1.528 934	.002 969	380
4.3	−24.227 380	1.649 979	.002 633	336
4.4	−26.008 973	1.781 593	.002 336	297
4.5	−27.933 697	1.924 723	.002 073	263
4.6	−30.014 099	2.080 403	.001 841	232
4.7	−32.263 860	2.249 760	.001 635	206
4.8	−34.697 890	2.434 030	.001 453	182
4.9	−37.332 451	2.634 561	.001 291	162
5.0	−40.185 275	2.852 825	+0.001 148	143

TABLE V

Col x, sil x, expl x, and expl $(-x)$ from 0 to 15

x	col x	sil x	expl x	expl $(-x)$	x	$x°$
0	+1.570796	∞	∞	∞	0	0
1	+ .624713	−.337404	− 1.895118	+.219384	1	57.2958
2	− .034617	−.422981	− 4.954234	.048901	2	114.5916
3	− .277856	−.119630	− 9.933833	.013048	3	171.8874
4	− .187407	+.140982	− 19.630874	.003779	4	229.1832
5	+ .020865	+.190093	− 40.185275	.001148	5	286.479
6	+ .146109	+.068057	− 85.990	.000360082	6	343.775
7	+ .116200	−.076695	− 191.505	.000115482	7	401.071
8	− .003391	−.122434	− 440.380	.000 37666	8	458.366
9	− .094244	−.055348	− 1037.878	.000012447	9	515.662
10	− .087551	+.045456	− 2492.229	.000004157	10	572.958
11	− .007511	+.089563	− 6071.406	.000001400	11	630.254
12	+ .065825	+.049780	− 14959	.000000475	12	687.550
13	+ .071435	−.026764	− 37198	.000000162	13	744.846
14	+ .014585	−.069396	− 93193	.000000056	14	802.142
15	− .047398	−.046279	−234956	+.000000019	15	859.438

TABLE VI

col x and sil x

x	co x	sil x	x	col x	sil x
0	+1.570 796	∞	150	+.004 630	+.004 796
5	+ .020 865	+.190 093	160	−.006 089	−.001 409
10	− .087 551	+.045 456	170	+.005 529	−.002 006
15	− .047 398	−.046 279	180	−.003 349	+.004 432
20	+ .022 555	−.044 420	190	+.000 377	−.005 249
25	+ .039 314	+.006 849	200	+.002 414	+.004 378
30	+ .004 040	+.033 032	300	−.000 085	+.003 332
35	− .026 126	+.011 480	400	−.001 319	+.002 124
40	− .016 189	−.019 020	500	−.001 770	+.000 932
45	+ .012 081	−.018 632	600	−.001 665	−.000 076
50	+ .019 179	+.005 628	700	−.001 198	−.000 779
55	+ .000 072	+.018 173	800	−.000 559	−.001 118
60	− .015 949	+.004 813	900	+.000 075	−.001 109
65	− .006 675	−.012 847	1 000	+.000 563	−.000 826
70	+ .009 201	−.010 922	2 000	−.000 183	−.000 465
75	+ .012 217	+.005 332	3 000	−.000 325	−.000 073
80	− .001 535	+.012 402	4 000	−.000 182	+.000 171
85	− .011 602	+.001 935	5 000	+.000 031	+.000 198
90	− .004 867	−.009 986	6 000	+.000 151	+.000 071
95	+ .007 760	−.007 110	7 000	+.000 123	−.000 072
100	+ .008 571	+.005 149	8 000	+.000 008	−.000 125
110	− .009 084	+.000 320	9 000	−.000 088	−.000 068
120	+ .007 824	−.004 781	10 000	−.000 095	+.000 031
130	− .002 880	+.007 132	11 000	−.000 026	+.000 087
140	− .001 363	−.007 011	100 000	−.000 010	−.000 000
150	+ .004 630	+.004 796	1 000 000	+.000 001	+.000 000

TABLE VII

Maxima and Minima of $\operatorname{col}\frac{\pi}{2}x$ and $\operatorname{sil}\frac{\pi}{2}x$

x	$\operatorname{col}\frac{\pi}{2}x$	$\frac{2}{\pi x}-\operatorname{col}\frac{\pi}{2}x$
0	$+1.570\ 796$	
2	$-\ .281\ 141$	
4	$+\ .152\ 645$	
6	$-\ .103\ 966$	
8	$+\ .078\ 635$	$x(-1)^{\frac{x}{2}}$
10	$-\ .063\ 168$	494
12	$+\ .052\ 762$	290
14	$-\ .045\ 289$	184
16	$+\ .039\ 665$	123
18	$-\ .035\ 281$	94
20	$+\ .031\ 767$	64
22	$-\ .028\ 889$	48
24	$+\ .026\ 489$	37
26	$-\ .024\ 456$	29
28	$+\ .022\ 713$	23
30	$-\ .021\ 202$	19
32	$+\ .019\ 879$	15
34	$-\ .018\ 711$	13
36	$+\ .017\ 673$	11
38	$-\ .016\ 744$	10
40	$+\ .015\ 907$	9
42	$-\ .015\ 151$	8
44	$+\ .014\ 462$	7
46	$-\ .013\ 834$	6
48	$+\ .013\ 258$	5
50	$-\ .012\ 728$	4
52	$+\ .012\ 239$	3
54	$-\ .011\ 786$	3
56	$+\ .011\ 365$	3
58	$-\ .010\ 974$	2
60	$+\ .010\ 608$	2
62	$-\ .010\ 266$	2
64	$+\ .009\ 945$	2
66	$-\ .009\ 644$	2
68	$+\ .009\ 360$	2
70	$-\ .009\ 093$	2
72	$+\ .008\ 841$	1
74	$-\ .008\ 602$	1
76	$+\ .008\ 375$	1
78	$-\ .008\ 161$	1
80	$+\ .007\ 957$	1

$x>80$

$$\operatorname{col}\frac{\pi}{2}x=(-1)^{\frac{x}{2}}\frac{2}{\pi x}$$

x	$\operatorname{sil}\frac{\pi}{2}x$	$\frac{2}{\pi x}-\operatorname{sil}\frac{\pi}{2}x$
1	$-\ .472\ 001$	
3	$+\ .198\ 408$	
5	$-\ .123\ 772$	
7	$+\ .089\ 564$	
9	$-\ .070\ 065$	$x(-1)^{\frac{x+1}{2}}$
11	$+\ .057\ 501$	374
13	$-\ .048\ 742$	229
15	$+\ .042\ 292$	149
17	$-\ .037\ 345$	103
19	$+\ .033\ 432$	74
21	$-\ .030\ 260$	55
23	$+\ .027\ 637$	42
25	$-\ .025\ 432$	33
27	$+\ .023\ 552$	26
29	$-\ .021\ 931$	21
31	$+\ .020\ 519$	17
33	$-\ .019\ 277$	14
35	$+\ .018\ 177$	12
37	$-\ .017\ 196$	10
39	$+\ .016\ 315$	9
41	$-\ .015\ 520$	8
43	$+\ .014\ 799$	7
45	$-\ .014\ 141$	6
47	$+\ .013\ 540$	5
49	$-\ .012\ 988$	4
51	$+\ .012\ 480$	3
53	$-\ .012\ 008$	3
55	$+\ .011\ 572$	3
57	$-\ .011\ 166$	3
59	$+\ .010\ 788$	2
61	$-\ .010\ 434$	2
63	$+\ .010\ 103$	2
65	$-\ .009\ 792$	2
67	$+\ .009\ 500$	2
69	$-\ .009\ 225$	2
71	$+\ .008\ 965$	1
73	$-\ .008\ 719$	1
75	$+\ .008\ 487$	1
77	$-\ .008\ 267$	1
79	$+\ .008\ 057$	1
81	$-\ .007\ 858$	0

$x>79$

$$\operatorname{sil}\frac{\pi}{2}x=(-1)^{\frac{x+1}{2}}\frac{2}{\pi x}$$

INDEX TO VOLUME III

SOME DOVER SCIENCE BOOKS

SOME DOVER SCIENCE BOOKS

WHAT IS SCIENCE?,
Norman Campbell

This excellent introduction explains scientific method, role of mathematics, types of scientific laws. Contents: 2 aspects of science, science & nature, laws of science, discovery of laws, explanation of laws, measurement & numerical laws, applications of science. 192pp. 5⅜ x 8. 60043-2 Paperbound $1.25

FADS AND FALLACIES IN THE NAME OF SCIENCE,
Martin Gardner

Examines various cults, quack systems, frauds, delusions which at various times have masqueraded as science. Accounts of hollow-earth fanatics like Symmes; Velikovsky and wandering planets; Hoerbiger; Bellamy and the theory of multiple moons; Charles Fort; dowsing, pseudoscientific methods for finding water, ores, oil. Sections on naturopathy, iridiagnosis, zone therapy, food fads, etc. Analytical accounts of Wilhelm Reich and orgone sex energy; L. Ron Hubbard and Dianetics; A. Korzybski and General Semantics; many others. Brought up to date to include Bridey Murphy, others. Not just a collection of anecdotes, but a fair, reasoned appraisal of eccentric theory. Formerly titled *In the Name of Science*. Preface. Index. x + 384pp. 5⅜ x 8.
20394-8 Paperbound $2.00

PHYSICS, THE PIONEER SCIENCE,
L. W. Taylor

First thorough text to place all important physical phenomena in cultural-historical framework; remains best work of its kind. Exposition of physical laws, theories developed chronologically, with great historical, illustrative experiments diagrammed, described, worked out mathematically. Excellent physics text for self-study as well as class work. Vol. 1: Heat, Sound: motion, acceleration, gravitation, conservation of energy, heat engines, rotation, heat, mechanical energy, etc. 211 illus. 407pp. 5⅜ x 8. Vol. 2: Light, Electricity: images, lenses, prisms, magnetism, Ohm's law, dynamos, telegraph, quantum theory, decline of mechanical view of nature, etc. Bibliography. 13 table appendix. Index. 551 illus. 2 color plates. 508pp. 5⅜ x 8.
60565-5, 60566-3 Two volume set, paperbound $5.50

THE EVOLUTION OF SCIENTIFIC THOUGHT FROM NEWTON TO EINSTEIN,
A. d'Abro

Einstein's special and general theories of relativity, with their historical implications, are analyzed in non-technical terms. Excellent accounts of the contributions of Newton, Riemann, Weyl, Planck, Eddington, Maxwell, Lorentz and others are treated in terms of space and time, equations of electromagnetics, finiteness of the universe, methodology of science. 21 diagrams. 482pp. 5⅜ x 8.
20002-7 Paperbound $2.50

CHANCE, LUCK AND STATISTICS: THE SCIENCE OF CHANCE,
Horace C. Levinson

Theory of probability and science of statistics in simple, non-technical language. Part I deals with theory of probability, covering odd superstitions in regard to "luck," the meaning of betting odds, the law of mathematical expectation, gambling, and applications in poker, roulette, lotteries, dice, bridge, and other games of chance. Part II discusses the misuse of statistics, the concept of statistical probabilities, normal and skew frequency distributions, and statistics applied to various fields—birth rates, stock speculation, insurance rates, advertising, etc. "Presented in an easy humorous style which I consider the best kind of expository writing," Prof. A. C. Cohen, Industry Quality Control. Enlarged revised edition. Formerly titled *The Science of Chance*. Preface and two new appendices by the author. xiv + 365pp. 5⅜ x 8. 21007-3 Paperbound $2.00

BASIC ELECTRONICS,
prepared by the U.S. Navy Training Publications Center

A thorough and comprehensive manual on the fundamentals of electronics. Written clearly, it is equally useful for self-study or course work for those with a knowledge of the principles of basic electricity. Partial contents: Operating Principles of the Electron Tube; Introduction to Transistors; Power Supplies for Electronic Equipment; Tuned Circuits; Electron-Tube Amplifiers; Audio Power Amplifiers; Oscillators; Transmitters; Transmission Lines; Antennas and Propagation; Introduction to Computers; and related topics. Appendix. Index. Hundreds of illustrations and diagrams. vi + 471pp. 6½ x 9¼.
61076-4 Paperbound $2.95

BASIC THEORY AND APPLICATION OF TRANSISTORS,
prepared by the U.S. Department of the Army

An introductory manual prepared for an army training program. One of the finest available surveys of theory and application of transistor design and operation. Minimal knowledge of physics and theory of electron tubes required. Suitable for textbook use, course supplement, or home study. Chapters: Introduction; fundamental theory of transistors; transistor amplifier fundamentals; parameters, equivalent circuits, and characteristic curves; bias stabilization; transistor analysis and comparison using characteristic curves and charts; audio amplifiers; tuned amplifiers; wide-band amplifiers; oscillators; pulse and switching circuits; modulation, mixing, and demodulation; and additional semiconductor devices. Unabridged, corrected edition. 240 schematic drawings, photographs, wiring diagrams, etc. 2 Appendices. Glossary. Index. 263pp. 6½ x 9¼. 60380-6 Paperbound $1.75

GUIDE TO THE LITERATURE OF MATHEMATICS AND PHYSICS,
N. G. Parke III

Over 5000 entries included under approximately 120 major subject headings of selected most important books, monographs, periodicals, articles in English, plus important works in German, French, Italian, Spanish, Russian (many recently available works). Covers every branch of physics, math, related engineering. Includes author, title, edition, publisher, place, date, number of volumes, number of pages. A 40-page introduction on the basic problems of research and study provides useful information on the organization and use of libraries, the psychology of learning, etc. This reference work will save you hours of time. 2nd revised edition. Indices of authors, subjects, 464pp. 5⅜ x 8.
60447-0 Paperbound $2.75

THE RISE OF THE NEW PHYSICS (formerly THE DECLINE OF MECHANISM), *A. d'Abro*

This authoritative and comprehensive 2-volume exposition is unique in scientific publishing. Written for intelligent readers not familiar with higher mathematics, it is the only thorough explanation in non-technical language of modern mathematical-physical theory. Combining both history and exposition, it ranges from classical Newtonian concepts up through the electronic theories of Dirac and Heisenberg, the statistical mechanics of Fermi, and Einstein's relativity theories. "A must for anyone doing serious study in the physical sciences," *J. of Franklin Inst.* 97 illustrations. 991pp. 2 volumes.

20003-5, 20004-3 Two volume set, paperbound $5.50

THE STRANGE STORY OF THE QUANTUM, AN ACCOUNT FOR THE GENERAL READER OF THE GROWTH OF IDEAS UNDERLYING OUR PRESENT ATOMIC KNOWLEDGE, *B. Hoffmann*

Presents lucidly and expertly, with barest amount of mathematics, the problems and theories which led to modern quantum physics. Dr. Hoffmann begins with the closing years of the 19th century, when certain trifling discrepancies were noticed, and with illuminating analogies and examples takes you through the brilliant concepts of Planck, Einstein, Pauli, de Broglie, Bohr, Schroedinger, Heisenberg, Dirac, Sommerfeld, Feynman, etc. This edition includes a new, long postscript carrying the story through 1958. "Of the books attempting an account of the history and contents of our modern atomic physics which have come to my attention, this is the best," H. Margenau, Yale University, in *American Journal of Physics.* 32 tables and line illustrations. Index. 275pp. 5⅜ x 8.

20518-5 Paperbound $2.00

GREAT IDEAS AND THEORIES OF MODERN COSMOLOGY, *Jagjit Singh*

The theories of Jeans, Eddington, Milne, Kant, Bondi, Gold, Newton, Einstein, Gamow, Hoyle, Dirac, Kuiper, Hubble, Weizsäcker and many others on such cosmological questions as the origin of the universe, space and time, planet formation, "continuous creation," the birth, life, and death of the stars, the origin of the galaxies, etc. By the author of the popular *Great Ideas of Modern Mathematics.* A gifted popularizer of science, he makes the most difficult abstractions crystal-clear even to the most non-mathematical reader. Index. xii + 276pp. 5⅜ x 8½. 20925-3 Paperbound $2.50

GREAT IDEAS OF MODERN MATHEMATICS: THEIR NATURE AND USE, *Jagjit Singh*

Reader with only high school math will understand main mathematical ideas of modern physics, astronomy, genetics, psychology, evolution, etc., better than many who use them as tools, but comprehend little of their basic structure. Author uses his wide knowledge of non-mathematical fields in brilliant exposition of differential equations, matrices, group theory, logic, statistics, problems of mathematical foundations, imaginary numbers, vectors, etc. Original publications, appendices. indexes. 65 illustr. 322pp. 5⅜ x 8. 20587-8 Paperbound $2.25

THE MATHEMATICS OF GREAT AMATEURS, *Julian L. Coolidge*

Great discoveries made by poets, theologians, philosophers, artists and other non-mathematicians: Omar Khayyam, Leonardo da Vinci, Albrecht Dürer, John Napier, Pascal, Diderot, Bolzano, etc. Surprising accounts of what can result from a non-professional preoccupation with the oldest of sciences. 56 figures. viii + 211pp. 5⅜ x 8½. 61009-8 Paperbound $2.00

College Algebra, *H. B. Fine*
Standard college text that gives a systematic and deductive structure to algebra; comprehensive, connected, with emphasis on theory. Discusses the commutative, associative, and distributive laws of number in unusual detail, and goes on with undetermined coefficients, quadratic equations, progressions, logarithms, permutations, probability, power series, and much more. Still most valuable elementary-intermediate text on the science and structure of algebra. Index. 1560 problems, all with answers. x + 631pp. 5⅜ x 8. 60211-7 Paperbound $2.75

Higher Mathematics for Students of Chemistry and Physics, *J. W. Mellor*
Not abstract, but practical, building its problems out of familiar laboratory material, this covers differential calculus, coordinate, analytical geometry, functions, integral calculus, infinite series, numerical equations, differential equations, Fourier's theorem, probability, theory of errors, calculus of variations, determinants. "If the reader is not familiar with this book, it will repay him to examine it," *Chem. & Engineering News.* 800 problems. 189 figures. Bibliography. xxi + 641pp. 5⅜ x 8. 60193-5 Paperbound $3.50

Trigonometry Refresher for Technical Men, *A. A. Klaf*
A modern question and answer text on plane and spherical trigonometry. Part I covers plane trigonometry: angles, quadrants, trigonometrical functions, graphical representation, interpolation, equations, logarithms, solution of triangles, slide rules, etc. Part II discusses applications to navigation, surveying, elasticity, architecture, and engineering. Small angles, periodic functions, vectors, polar coordinates, De Moivre's theorem, fully covered. Part III is devoted to spherical trigonometry and the solution of spherical triangles, with applications to terrestrial and astronomical problems. Special time-savers for numerical calculation. 913 questions answered for you! 1738 problems; answers to odd numbers. 494 figures. 14 pages of functions, formulae. Index. x + 629pp. 5⅜ x 8.
20371-9 Paperbound $3.00

Calculus Refresher for Technical Men, *A. A. Klaf*
Not an ordinary textbook but a unique refresher for engineers, technicians, and students. An examination of the most important aspects of differential and integral calculus by means of 756 key questions. Part I covers simple differential calculus: constants, variables, functions, increments, derivatives, logarithms, curvature, etc. Part II treats fundamental concepts of integration: inspection, substitution, transformation, reduction, areas and volumes, mean value, successive and partial integration, double and triple integration. Stresses practical aspects! A 50 page section gives applications to civil and nautical engineering, electricity, stress and strain, elasticity, industrial engineering, and similar fields. 756 questions answered. 556 problems; solutions to odd numbers. 36 pages of constants, formulae. Index. v + 431pp. 5⅜ x 8. 20370-0 Paperbound $2.25

Introduction to the Theory of Groups of Finite Order, *R. Carmichael*
Examines fundamental theorems and their application. Beginning with sets, systems, permutations, etc., it progresses in easy stages through important types of groups: Abelian, prime power, permutation, etc. Except 1 chapter where matrices are desirable, no higher math needed. 783 exercises, problems. Index. xvi + 447pp. 5⅜ x 8. 60300-8 Paperbound $3.00

FIVE VOLUME "THEORY OF FUNCTIONS" SET BY KONRAD KNOPP

This five-volume set, prepared by Konrad Knopp, provides a complete and readily followed account of theory of functions. Proofs are given concisely, yet without sacrifice of completeness or rigor. These volumes are used as texts by such universities as M.I.T., University of Chicago, N. Y. City College, and many others. "Excellent introduction . . . remarkably readable, concise, clear, rigorous," *Journal of the American Statistical Association.*

ELEMENTS OF THE THEORY OF FUNCTIONS,
Konrad Knopp
This book provides the student with background for further volumes in this set, or texts on a similar level. Partial contents: foundations, system of complex numbers and the Gaussian plane of numbers, Riemann sphere of numbers, mapping by linear functions, normal forms, the logarithm, the cyclometric functions and binomial series. "Not only for the young student, but also for the student who knows all about what is in it," *Mathematical Journal.* Bibliography. Index. 140pp. 5⅜ x 8. 60154-4 Paperbound $1.50

THEORY OF FUNCTIONS, PART I,
Konrad Knopp
With volume II, this book provides coverage of basic concepts and theorems. Partial contents: numbers and points, functions of a complex variable, integral of a continuous function, Cauchy's integral theorem, Cauchy's integral formulae, series with variable terms, expansion of analytic functions in power series, analytic continuation and complete definition of analytic functions, entire transcendental functions, Laurent expansion, types of singularities. Bibliography. Index. vii + 146pp. 5⅜ x 8. 60156-0 Paperbound $1.50

THEORY OF FUNCTIONS, PART II,
Konrad Knopp
Application and further development of general theory, special topics. Single valued functions. Entire, Weierstrass, Meromorphic functions. Riemann surfaces. Algebraic functions. Analytical configuration, Riemann surface. Bibliography. Index. x + 150pp. 5⅜ x 8. 60157-9 Paperbound $1.50

PROBLEM BOOK IN THE THEORY OF FUNCTIONS, VOLUME 1.
Konrad Knopp
Problems in elementary theory, for use with Knopp's *Theory of Functions,* or any other text, arranged according to increasing difficulty. Fundamental concepts, sequences of numbers and infinite series, complex variable, integral theorems, development in series, conformal mapping. 182 problems. Answers. viii + 126pp. 5⅜ x 8. 60158-7 Paperbound $1.50

PROBLEM BOOK IN THE THEORY OF FUNCTIONS, VOLUME 2,
Konrad Knopp
Advanced theory of functions, to be used either with Knopp's *Theory of Functions,* or any other comparable text. Singularities, entire & meromorphic functions, periodic, analytic, continuation, multiple-valued functions, Riemann surfaces, conformal mapping. Includes a section of additional elementary problems. "The difficult task of selecting from the immense material of the modern theory of functions the problems just within the reach of the beginner is here masterfully accomplished," *Am. Math. Soc.* Answers. 138pp. 5⅜ x 8.
60159-5 Paperbound $1.50

NUMERICAL SOLUTIONS OF DIFFERENTIAL EQUATIONS, *H. Levy & E. A. Baggott*

Comprehensive collection of methods for solving ordinary differential equations of first and higher order. All must pass 2 requirements: easy to grasp and practical, more rapid than school methods. Partial contents: graphical integration of differential equations, graphical methods for detailed solution. Numerical solution. Simultaneous equations and equations of 2nd and higher orders. "Should be in the hands of all in research in applied mathematics, teaching," *Nature*. 21 figures. viii + 238pp. 5⅜ x 8. 60168-4 Paperbound $1.85

ELEMENTARY STATISTICS, WITH APPLICATIONS IN MEDICINE AND THE BIOLOGICAL SCIENCES, *F. E. Croxton*

A sound introduction to statistics for anyone in the physical sciences, assuming no prior acquaintance and requiring only a modest knowledge of math. All basic formulas carefully explained and illustrated; all necessary reference tables included. From basic terms and concepts, the study proceeds to frequency distribution, linear, non-linear, and multiple correlation, skewness, kurtosis, etc. A large section deals with reliability and significance of statistical methods. Containing concrete examples from medicine and biology, this book will prove unusually helpful to workers in those fields who increasingly must evaluate, check, and interpret statistics. Formerly titled "Elementary Statistics with Applications in Medicine." 101 charts. 57 tables. 14 appendices. Index. vi + 376pp. 5⅜ x 8. 60506-X Paperbound $2.25

INTRODUCTION TO SYMBOLIC LOGIC, *S. Langer*

No special knowledge of math required—probably the clearest book ever written on symbolic logic, suitable for the layman, general scientist, and philosopher. You start with simple symbols and advance to a knowledge of the Boole-Schroeder and Russell-Whitehead systems. Forms, logical structure, classes, the calculus of propositions, logic of the syllogism, etc. are all covered. "One of the clearest and simplest introductions," *Mathematics Gazette*. Second enlarged, revised edition. 368pp. 5⅜ x 8. 60164-1 Paperbound $2.25

A SHORT ACCOUNT OF THE HISTORY OF MATHEMATICS, *W. W. R. Ball*

Most readable non-technical history of mathematics treats lives, discoveries of every important figure from Egyptian, Phoenician, mathematicians to late 19th century. Discusses schools of Ionia, Pythagoras, Athens, Cyzicus, Alexandria, Byzantium, systems of numeration; primitive arithmetic; Middle Ages, Renaissance, including Arabs, Bacon, Regiomontanus, Tartaglia, Cardan, Stevinus, Galileo, Kepler; modern mathematics of Descartes, Pascal, Wallis, Huygens, Newton, Leibnitz, d'Alembert, Euler, Lambert, Laplace, Legendre, Gauss, Hermite, Weierstrass, scores more. Index. 25 figures. 546pp. 5⅜ x 8. 20630-0 Paperbound $2.75

INTRODUCTION TO NONLINEAR DIFFERENTIAL AND INTEGRAL EQUATIONS, *Harold T. Davis*

Aspects of the problem of nonlinear equations, transformations that lead to equations solvable by classical means, results in special cases, and useful generalizations. Thorough, but easily followed by mathematically sophisticated reader who knows little about non-linear equations. 137 problems for student to solve. xv + 566pp. 5⅜ x 8½. 60971-5 Paperbound $2.75

AN INTRODUCTION TO THE GEOMETRY OF N DIMENSIONS, *D. H. Y. Sommerville*

An introduction presupposing no prior knowledge of the field, the only book in English devoted exclusively to higher dimensional geometry. Discusses fundamental ideas of incidence, parallelism, perpendicularity, angles between linear space; enumerative geometry; analytical geometry from projective and metric points of view; polytopes; elementary ideas in analysis situs; content of hyper-spacial figures. Bibliography. Index. 60 diagrams. 196pp. 5⅜ x 8.
60494-2 Paperbound $1.50

ELEMENTARY CONCEPTS OF TOPOLOGY, *P. Alexandroff*

First English translation of the famous brief introduction to topology for the beginner or for the mathematician not undertaking extensive study. This unusually useful intuitive approach deals primarily with the concepts of complex, cycle, and homology, and is wholly consistent with current investigations. Ranges from basic concepts of set-theoretic topology to the concept of Betti groups. "Glowing example of harmony between intuition and thought," David Hilbert. Translated by A. E. Farley. Introduction by D. Hilbert. Index. 25 figures. 73pp. 5⅜ x 8.
60747-X Paperbound $1.25

ELEMENTS OF NON-EUCLIDEAN GEOMETRY, *D. M. Y. Sommerville*

Unique in proceeding step-by-step, in the manner of traditional geometry. Enables the student with only a good knowledge of high school algebra and geometry to grasp elementary hyperbolic, elliptic, analytic non-Euclidean geometries; space curvature and its philosophical implications; theory of radical axes; homothetic centres and systems of circles; parataxy and parallelism; absolute measure; Gauss' proof of the defect area theorem; geodesic representation; much more, all with exceptional clarity. 126 problems at chapter endings provide progressive practice and familiarity. 133 figures. Index. xvi + 274pp. 5⅜ x 8.
60460-8 Paperbound $2.00

INTRODUCTION TO THE THEORY OF NUMBERS, *L. E. Dickson*

Thorough, comprehensive approach with adequate coverage of classical literature, an introductory volume beginners can follow. Chapters on divisibility, congruences, quadratic residues & reciprocity. Diophantine equations, etc. Full treatment of binary quadratic forms without usual restriction to integral coefficients. Covers infinitude of primes, least residues. Fermat's theorem. Euler's phi function, Legendre's symbol, Gauss's lemma, automorphs, reduced forms, recent theorems of Thue & Siegel, many more. Much material not readily available elsewhere. 239 problems. Index. I figure. viii + 183pp. 5⅜ x 8.
60342-3 Paperbound $1.75

MATHEMATICAL TABLES AND FORMULAS, *compiled by Robert D. Carmichael and Edwin R. Smith*

Valuable collection for students, etc. Contains all tables necessary in college algebra and trigonometry, such as five-place common logarithms, logarithmic sines and tangents of small angles, logarithmic trigonometric functions, natural trigonometric functions, four-place antilogarithms, tables for changing from sexagesimal to circular and from circular to sexagesimal measure of angles, etc. Also many tables and formulas not ordinarily accessible, including powers, roots, and reciprocals, exponential and hyperbolic functions, ten-place logarithms of prime numbers, and formulas and theorems from analytical and elementary geometry and from calculus. Explanatory introduction. viii + 269pp. 5⅜ x 8½.
60111-0 Paperbound $1.50

A SOURCE BOOK IN MATHEMATICS,
D. E. Smith

Great discoveries in math, from Renaissance to end of 19th century, in English translation. Read announcements by Dedekind, Gauss, Delamain, Pascal, Fermat, Newton, Abel, Lobachevsky, Bolyai, Riemann, De Moivre, Legendre, Laplace, others of discoveries about imaginary numbers, number congruence, slide rule, equations, symbolism, cubic algebraic equations, non-Euclidean forms of geometry, calculus, function theory, quaternions, etc. Succinct selections from 125 different treatises, articles, most unavailable elsewhere in English. Each article preceded by biographical introduction. Vol. I: Fields of Number, Algebra. Index. 32 illus. 338pp. 5⅜ x 8. Vol. II: Fields of Geometry, Probability, Calculus, Functions, Quaternions. 83 illus. 432pp. 5⅜ x 8.
60552-3, 60553-1 Two volume set, paperbound $5.00

FOUNDATIONS OF PHYSICS,
R. B. Lindsay & H. Margenau

Excellent bridge between semi-popular works & technical treatises. A discussion of methods of physical description, construction of theory; valuable for physicist with elementary calculus who is interested in ideas that give meaning to data, tools of modern physics. Contents include symbolism; mathematical equations; space & time foundations of mechanics; probability; physics & continua; electron theory; special & general relativity; quantum mechanics; causality. "Thorough and yet not overdetailed. Unreservedly recommended," *Nature* (London). Unabridged, corrected edition. List of recommended readings. 35 illustrations. xi + 537pp. 5⅜ x 8. 60377-6 Paperbound $3.50

FUNDAMENTAL FORMULAS OF PHYSICS,
ed. by D. H. Menzel

High useful, full, inexpensive reference and study text, ranging from simple to highly sophisticated operations. Mathematics integrated into text—each chapter stands as short textbook of field represented. Vol. 1: Statistics, Physical Constants, Special Theory of Relativity, Hydrodynamics, Aerodynamics, Boundary Value Problems in Math, Physics, Viscosity, Electromagnetic Theory, etc. Vol. 2: Sound, Acoustics, Geometrical Optics, Electron Optics, High-Energy Phenomena, Magnetism, Biophysics, much more. Index. Total of 800pp. 5⅜ x 8.
60595-7, 60596-5 Two volume set, paperbound $4.75

THEORETICAL PHYSICS,
A. S. Kompaneyets

One of the very few thorough studies of the subject in this price range. Provides advanced students with a comprehensive theoretical background. Especially strong on recent experimentation and developments in quantum theory. Contents: Mechanics (Generalized Coordinates, Lagrange's Equation, Collision of Particles, etc.), Electrodynamics (Vector Analysis, Maxwell's equations, Transmission of Signals, Theory of Relativity, etc.), Quantum Mechanics (the Inadequacy of Classical Mechanics, the Wave Equation, Motion in a Central Field, Quantum Theory of Radiation, Quantum Theories of Dispersion and Scattering, etc.), and Statistical Physics (Equilibrium Distribution of Molecules in an Ideal Gas, Boltzmann Statistics, Bose and Fermi Distribution. Thermodynamic Quantities, etc.). Revised to 1961. Translated by George Yankovsky, authorized by Kompaneyets. 137 exercises. 56 figures. 529pp. 5⅜ x 8½.
60972-3 Paperbound $3.50

MATHEMATICAL PHYSICS, *D. H. Menzel*
Thorough one-volume treatment of the mathematical techniques vital for classical mechanics, electromagnetic theory, quantum theory, and relativity. Written by the Harvard Professor of Astrophysics for junior, senior, and graduate courses, it gives clear explanations of all those aspects of function theory, vectors, matrices, dyadics, tensors, partial differential equations, etc., necessary for the understanding of the various physical theories. Electron theory, relativity, and other topics seldom presented appear here in considerable detail. Scores of definition, conversion factors, dimensional constants, etc. "More detailed than normal for an advanced text . . . excellent set of sections on Dyadics, Matrices, and Tensors," *Journal of the Franklin Institute*. Index. 193 problems, with answers. x + 412pp. 5⅜ x 8. 60056-4 Paperbound $2.50

THE THEORY OF SOUND, *Lord Rayleigh*
Most vibrating systems likely to be encountered in practice can be tackled successfully by the methods set forth by the great Nobel laureate, Lord Rayleigh. Complete coverage of experimental, mathematical aspects of sound theory. Partial contents: Harmonic motions, vibrating systems in general, lateral vibrations of bars, curved plates or shells, applications of Laplace's functions to acoustical problems, fluid friction, plane vortex-sheet, vibrations of solid bodies, etc. This is the first inexpensive edition of this great reference and study work. Bibliography, Historical introduction by R. B. Lindsay. Total of 1040pp. 97 figures. 5⅜ x 8. 60292-3, 60293-1 Two volume set, paperbound $6.00

HYDRODYNAMICS, *Horace Lamb*
Internationally famous complete coverage of standard reference work on dynamics of liquids & gases. Fundamental theorems, equations, methods, solutions, background, for classical hydrodynamics. Chapters include Equations of Motion, Integration of Equations in Special Gases, Irrotational Motion, Motion of Liquid in 2 Dimensions, Motion of Solids through Liquid-Dynamical Theory, Vortex Motion, Tidal Waves, Surface Waves, Waves of Expansion, Viscosity, Rotating Masses of Liquids. Excellently planned, arranged; clear, lucid presentation. 6th enlarged, revised edition. Index. Over 900 footnotes, mostly bibliographical. 119 figures. xv + 738pp. 6⅛ x 9¼. 60256-7 Paperbound $4.00

DYNAMICAL THEORY OF GASES, *James Jeans*
Divided into mathematical and physical chapters for the convenience of those not expert in mathematics, this volume discusses the mathematical theory of gas in a steady state, thermodynamics, Boltzmann and Maxwell, kinetic theory, quantum theory, exponentials, etc. 4th enlarged edition, with new material on quantum theory, quantum dynamics, etc. Indexes. 28 figures. 444pp. 6⅛ x 9¼. 60136-6 Paperbound $2.75

THERMODYNAMICS, *Enrico Fermi*
Unabridged reproduction of 1937 edition. Elementary in treatment; remarkable for clarity, organization. Requires no knowledge of advanced math beyond calculus, only familiarity with fundamentals of thermometry, calorimetry. Partial Contents: Thermodynamic systems; First & Second laws of thermodynamics; Entropy; Thermodynamic potentials: phase rule, reversible electric cell; Gaseous reactions: van't Hoff reaction box, principle of LeChatelier; Thermodynamics of dilute solutions: osmotic & vapor pressures, boiling & freezing points; Entropy constant. Index. 25 problems. 24 illustrations. x + 160pp. 5⅜ x 8. 60361-X Paperbound $2.00

CELESTIAL OBJECTS FOR COMMON TELESCOPES,
Rev. T. W. Webb
Classic handbook for the use and pleasure of the amateur astronomer. Of inestimable aid in locating and identifying thousands of celestial objects. Vol I, The Solar System: discussions of the principle and operation of the telescope, procedures of observations and telescope-photography, spectroscopy, etc., precise location information of sun, moon, planets, meteors. Vol. II, The Stars: alphabetical listing of constellations, information on double stars, clusters, stars with unusual spectra, variables, and nebulae, etc. Nearly 4,000 objects noted. Edited and extensively revised by Margaret W. Mayall, director of the American Assn. of Variable Star Observers. New Index by Mrs. Mayall giving the location of all objects mentioned in the text for Epoch 2000. New Precession Table added. New appendices on the planetary satellites, constellation names and abbreviations, and solar system data. Total of 46 illustrations. Total of xxxix + 606pp. 5⅜ x 8. 20917-2, 20918-0 Two volume set, paperbound $5.00

PLANETARY THEORY,
E. W. Brown and C. A. Shook
Provides a clear presentation of basic methods for calculating planetary orbits for today's astronomer. Begins with a careful exposition of specialized mathematical topics essential for handling perturbation theory and then goes on to indicate how most of the previous methods reduce ultimately to two general calculation methods: obtaining expressions either for the coordinates of planetary positions or for the elements which determine the perturbed paths. An example of each is given and worked in detail. Corrected edition. Preface. Appendix. Index. xii + 302pp. 5⅜ x 8½. 61133-7 Paperbound $2.25

STAR NAMES AND THEIR MEANINGS,
Richard Hinckley Allen
An unusual book documenting the various attributions of names to the individual stars over the centuries. Here is a treasure-house of information on a topic not normally delved into even by professional astronomers; provides a fascinating background to the stars in folk-lore, literary references, ancient writings, star catalogs and maps over the centuries. Constellation-by-constellation analysis covers hundreds of stars and other asterisms, including the Pleiades, Hyades, Andromedan Nebula, etc. Introduction. Indices. List of authors and authorities. xx + 563pp. 5⅜ x 8½. 21079-0 Paperbound $3.00

A SHORT HISTORY OF ASTRONOMY, *A. Berry*
Popular standard work for over 50 years, this thorough and accurate volume covers the science from primitive times to the end of the 19th century. After the Greeks and the Middle Ages, individual chapters analyze Copernicus, Brahe, Galileo, Kepler, and Newton, and the mixed reception of their discoveries. Post-Newtonian achievements are then discussed in unusual detail: Halley, Bradley, Lagrange, Laplace, Herschel, Bessel, etc. 2 Indexes. 104 illustrations, 9 portraits. xxxi + 440pp. 5⅜ x 8. 20210-0 Paperbound $2.75

SOME THEORY OF SAMPLING, *W. E. Deming*
The purpose of this book is to make sampling techniques understandable to and useable by social scientists, industrial managers, and natural scientists who are finding statistics increasingly part of their work. Over 200 exercises, plus dozens of actual applications. 61 tables. 90 figs. xix + 602pp. 5⅜ x 8½.
61755-6 Paperbound $3.50

PRINCIPLES OF STRATIGRAPHY,
A. W. Grabau
Classic of 20th century geology, unmatched in scope and comprehensiveness. Nearly 600 pages cover the structure and origins of every kind of sedimentary, hydrogenic, oceanic, pyroclastic, atmoclastic, hydroclastic, marine hydroclastic, and bioclastic rock; metamorphism; erosion; etc. Includes also the constitution of the atmosphere; morphology of oceans, rivers, glaciers; volcanic activities; faults and earthquakes; and fundamental principles of paleontology (nearly 200 pages). New introduction by Prof. M. Kay, Columbia U. 1277 bibliographical entries. 264 diagrams. Tables, maps, etc. Two volume set. Total of xxxii + 1185pp. 5⅜ x 8. 60686-4, 60687-2 Two volume set, paperbound $6.25

SNOW CRYSTALS, *W. A. Bentley and W. J. Humphreys*
Over 200 pages of Bentley's famous microphotographs of snow flakes—the product of painstaking, methodical work at his Jericho, Vermont studio. The pictures, which also include plates of frost, glaze and dew on vegetation, spider webs, windowpanes; sleet; graupel or soft hail, were chosen both for their scientific interest and their aesthetic qualities. The wonder of nature's diversity is exhibited in the intricate, beautiful patterns of the snow flakes. Introductory text by W. J. Humphreys. Selected bibliography. 2,453 illustrations. 224pp. 8 x 10¼. 20287-9 Paperbound $3.25

THE BIRTH AND DEVELOPMENT OF THE GEOLOGICAL SCIENCES,
F. D. Adams
Most thorough history of the earth sciences ever written. Geological thought from earliest times to the end of the 19th century, covering over 300 early thinkers & systems: fossils & their explanation, vulcanists vs. neptunists, figured stones & paleontology, generation of stones, dozens of similar topics. 91 illustrations, including medieval, renaissance woodcuts, etc. Index. 632 footnotes, mostly bibliographical. 511pp. 5⅜ x 8. 20005-1 Paperbound $2.75

ORGANIC CHEMISTRY, *F. C. Whitmore*
The entire subject of organic chemistry for the practicing chemist and the advanced student. Storehouse of facts, theories, processes found elsewhere only in specialized journals. Covers aliphatic compounds (500 pages on the properties and synthetic preparation of hydrocarbons, halides, proteins, ketones, etc.), alicyclic compounds, aromatic compounds, heterocyclic compounds, organophosphorus and organometallic compounds. Methods of synthetic preparation analyzed critically throughout. Includes much of biochemical interest. "The scope of this volume is astonishing," *Industrial and Engineering Chemistry*. 12,000-reference index. 2387-item bibliography. Total of x + 1005pp. 5⅜ x 8. 60700-3, 60701-1 Two volume set, paperbound $4.50

THE PHASE RULE AND ITS APPLICATION,
Alexander Findlay
Covering chemical phenomena of 1, 2, 3, 4, and multiple component systems, this "standard work on the subject" (*Nature*, London), has been completely revised and brought up to date by A. N. Campbell and N. O. Smith. Brand new material has been added on such matters as binary, tertiary liquid equilibria, solid solutions in ternary systems, quinary systems of salts and water. Completely revised to triangular coordinates in ternary systems, clarified graphic representation, solid models, etc. 9th revised edition. Author, subject indexes. 236 figures. 505 footnotes, mostly bibliographic. xii + 494pp. 5⅜ x 8. 60091-2 Paperbound $2.75

A COURSE IN MATHEMATICAL ANALYSIS,
Edouard Goursat

Trans. by E. R. Hedrick, O. Dunkel, H. G. Bergmann. Classic study of fundamental material thoroughly treated. Extremely lucid exposition of wide range of subject matter for student with one year of calculus. Vol. 1: Derivatives and differentials, definite integrals, expansions in series, applications to geometry. 52 figures, 556pp. 60554-X Paperbound $3.00. Vol. 2, Part I: Functions of a complex variable, conformal representations, doubly periodic functions, natural boundaries, etc. 38 figures, 269pp. 60555-8 Paperbound $2.25. Vol. 2, Part II: Differential equations, Cauchy-Lipschitz method, nonlinear differential equations, simultaneous equations, etc. 308pp. 60556-6 Paperbound $2.50. Vol. 3, Part I: Variation of solutions, partial differential equations of the second order. 15 figures, 339pp. 61176-0 Paperbound $3.00. Vol. 3, Part II: Integral equations, calculus of variations. 13 figures, 389pp. 61177-9 Paperbound $3.00 60554-X, 60555-8, 60556-6 61176-0, 61177-9 Six volume set,
paperbound $13.75

PLANETS, STARS AND GALAXIES,
A. E. Fanning

Descriptive astronomy for beginners: the solar system; neighboring galaxies; seasons; quasars; fly-by results from Mars, Venus, Moon; radio astronomy; etc. all simply explained. Revised up to 1966 by author and Prof. D. H. Menzel, former Director, Harvard College Observatory. 29 photos, 16 figures. 189pp. 5⅜ x 8½. 21680-2 Paperbound $1.50

GREAT IDEAS IN INFORMATION THEORY, LANGUAGE AND CYBERNETICS,
Jagjit Singh

Winner of Unesco's Kalinga Prize covers language, metalanguages, analog and digital computers, neural systems, work of McCulloch, Pitts, von Neumann, Turing, other important topics. No advanced mathematics needed, yet a full discussion without compromise or distortion. 118 figures. ix + 338pp. 5⅜ x 8½.
21694-2 Paperbound $2.25

GEOMETRIC EXERCISES IN PAPER FOLDING,
T. Sundara Row

Regular polygons, circles and other curves can be folded or pricked on paper, then used to demonstrate geometric propositions, work out proofs, set up well-known problems. 89 illustrations, photographs of actually folded sheets. xii + 148pp. 5⅜ x 8½. 21594-6 Paperbound $1.00

VISUAL ILLUSIONS, THEIR CAUSES, CHARACTERISTICS AND APPLICATIONS,
M. Luckiesh

The visual process, the structure of the eye, geometric, perspective illusions, influence of angles, illusions of depth and distance, color illusions, lighting effects, illusions in nature, special uses in painting, decoration, architecture, magic, camouflage. New introduction by W. H. Ittleson covers modern developments in this area. 100 illustrations. xxi + 252pp. 5⅜ x 8.
21530-X Paperbound $1.50

ATOMS AND MOLECULES SIMPLY EXPLAINED,
B. C. Saunders and R. E. D. Clark

Introduction to chemical phenomena and their applications: cohesion, particles, crystals, tailoring big molecules, chemist as architect, with applications in radioactivity, color photography, synthetics, biochemistry, polymers, and many other important areas. Non technical. 95 figures. x + 299pp. 5⅜ x 8½.
21282-3 Paperbound $1.50

THE PRINCIPLES OF ELECTROCHEMISTRY,
D. A. MacInnes

Basic equations for almost every subfield of electrochemistry from first principles, referring at all times to the soundest and most recent theories and results; unusually useful as text or as reference. Covers coulometers and Faraday's Law, electrolytic conductance, the Debye-Hueckel method for the theoretical calculation of activity coefficients, concentration cells, standard electrode potentials, thermodynamic ionization constants, pH, potentiometric titrations, irreversible phenomena. Planck's equation, and much more. 2 indices. Appendix. 585-item bibliography. 137 figures. 94 tables. ii + 478pp. 5⅝ x 8⅜.
60052-1 Paperbound $3.00

MATHEMATICS OF MODERN ENGINEERING,
E. G. Keller and R. E. Doherty

Written for the Advanced Course in Engineering of the General Electric Corporation, deals with the engineering use of determinants, tensors, the Heaviside operational calculus, dyadics, the calculus of variations, etc. Presents underlying principles fully, but emphasis is on the perennial engineering attack of set-up and solve. Indexes. Over 185 figures and tables. Hundreds of exercises, problems, and worked-out examples. References. Total of xxxiii + 623pp. 5⅜ x 8. 60734-8, 60735-6 Two volume set, paperbound $3.70

AERODYNAMIC THEORY: A GENERAL REVIEW OF PROGRESS,
William F. Durand, editor-in-chief

A monumental joint effort by the world's leading authorities prepared under a grant of the Guggenheim Fund for the Promotion of Aeronautics. Never equalled for breadth, depth, reliability. Contains discussions of special mathematical topics not usually taught in the engineering or technical courses. Also: an extended two-part treatise on Fluid Mechanics, discussions of aerodynamics of perfect fluids, analyses of experiments with wind tunnels, applied airfoil theory, the nonlifting system of the airplane, the air propeller, hydrodynamics of boats and floats, the aerodynamics of cooling, etc. Contributing experts include Munk, Giacomelli, Prandtl, Toussaint, Von Karman, Klemperer, among others. Unabridged republication. 6 volumes. Total of 1,012 figures, 12 plates, 2,186pp. Bibliographies. Notes. Indices. 5⅜ x 8½. 61709-2, 61710-6, 61711-4, 61712-2, 61713-0, 61715-9 Six volume set, paperbound $13.50

FUNDAMENTALS OF HYDRO- AND AEROMECHANICS,
L. Prandtl and O. G. Tietjens

The well-known standard work based upon Prandtl's lectures at Goettingen. Wherever possible hydrodynamics theory is referred to practical considerations in hydraulics, with the view of unifying theory and experience. Presentation is extremely clear and though primarily physical, mathematical proofs are rigorous and use vector analysis to a considerable extent. An Engineering Society Monograph, 1934. 186 figures. Index. xvi + 270pp. 5⅜ x 8.
60374-1 Paperbound $2.25

APPLIED HYDRO- AND AEROMECHANICS,
L. Prandtl and O. G. Tietjens

Presents for the most part methods which will be valuable to engineers. Covers flow in pipes, boundary layers, airfoil theory, entry conditions, turbulent flow in pipes, and the boundary layer, determining drag from measurements of pressure and velocity, etc. Unabridged, unaltered. An Engineering Society Monograph. 1934. Index. 226 figures, 28 photographic plates illustrating flow patterns. xvi + 311pp. 5⅜ x 8. 60375-X Paperbound $2.50